Trends in Fisheries and Aquatic Animal Health

Edited by

Panagiotis Berillis

Department of Ichthyology and Aquatic Environment, School of Agricultural Sciences, University of Thessaly, Volos, Hellas, Greece

Trends in Fisheries and Aquatic Animal Health

Editor: Panagiotis Berillis

ISBN (Online): 978-1-68108-580-7

ISBN (Print): 978-1-68108-581-4

First published in 2017.

General:

1. Any dispute or claim arising out of or in connection with this License Agreement or the Work (including non-contractual disputes or claims) will be governed by and construed in accordance with the laws of the U.A.E. as applied in the Emirate of Dubai. Each party agrees that the courts of the Emirate of Dubai shall have exclusive jurisdiction to settle any dispute or claim arising out of or in connection with this License Agreement or the Work (including non-contractual disputes or claims).
2. Your rights under this License Agreement will automatically terminate without notice and without the need for a court order if at any point you breach any terms of this License Agreement. In no event will any delay or failure by Bentham Science Publishers in enforcing your compliance with this License Agreement constitute a waiver of any of its rights.
3. You acknowledge that you have read this License Agreement, and agree to be bound by its terms and conditions. To the extent that any other terms and conditions presented on any website of Bentham Science Publishers conflict with, or are inconsistent with, the terms and conditions set out in this License Agreement, you acknowledge that the terms and conditions set out in this License Agreement shall prevail.

Bentham Science Publishers Ltd.
Executive Suite Y - 2
PO Box 7917, Saif Zone
Sharjah, U.A.E.
Email: subscriptions@benthamscience.org

BENTHAM SCIENCE

CONTENTS

FOREWORD

"Trends in Fisheries and Aquatic Animal Health" is an effort to cover some main topics about fisheries, aquaculture and aquatic animal health aspects. The book is a combination of reviews and original research articles.

The first two chapters examine the current state and the potential future scenarios of fish and fisheries of the eastern Adriatic Sea under the light of climate changes and the fresh water fishes of Croatia. The third and the fourth chapters focus on the challenges faced by marine aquaculture and Mediterranean fisheries in the framework of the new common fisheries policy. The fifth chapter is an introduction to the relationship between the use of antibiotics and the presence of antibiotic resistant strains of both pathogenic and commensal bacteria isolated from fish and aquatic environment. The topic of the antibiotic residues in fish due to irregular use of antibiotics is also examined. The sixth chapter deals with the stress and fish health towards an understanding of allostatic load. Chapter 7 is an excellent introduction to fish histopathology as a biomarker in ecotoxicology, while chapter 8 examines the effect of cyanobacteria and their toxins on fishes. The environmental and technical issues on developing rainbow trout farming in desert underground brackish water, focusing on the Iran area, is the theme of chapter 9. The last of the review chapters focuses on the fishing landings of crustacean decapods and the problems associated with their culture diseases.

The next six chapters are selective research papers mainly presented at the "HydroMediT 2016" international conference, held at Messolonghi, Greece, on 10 - 12 November 2016. A histological method to assess the effect of diet and a single meal on the liver and intestine of rainbow trout is presented in chapter 11. Chapter 12 deals with an innovative methodology for the mariculture of bath sponges in Greece, while examining the establishment of blue crab *Callinectes sapidus* in the Pogonitsa lagoon of the Amvrakikos gulf in western Greece is discussed in chapter 13. The CyHV-2 outbreak associated with *Aeromonas spp.* in crucian carp and the epidemiology of infectious hematopoietic necrosis (IHN) and viral haemorrhagic septicaemia (VHS) in salmonid fish in the Piedmont region of Italy are discussed in chapters 14 & 15. The final two chapters are focused in the area of Turkey dealing with the effects of chronic low level zinc exposure on the hematological profile of *Tinca tinca* and with the introduce of acoustic telemetry as a fishery management tool in artificial reefs.

The whole book is serving as an excellent introduction to the topics of fisheries, aquaculture and aquatic animal health. It is appropriate for the general public, students, or scientists from outside the field of marine sciences.

Dimitris Vafidis
Depatrment of Ichthyology and aquatic environment,
School of agricultural sciences,
Univesrity of Thessaly
Greece

PREFACE

We live in a world where the global population is expected to grow by another 2 billion to reach 9.6 billion people by 2050. Current food resources will not be enough and feeding the planet will be a huge challenge. Food coming from marine or inland waters could be the key. Fish and seafood have always been considered as an important part of human diet and it has long been recognized as a health-promoting food for human nutrition.

The Earth's surface is 71% water-covered, and oceans hold about 96.5 % of all Earth's water. Oceans contain almost the 90% of the world's living biomass. Globally, the fish stocks are over-exploited, depleted, fully exploited or in recovery from exploitation in a portion of 85%. We may be the last generation able to catch food from the oceans and seas. Aquaculture is the alternative solution for sea and inland waters food production. According to FAO, aquaculture is probably the fastest growing food-producing sector and can be a sustainable solution for fish production. The latest FAO 2014 report highlights that fact. In 2014, world aquaculture production attained another all-time high at 66.6 million tones. The global capture fishery production that year was 86.6 million tonnes (7.9 million tonnes were discard) and the global aquaculture production was more than 80% of the fishery production (66.6 million tonnes).

In order to maximize production and also having sustainable management of the aquatic environment, knowledge about aspects of fisheries and aquatic animal health is very important. This book comes to cover part of these aspects. The need to exchange reliable information on all related subjects is becoming a key issue.

The editor would like to thank Bentham Science Publishers for providing this opportunity to bring this book to the general public.

Panagiotis Berillis
Department of Ichthyology and Aquatic Environment,
School of Agricultural Sciences,
University of Thessaly, Volos, Hellas,
Greece

List of Contributors

Dragana B. Ljubojević	Scientific Veterinary Institute "Novi Sad", Novi Sad, Republic of Serbia
Perica Mustafić	University of Zagreb, Faculty of Science, Zagreb, Croatia
Milorad Mrakovčić	University of Zagreb, Faculty of Science, Zagreb, Croatia
Ivan Katavic	Institute of Oceanography and Fisheries, Split, Croatia
Claudio Caruso	Department of Virology, Istituto Zooprofilattico Sperimentale del Piemonte, Liguria e Valle d'Aosta, *via* Bologna 148, 10154 Turin, Italy
Maria Chistina Bona	Fish Diseases Laboratory, Istituto Zooprofilattico Sperimentale del Piemonte, Liguria e Valle d'Aosta, *via* Bologna 148, 10154 , Turin, Italy
Paolo Pastorino	Fish Diseases Laboratory, Istituto Zooprofilattico Sperimentale del Piemonte, Liguria e Valle d'Aosta, *via* Bologna 148, 10154 , Turin, Italy
Riccarco Prato	Department of Virology, Istituto Zooprofilattico Sperimentale del Piemonte, Liguria e Valle d'Aosta, *via* Bologna 148, 10154 Turin, Italy
Erika Astrid Virginie Burioli	Fish Diseases Laboratory, Istituto Zooprofilattico Sperimentale del Piemonte, Liguria e Valle d'Aosta, *via* Bologna 148, 10154 , Turin, Italy
Simone Peletto	Department of Genetics and Immunobiochemical, Istituto Zooprofilattico Sperimentale del Piemonte, Liguria e Valle d'Aosta, *via* Bologna 148, 10154, Turin, Italy
Marzia Righetti	Fish Diseases Laboratory, Istituto Zooprofilattico Sperimentale del Piemonte, Liguria e Valle d'Aosta, *via* Bologna 148, 10154 , Turin, Italy
Claudio Foglini	Fish Diseases Laboratory, Istituto Zooprofilattico Sperimentale del Piemonte, Liguria e Valle d'Aosta, *via* Bologna 148, 10154 , Turin, Italy
Pier Luigi Acutis	Department of Genetics and Immunobiochemical, Istituto Zooprofilattico Sperimentale del Piemonte, Liguria e Valle d'Aosta, *via* Bologna 148, 10154, Turin, Italy
Paola Arsieni	Fish Diseases Laboratory, Istituto Zooprofilattico Sperimentale del Piemonte, Liguria e Valle d'Aosta, *via* Bologna 148, 10154 , Turin, Italy
Loretta Masoero	Department of Genetics and Immunobiochemical, Istituto Zooprofilattico Sperimentale del Piemonte, Liguria e Valle d'Aosta, *via* Bologna 148, 10154, Turin, Italy
Marino Prearo	Fish Diseases Laboratory, Istituto Zooprofilattico Sperimentale del Piemonte, Liguria e Valle d'Aosta, *via* Bologna 148, 10154 , Turin, Italy
Jane Castritsi – Catharios	Department of Ichthyology and Aquatic Environment, School of Agricultural Sciences, University of Thessaly, Volos, Hellas, Greece
Constantin Vamvakas	Laboratory of Aquaculture & Artemia Reference Center, Faculty of Bioscience Engineering, University of Ghent, Ghent, Belgium
Ilias Baras	Selonda Aquaculture SA, Navarchou Nikodimou 26-30, 10556 Athens, Hellas, Greece
George Ekonomou	Department of Ichthyology and Aquatic Environment, School of Agricultural Sciences, University of Thessaly, Volos, Hellas, Greece

Branko Dragičević	Institute of Oceanography and Fisheries, Split, Croatia
Sanja Matić-Skoko	Institute of Oceanography and Fisheries, Split, Croatia
Jakov Dulčić	Institute of Oceanography and Fisheries, Split, Croatia
Božidar Rašković	University of Belgrade - Faculty of Agriculture, Institute of Animal Science, Belgrade - Zemun, Serbia
Vesna Poleksić	University of Belgrade - Faculty of Agriculture, Institute of Animal Science, Belgrade - Zemun, Serbia
Panagiotis Berillis	Department of Ichthyology and Aquatic Environment, School of Agricultural Sciences, University of Thessaly, Volos, Hellas, Greece
Eleni Mente	Department of Ichthyology and Aquatic Environment, School of Agricultural Sciences, University of Thessaly, Volos, Hellas, Greece
Sam Martin	School of Biological Sciences, University of Aberdeen, Tillydrone Avenue, AB24 2TZ Aberdenn, UK
Morteza Alizadeh	Iranian Fisheries Science Research Institute, Agricultural Research, Education and Extension Organization, Tehran, Iran
Shahram Dadgar	Iranian Fisheries Science Research Institute, Agricultural Research, Education and Extension Organization, Tehran, Iran
Shohre Masaeli	Fisheries office of Isfahan agriculture organization, Isfahan, Iran
Angeles Jos	Area of Toxicology, Faculty of Pharmacy, University of Sevilla, Spain
Ana M. Cameán	Area of Toxicology, Faculty of Pharmacy, University of Sevilla, Spain
Reyhan Akcaalan	Istanbul University, Fisheries Faculty, Turkey
Meriç Albay	Istanbul University, Fisheries Faculty, Turkey
George N. Katselis	Department of Fisheries and aquaculture Technology, TEI of Western Greece, Greece, Messolonghi
Constantin Koutsikopoulos	Department of Biology, University of Patras, Greece, Patra
Mehmet B. Ergönül	Ankara University, Faculty of Science, Department of Biology, 06100-Tandoğan, Ankara, Turkey
Sibel Atasagun	Ankara University, Faculty of Science, Department of Biology, 06100-Tandoğan, Ankara, Turkey
Kostas Kapiris	Hellenic Centre of Marine Research, Institute of Marine Biological Resources and Inland Waters, 46,7 km Athens-Sounio, P.O. Box 712, 19013, Anavissos, Athens, Hellas
Hijran Yavuzcan Yıldız	Ankara University, Faculty of Agriculture, Department of Fisheries and Aquaculture, 06110, Ankara, Turkey
F. Sertel Seçer	Ankara University, Faculty of Agriculture, Department of Fisheries and Aquaculture, 06110, Ankara, Turkey
Aytaç Özgül	Ege University, Faculty of Fisheries, 35100 İzmir, Turkey
Altan Lök	Ege University, Faculty of Fisheries, 35100 İzmir, Turkey

CHAPTER 1

Fish and Fisheries of The Eastern Adriatic Sea in The Light of Climate Change

Branko Dragičević[*], **Sanja Matić-Skoko** and **Jakov Dulčić**

Institute of oceanography and Fisheries, Split, Croatia

Abstract: Climate change is affecting biodiversity and overall system functioning of all the worlds' oceans. Over the past decades, significant amount of evidence has accumulated which allow us to decipher its effects on the Adriatic Sea. Investigations of fish communities in its eastern part showed us that it has undergone a significant biodiversity changes which are dominated by the influx of alien fishes, distributional shifts of native fishes and community-level changes of native populations of fishes. These changes have profound influence on fishery and aquaculture of the eastern Adriatic Sea. Beside current state, potential future scenarios which will affect its fish and fisheries are reviewed.

Keywords: Exotic fishes, Fisheries, Lessepsian, Northward spreading, Thermophilic.

INTRODUCTION

Adriatic Sea is considered as a hydrologicaly distinct Mediterranean sub-system. It is the northernmost area of the Mediterranean (excluding the Black Sea) and is generally divided into three subdivisions, North, Middle and South Adriatic. Also, the eastern and western coasts are different; the former is high, rocky and articulated with many islands while the western coast is flat and alluvial with raised terraces in certain areas [1]. Due to the presence of numerous islands, eastern coast is characterized by the abundance of coastal habitats. The depth gradually decreases from south to north and depths in the North Adriatic never exceed 100 meters. Greatest depth in the area of Middle Adriatic is 273 m in the Jabuka/Pomo Pit while the South Adriatic hosts the deepest point of the Adriatic Sea (South Adriatic Pit) with the depth of 1233 m. Mean depth of the Adriatic is around 252 m [2]. Its thermohaline properties are mainly determined by the interactions between sea and air, river discharge, mixing, currents, water exchange

* **Corresponding author Branko Dragicevic:** Institute of Oceanography and Fisheries, Split, Croatia; Tel: +38521408013; E-mail: brankod@izor.hr

Panagiotis Berillis (Ed)

through the Otranto Strait and general topography of the basin. The temperature range of the surface water is around 18°C in the South, but can be as high as 25°C in the North Adriatic. The extremes of surface temperature can range from 6°C to 29°C. Mean temperatures of even the deepest layers are for the most part above 10°C. During winter, the South Adriatic is 8-10°C warmer than its central and northern parts. The horizontal temperature profile is more uniform in other seasons. In general, the open sea is warmer than the coastal waters [3]. Salinity is relatively high with significant ranges and highest salinity occurs in the area of South Adriatic (38.4-38.9) especially in the intermediate layer. Salinity decreases from south to north and from open sea to the coast. The North and western Adriatic are more influenced by river floods (mainly Po River) which affect water circulation through buoyancy input and the ecosystem dynamics by large nutrient influx [4].

Currents in the Adriatic Sea are rather complex and layered (surface, intermediate and bottom layer) and are influenced by the general Mediterranean circulation regimes. In essence, surface circulation of the Adriatic Sea can be described as a large-scale meander with the northerly flow along the eastern coast and a southerly (return flow) along the western coast [5]. However, specific geomorphology and significant seasonal variations elevate complexity of the Adriatic hydro-biological system making it rather difficult but interesting to study.

One of the most important hydrological features connecting the Adriatic Sea with the rest of the Mediterranean through the Ionian Sea is the Bimodal Oscillating System (BiOS) mechanism that changes the circulation of the North Ionian Gyre (NIG) from cyclonic to anticyclonic and *vice versa*, on decadal time scale [6]. This mechanism allows for the influx of water from the Ionian Sea and/or central Mediterranean into the Adriatic Sea depending on the cyclonic or anticyclonic regime, respectively. Beside influence on hydrology, this mechanism is also influencing biodiversity of the Adriatic Sea. It has been shown that increased abundance or first appearance of some organisms in the Adriatic Sea indicate BiOS regimes (inflow of Western Mediterranean/Atlantic or Eastern Mediterranean water) [6, 7]. For a long period, some features influenced by BiOS mechanism have been attributed to a phenomenon of “Adriatic Ingressions”, a theory which got updated with the description of BiOS [6, 8]. Presence of various thermophilic species has been previously attributed to the phenomenon of “Adriatic Ingressions” [9, 10]. In any event, influx of more warmer, nutrient rich and more saline Ionian waters has not only contributed to the presence of some rare and alien species in the Adriatic waters, but has a profound influence on the Adriatic biodiversity in general.

The present-day flora and fauna of the Adriatic Sea is a result of the numerous geological, geographical, climatic and biological processes occurring during its formation. Nowadays, it seems that climatic processes again play a significant role in shaping of the Adriatic ecosystem. Although it is a part of the Mediterranean, Adriatic Sea is also an independent biogeographical and ecological subunit, which is manifested in the composition and properties of its life communities. The marine biodiversity of the Mediterranean Sea is nowadays facing substantial structural changes in flora and fauna [11]. It is rapidly changing due to the increasing arrival of non-indigenous species [12]. Such changes were recorded in the Adriatic Sea, as well. During the last few decades, various factors including climate change, anthropogenic activity and "lessepsian migration" have altered the composition of Adriatic ecosystem, ichthyofauna in particular [13 - 16].

INFLUENCE OF CLIMATE CHANGE

Although discussion on the origin and mechanism of climate change can be lead from different angles and is a source of controversy (scientific evidence, however, leaves little maneuvering space for climate change denialists), the fact that climate change is actually happening stands stronger than ever before. It is usually perceived through a rise in mean atmospheric temperature, but several features of the earth's climate such as atmospheric circulation, ocean patterns and mixing, stratification, hydrological and seasonal patterns are influenced by the build-up of carbon dioxide and other greenhouse gases in the atmosphere [17].

Climate is one of the most important abiotic drivers giving shape to the distribution of plants and animals all over the planet through the combination of its direct and indirect effects. In organisms whose body temperature varies according to the environmental temperature (poikilotherms) such as fishes, the temperature may shape population and community structures through its influence on the survival, reproduction and patterns of resource use of single individuals [18]. Indirect effects are exhibited for example through modification in water circulation which can influence larval dispersal and recruitment [13]. Fishes have been used as indicators of environmental changes for a long time [19, 20]. Their high dispersal potential, ecological differentiation, sensitivity to temperature, usually large size and ease of identification, make them excellent indicators of the effects of climate variability. Beside this, climatic changes have been known for their damaging impact on the biology, fecundity, growth and biodiversity of aquatic, terrestrial and aerial animals [21]. It is suggested that climate change has an effect on individual organisms during all life stages hence affecting populations of a species, communities and the functioning of the ecosystem [22].

It seems that detection of influence of climatic changes on the Mediterranean hydrology and hydrobiology is rather difficult due to the significant variability in local climate which masks trends in the "noise" of natural fluctuations [23]. However, in the recent decade, significant progress has been made to tackle this issue. This is especially relevant for the Adriatic basin where local athmospherical conditions significantly influence hydrological properties. Ferrarese *et al.* [24] pointed out that the Adriatic Sea circulation and temperature patterns change abruptly at onset and growth of strong winds (*i.e.* Bora and Scirocco). Indeed, highest seasonal variability of sea surface temperature (SST) in the Mediterranean has been documented for the area of North Adriatic [25]. Vilibić *et al.* [26] pointed out that weakening of thermohaline circulation in the Adriatic Sea due to climate change can potentially affect deep pelagic and Benthic Organisms especially biodiversity in niches such as those found in the nearby Jabuka Pit. In the last decade a significant amount of papers acknowledged influence of climate change on hydrological and biological processes in the Mediterranean and Adriatic Sea, as well.

Potential responses to climate warming include a wide array of taxa seeking cooler environments by shifting toward poles and higher latitudes, higher global extinction rates and reorganization of local communities resulting from local extinctions and expansion of thermophilic species [27]. In case of fishes, climate change can influence their distribution and abundance through changes in growth, reproduction, survival or response to changes at other trophic levels [28]. Other authors predict the species niche reductions by the middle and the end of the 21st century that might result from sea temperature increase [27].

There are few theoretical consequences that can be experienced by the mobile fauna (especially fishes) of the Adriatic Sea which can be perceived as an increase of abundance of certain species but due to different mechanism. One is the extension of the northern limit of species distributions which usually affects thermophillic species (northward expansion), and the other is the reduction of distribution of species of cold water affinity with subsequent northward shift of center of population distribution (seeking refuge in northern areas). Thermophilic fishes have evolved in tropical or subtropical marine environments and are therefore well adapted to warm waters. In the Mediterranean (including Adriatic Sea), these species can broadly be categorized into two groups distinguished by different histories: a) native species (with tropical or subtropical affinity and origin), and b) exotic species recently entered in [18]. Beside phenomenon of shift in population distribution (usually northward) by native Mediterranean species which is usually termed "meridionalization", a process of "tropicalization" (arrival of alien species of tropical origin) also plays an important role in carving of the faunal assemblages of the Mediterranean and Adriatic Sea.

It is presumed that coldest parts of the Mediterranean Sea (Gulf of Lyon and North Adriatic) could initially serve as a sanctuary for cold-temperate species, but with the continuation of warming those areas could become "cul-de-sac" for those species. This is especially important for endemic species which could go extinct due to the trapping effect [27].

CHANGES IN THE ADRIATIC ICHTHYOFAUNA

As previously mentioned, one of the main consequences of the climate change on the biodiversity, or at least easily observable, is the shift in species geographical distribution. There are numerous examples of such shifts in the world oceans including the Mediterranean and Adriatic Sea. In the last two decades, the advance of thermophilic species has been a largely cited evidence of the linkage between climate change and distribution patterns of Mediterranean Sea biodiversity. Earlier, when consequences of climate warming still had a hypothetical character, these unusual occurrences provided the first indication of changes that followed [29, 30]. Numerous northward shift records came from the Adriatic Sea [13, 31], where even juveniles of some previously rare or absent, thermophilic fishes have been recently registered.

Presence or increase in abundance of certain thermophilic species in the Adriatic Sea is usually attributed to the water warming, but it is a very complex task to distinguish it from other potential drivers. However, it is most likely that majority of possible drivers are essentially related to the increased average sea temperature. In the last 20 years numerous thermophilous fish species have been recorded for the first time in the Adriatic Sea and their presence might be related to climate change. There is also an interesting issue of the presence of lessepsian fish species whose presence cannot easily be attributed to climate change but is certainly facilitated by it.

Even if you are just a recreational fisherman, you can actually witness effects of climate change in the marine environment. It is a gradual change and its effects are observable at least on a decadal scale, however, certain phenomenons are represented by rapid changes, *i.e.* population explosions. If you are living on the eastern coast of the South Adriatic Sea, one thing you might notice is a triggerfish (*Balistes carolinensis*) eating hard shelled animals of the bridge pier, beautifully colored ornate wrasse (*Thalassoma pavo*) graciously moving among the weedy rocks and, if you are lucky enough, you might witness an elaborate mating of parrotfish (*Sparisoma cretense*) somewhere in the waters off Korcula island. These fishes are now becoming increasingly abundant and slowly progressing toward northern areas. Any regular visitor of the fishmarket in Split (for example) can witness increasing occurrence of previously unknown fish species in the last decade. Bluefish (*Pomatomus saltatrix*), dolphinfish (*Coryphaena hippurus*), atlantic lizardfish (*Synodus saurus*), yellowmouth barracuda (*Sphyraena viridensis*) are now frequent fishmarket items, although only ten years ago they were either rare or totally absent from fishmarkets. Even native sea bream (*Sparus aurata*) has become exceptionally abundant in the last few years. Are effects of climate change the only factor that influences these changes? Probably not. But in any event, climate change might still be a significant if not major culprit for these changes. The reason for uncertainty is quite a large number of variables involved in the Adriatic ecosystem equation. Brander [32] stated that an important issue in assessing the influence of climate change on fish populations is the disentanglement of its effects from those of other drivers (*i.e.* fishing). First of all, overfishing of commercial fish stocks can be considered as the primary source of fluctuations in this complex web of ecological interactions. Heavy fishing can influence success of non-commercial species by reducing competition by the commercial species for the same resources hence allowing previously rare species to establish more numerous and resilient populations. Secondly, since Adriatic Sea is not isolated from the rest of the Mediterranean, changes occurring in other areas have consequences for the Adriatic ecosystem. This is especially important for the presence of lessepsian fish species whose arrival in the Adriatic is not only facilitated by climate change but also by the presence of already established populations in the southern regions which probably act as recruitment areas for subsequent northward spreading.

There has been numerous studies connecting spreading and arrival of species of tropical affinity into new areas as a consequence of water warming. Beside arrival and spreading of non-native species there are also certain changes affecting populations of native fishes. This is evident through either increased abundances, northward extension or decline in occurrence of some species. If latter case is concerned, it is very difficult to resolve whether decline of particular fish species is due to unsuitable hydrological conditions or is overfishing, food chain instability, disease or some other factor affecting its population. However, it is very likely that some cold water species will be negatively affected by the water warming while thermophilious species will benefit from it. This issue is of particular importance for the Adriatic Sea since the impacts of the global warming are particularly critical in semi-enclosed seas [33]. There are already indications that some cold-water fish species, particularly *Sprattus sprattus*, are in decline in the last few decades [34].

Effects of water warming probably influenced shifts in distributions of some native species which is usually being perceived as an increase in abundance (or *vice versa*) of such species. Scarcity of historical data usually don't allow for more elaborate investigation of some of such cases, but there is a growing number of reports on the unusual occurrences of certain species. It is much easier to track changes in the distribution and dynamics of colonization of the species whose presence is recently reported for the first time as such records usually provide a baseline for the tracking based on subsequent occurrences. Due to increased research interest, advance of social media and awareness raising campaigns, a possibility of drawing biased conclusions based solely on the occurrence reports is very much possible. In the future we'll have to rely on local ecological knowledge (LEK) as a source of needed data and such attempts have already been undertaken for the Mediterranean Sea [35, 36].

Perhaps the most striking change in the fish communities is an increase in abundance of before mentioned bluefish (*P. saltatrix*). For the area of northwestern Mediterranean it is documented that this species reacts to the increased sea surface temperatures (SST) by shifting areas of distribution and reproduction northward in the northwestern Mediterranean [37]. We can assume this is also the case in the Adriatic waters, but due to lack of landing data on this species it is impossible to perform this type of study at present time. However, preliminary results originating from local ecological knowledge survey (LEK) indicate a rise in abundance of this species in the last decade (unpublished data). An unusually large catch (1.5 tonnes) of this species has been documented for the area of North Adriatic in winter of 2003 [38]. Additionally, there is evidence that this species causes significant negative effects on populations of native fishes such as mugils especially in the area of Neretva estuary [39]. It is therefore plausible to consider this fish as a danger to the communities of local fish populations but which could also be compensated through targeted fisheries of this species.

Presence of thermophillic dolphin fish *Coryphaena hippurus* is not new for the Adriatic Sea. This species regularly enters Adriatic waters in the warmer period of the year but was considered rare in the older literature [40]. However, results of the LEK survey among sports fishermen in Croatia indicated that abundance of this species is on the increase in the recent years (unpublished data). This is in agreement with results of the LEK study performed by Azzurro *et al.* [35]. It is also suggested that this species reproduces in the Adriatic waters which is indicated by the presence of larvae and early juveniles [41, 42]. Unfortunately, lack of detailed landing data on this species for the Adriatic Sea, as is the case with *P. saltatrix*, makes it particularly hard to establish a scientifically based connection between perceived changes in population abundance and water warming.

There are interesting cases which consider thermophillic species of Carangidae family, namely blue runner *Caranx crysos* and false scad *Caranx rhonchus*.

Adriatic Sea experienced very rapid, but somewhat overlooked colonization by these two species. Both appeared recently for the first time in the Adriatic waters, in 2009 and 2011, respectively [43, 44]. In the meantime, there were numerous occurrences of both species including a significant number of juvenile specimens of *C. crysos*. According to Psomadakis *et al.* [45], recent increase in the occurrence of this species in the Tyrrhenian Sea is probably indication of displacement of the northern limit in the Mediterranean and is attributable to the "tropicalization" phenomenon currently occurring in the Mediterranean. This is even more indicative for the Adriatic scenario since northernmost occurrence of this species in the Mediterranean was recorded in the area of North Adriatic. Similar is the case of *C. rhonchus* for which, after its first occurrence, few catches of significant quantities from the area of South Adriatic (Montenegrin coast) were reported.

Indications on the distributional shifts among some thermophillic native fishes are based mostly on increased number of reports in the recent decade. As previously mentioned, increased number of reports might not reflect real situation, but it can serve as a basis for focusing LEK investigations. Additionally, investigations undertaken in other areas in the Mediterranean can also provide a support for such theories. Among the most interesting examples are increases in abundances, possibly due to distributional shift, of species like *Sphyraena viridensis*, *Synodus saurus*, *Dactylopterus volitans*, *Thallasoma pavo*, *Balistes carolinensis, Pseudocaranx dentex*, *Sparisoma cretense* and *Enchelycore anatina.* Previously predominant in the South Adriatic, species of Serranidae family also experienced northward shift. This is especially indicative in the cases of white grouper *Epinephelus aeneus* and mottled grouper *Mycteroperca rubra*, which, after their first records from South Adriatic in 1999 and 2000, respectively, experienced a northward expansion and are now occasionally reported from the areas of South and Middle Adriatic Sea [46, 47].

As previously mentioned, presence of some rare, alien or theremophillic fish species was usually attributed to the phenomenon of "Adriatic ingressions". This is probably true, although the phenomenon itself is now incorporated in the mechanism of BiOS. It is possible that, depending on the regime of BiOS, influx of species originating from central Mediterranean or western Ionian sea (Atlantic/Wester Mediterranean species) is facilitated during anticyclonic period and those from eastern Ionian sea (Lessepsian species) during cyclonic period. Unfortunately, since the number of records of fish species whose presence might be connected with this mechanism is still relatively low and the fact that time of record doesn't necessary indicate the time of their arrival, ability to correlate these events is difficult but future investigations should shed some light on these processes. Investigations by Batistić *et al.* [7] already established such connection with zooplankton communities.

LESSEPSIAN MIGRATION

Since the opening of the Suez Canal in 1869 until now a tremendous number of Indo-Pacific species penetrated into the Mediterranean Sea significantly changing its ecosystem especially in its eastern part [48]. Since 1980's, lessepsian migrants started to occur in the waters of the South Adriatic as well. First encountered species were *Hemiramphus far* and *Paraexocoeus volitans*, two species whose occurrence was never subsequently documented for the Adriatic Sea [49]. The case of *Saurida lessepsianus* (previously determined as *S. undosquamis*) is interesting since due to similarity to native *Synodus saurus* might be an overlooked species. Two records from the South Adriatic were reported in 1995 and 2002, without subsequent documented records [50, 51]. Until around 2011, there were no evidences of established populations of lessepsian fish species in the Adriatic, but after that period subsequent records including occurrences of juveniles allowed us to presume that *Fistularia commersonii* and *Siganus luridus* established their populations in the South Adriatic [52]. Large schools of *S. luridus* which have recently been documented in the area of South Adriatic and additional record of juvenile *F. commersonii* in 2016. corroborate this opinion. Out of 14 lessepsian fish species encountered in the Adriatic, it seems that only these two species established their populations. Multiple occurrences of *Lagocephalus sceleratus* raised doubt whether this species also has self-sustaining population, but since only large specimens were found, its presence might be of periodical nature. This species also represents a threat to human health due to presence of potent neurotoxin tetrodotoxin which can be fatal if consumed. This problem triggered awareness-raising campaigns in some Adriatic countries.

In case of lessepsian migrants, it is difficult to detect whether range extension is caused by the natural movement or is dictated by the change in climate conditions. It seems, however, that in the last 20 years this process intensified, especially westward migration [53]. Presence of lessepsian species in the Adriatic Sea is probably facilitated if not a consequence of periodic influx of water originating from the eastern Mediterranean Sea (BiOS). Additionally, influence of climate change on the arrival of lessepsians is probably masked by inherent progression of the dynamics of colonization processes. Namely, invasion dynamics is usually characterized by the initial phase of low spread, followed by expansion phase by increasing spread rate [54]. In any event, future expansion of these species is expected in the Adriatic waters and additional research effort should be undertaken to evaluate their impact.

EXPECTED CHANGES IN FISHERIES AND AQUACULTURE

Any changes occurring in marine ecosystems have a definite impact on the users of marine bio-resources, primarily on fisheries. As previously stated, sea warming may directly influence marine populations through individual physiological adaptations but also indirectly through alterations in the abiotic environment or prey availability [55]. Adaptation strategies are largely dependent on the predicted impacts of the climate change on marine ecosystems, but these predictions are still blurred by immanent uncertainty.

Climate induced changes are expected to affect services provided by the ecosystem (*e.g.* aquaculture and fisheries) all around the world [56]. Particularly, the impact of climate change on the marine fisheries sector is complex, due to the fact that the effects can be both positive and negative in economic terms. The consequences of climate change on fishing communities will depend on their exposure to change, the sensitivity of important species and ecosystem in general to climate change but also on fishermen's ability to adapt to new situation [56]. Fisheries should be affected by both "meridionalization" and "tropicalization" of catch, *i.e.* an increase of warmer-water species in relation to colder-water ones, since shifts in distribution are expected to affect their availability to fisheries [57, 58]. This scenario is already taking place in the Adriatic waters where a significant number of thermophilic species occur in the catches more often, especially in the coastal areas (*Sphyraena viridensis, Synodus saurus, Balistes carolinensis, Pseudocaranx dentex etc.*). Landings may change in relation to global warming [59], and this may induce changes in the intensity and spatial distribution of fishing effort [60]. In particular, the vulnerability of a fishery to climate depends on previously induced changes in fish stocks that affect species composition and thus abundance in commercial catches [61]. As an example, the change in the mean temperature of the catch (MTC) of official and survey catches, parameter proposed as one method which can shed some light on changes in catch composition, in the Ionian and Aegean Seas indicated that the relative catch proportions of species preferring warmer waters and those preferring colder waters have changed in favor of the former and that this change is linked to sea surface temperature increase [62]. Generally, mean catches of temperate and polar species in the large marine ecosystems (LME) were higher in years of warm winters in the northern regions of the species distributional range and in years of cold winters in the southern regions of their ranges. Mean catches of subtropical species during cold years in LME were higher in lower latitudes and in warm years in higher latitude regions. These results indicate a poleward shift of fish species as a response to ocean warming, posing challenges for future fisheries management [56].

Climate change will provoke changes that will be reflected in all fishery sectors, professional and recreational. Both artisanal and industrial sectors may adapt to these changes mainly through expansion of fishing grounds that will consequently increase operation costs. Highly mobile and diversified fishery might have a better adaptational potential and therefore be less vulnerable to climate change. A practical implication of the reported changes in catch since 1970s is that vessels are now reportedly fishing further offshore, which has implications for fuel consumption and profitability of the fishery [63]. However, Belhabib *et al.* [64] highlighted that historical changes in target species are more common in industrial than artisanal fisheries which challenges the assumption that artisanal fisheries, with their limited movement capacity, would easily adapt to climate change by shifting target species and/or gear type.

The exposure of a fishing community will be greatest where other pressures, such as overfishing, are already causing problems in the socioecological system [65]. Also, overexploited fish stocks are more strongly affected by climate change. This is due to reduced age structure, limitations of geographic distribution, loss of diversity *etc.* [57, 66, 67]. Unfortunately, most of the Mediterranean fish stocks are currently overexploited [68] and it makes them particularly vulnerable to climate change, as observed, for instance, for the North Adriatic Sea [69]. There, commercial catch is almost entirely composed of species of cold water affinities that have decreased during the past decade as a consequence of climate change. As previously mentioned, North Adriatic can be considered as particularly vulnerable area because it hosts several species that are adapted to boreal climatic conditions and it is configured as a cul-de-sac that prevents the northward movement of such species [27, 70]. It is shown that the relatively healthy condition of fisheries is likely to assist successful adaptation on climate changes [71]. Unfortunately, fisheries resources of the Mediterranean and Black Sea are at risk of overexploitation [62] which makes this ecosystem highly vulnerable to effects of climate change.

Hare *et al.* [72] suggested that the overall climate vulnerability is high to very high for diadromous and benthic invertebrate species. In addition, the majority of assessed species have a high potential for a change in distributional range as a response to projected climate changes. Negative effects are expected for approximately half of the assessed species, but some are expected to be positively affected (*e.g.*, increase in productivity, migration into the new region).

Small pelagic fish are important ecological indicators of the state of the ecosystem due to their sensitivity to climate forcing that influences their distribution and abundance. It appears that the combined effects of climate change and fishing may significantly alter habitat characteristics so that both the distribution and productivity of a population are shifted [73]. This makes purse-seine fisheries particularly vulnerable to climate change.

The effect of sea warming on the faunal composition in marine ecosystem is already evident in the eastern Mediterranean Sea because of the invasion of alien species of indo-pacific origin through the Suez Canal (Lessepsian immigrants), which is facilitated by temperature increase [74]. Over 435 species have invaded the Mediterranean Sea through the Suez Canal and about 20% are fishes [75]. These invaders certainly alter the faunal composition but may also cause severe alterations to ecosystem structure and function [76, 77]. In certain areas of the eastern Mediterranean some alien fishes are commercially exploited [78, 79] but their catches are still not officially adequately recorded [80]. Similar scenarios are expected for the Adriatic due to the the fact that the presence of lessepsian species is on the increase. Some of them are top predators like the bluespotted cornetfish, *Fistularia commersonii* which can potentially affect populations of native fish species like sardine, anchovy, boque and picarel. Possible impacts don't have to necessarily be of ecological or economical nature. Health problems due to consumption of certain toxic species, like *Lagocephalus sceleratus* or other members of tetraodontidae family, should also be considered.

The most comprehensive catch records of alien species have been made for the Israel fishery. The catch of the lessepsian species has been estimated at approximately a third of the total landing since 1954 [81]. Nearly half of the trawl catches along the Israeli coast consists of lessepsian fish (Golani and Ben Tuvia, 1995). Similar scenario is practically following the path of succession of these species as new areas are being colonized, *i.e.* Aegean Sea [82]. This species is not targeted but if caught they are offered for sale by individual fishers directly on the market.

A potential displacement of native species by alien species is expectable due to the fact that alien species are usually more successful in competition for space, shelter and food. Displacement of native mullets (Mullus sp.) by the alien goatfishes (Upeneus spp.) [77], of salema, *Sarpa salpa* by spinefoot species (Siganus spp.) [79, 83] and/or anchovy, *Engraulis encrasicolus* by round-eye herring *E. golanii* [78] is on the way in certain areas and there is a growing evidence for increasing contribution of alien species of Indo-pacific origin in the Aegean Sea catches [12]. The alien fish *Upeneus moluccensis* and *U. pori*, formed 87% of the mullid catch off the coast of Israel at depth of 20 m, and 50% at 55 m, whereas the native mullids are more abundant in deeper waters [82]. Ismen (2002) stated that in the eastern Mediterranean almost total catch of *U. pori* was trawled in less than 50 m deep water and its market increased in the recent years. The percentage of the Erythrean mullids in the total mullid catch has increased steadily; from 30% in 1980, to close to 50% in 1989 [82].

A replacement of sardine *Sardina pilchardus* and sprat *Sprattus sprattus* with round sardinella *Sardinella aurita*, which is also distributed in the Greek Seas already occurs in the catches [84] because of round sardinella's expansion [37]. This species is also becoming increasingly abundant in the Adriatic Sea, especially in purse-seine fisheries. Although it is still being caught in lower quantities than anchovy and sardines, it's gaining importance in food and aquaculture industry (as fish food) [85].

Numerous findings of some new species which are economically important in the area of their former distribution could have a positive echo in recreational fishing and aquaculture. As an example, in the last decades, a 3 new species of groupers were recorded in Adriatic Sea: the orange-spotted grouper *Epinephelus coioides,* the white grouper, *E. aeneus*, and the mottled grouper, *Mycteroperca rubra.* These two latter species already show signs of established populations (especially *E. aeneus*) in the eastern Adriatic coast. This is not surprising since the native groupers also show positive response to sea warming. Namely, successful spawning of groupers in the South Adriatic resulted in better recruitment and new colonization of groupers in middle and northern Adriatic, particularly of adult dusky grouper, *E. marginatus* with still unconfirmed spawning in spreading areas. Those facts suggest a significant expansion of groupers in the last 10 years. In shallow waters, juvenile groupers represent the strongest carnivores and possible strong ecological impact through prey-predator relationship is expected [83].

Among economically important species that have potential to establish or have already established population in the Adriatic Sea are tripletail, *Lobotes surinamensis* and before mentioned spinefoots, *Siganus luridus* and *S. rivulatus.* The tripletail is a tasty fish and have economic importance along the western Atlantic coast. It is also becoming increasingly abundant in the area of Maltese islands and its presence in the Adriatic waters is showing similar trend but with lack of juvenile records [86, 87]. There was unsuccessful attempts of introducing *S. luridus* in the Mediterranean aquaculture but this species asks for special attention due to competitive behavior toward *S. salpa.* Moreover, the landings of S. *rivulatus* in the eastern Mediterranean are significant and point out the economic potential of this species as fishery resource in the spreading areas of its distribution.

Representatives of *Sphyraena* family are very interesting in socio-economic sense for purse- and beach-seine fisheries in the Adriatic Sea. Beside native European barracuda, *Sphyraena sphyraena,* two new species have been recorded in the recent decades: yellowstripe barracuda, *S. chrysotaenia* and yellowmouth barracuda, *S. viridensis*. The latter became quite abundant in the recent years but due to similarity of both species to native *S. sphyraena*, fishermen have difficulty in distinguishing those species which makes it difficult to track their spreading. The catch statistics of sphyraenids landings in Israel do not separate the Red Sea blunt barracuda from native Mediterranean species *S. sphyraena* and *S. viridensis*. However, the examination of the landed catch showed that the lessepsian barracuda had outnumbered the native sphyraenids in catches [88]. Similar scenario can be expected along the Mediterranean coast, although *S. chrysotaenia* is still very rare in the Adriatic Sea.

The brushtooth lizardfish, *Saurida undosquamis* also have certain fisheries potential since it has significant share in demersal catches along the Turkish coast where it is numerous and abundant. It was caught in Israel for the first time in 1952; only three years later 266 tons were landed by local trawlers, constituting almost 20% of the total trawler catch [89]. On the other side, this species is also in prey-predator relationships with native sardine and anchovy and thus can have possible negative influence on local communities and biomass of those pelagic species [83].

For alien species that are frequently being caught in commercial fisheries in the Mediterranean Sea, it is necessary to boost their value through public awareness-raising campaigns. These could be aimed at educating about their nutritional value, creating new processed products and promoting fresh and processed products in the market. The best way to reduce the pressure of the populations of alien species like *S. luridus*, *S. rivulatus* and *F. commersonii* on native species is to stimulate the commercial fishery to target these species. This shouldn't include only alien species but also other (thermophilic) species whose populations are experiencing significant increase like *Pomatomus saltatrix*, *Sphyraena viridensis* or *Balistes carolinensis*. The aim of such process should not only be a reduction of pressure induced by alien and other climate-driven species, but also reduction of fishing pressure on native, overexploited stocks. This would also contribute to the maintenance of the good environmental status (GES) of coastal marine ecosystems and the sustainable exploitation of species of fisheries concern. Eventually, this will lead to the improvement of the knowledge on alien species life history and their role in the marine ecosystem, as well as on the spatio-temporal variation of their landings.

Although the influence of climate change is easily seen through the emergence of new species in the areas where they were previously absent, it is necessary to simultaneously monitor populations of native species and track their adaptation to changing ecosystem. Groupers and bluefish are real examples of how changes in distribution of fish species may lead to changes in income of fisheries sector with both positive and negative consequences. Specifically, local fishermen still don't have effective fishing gears for catching bluefish (great damages on gillnets) and *P. saltatrix* is a predator of the flathead grey mullet, *Mugil cephalus* and it decimated populations of mullets from native habitats in Neretva Estuary. On the other side, groupers occupied ecological niches of overfished sparids, particularly of the white seabream, *Diplodus sargus* which is ecologically unacceptable but their catches might compensate for economic loss of decreasing abundance and biomass of sparids. Besides that, positive effects of sea warming on some physiological processes which influence spawning and reproductive success, namely better survivorship and recruitment, are observed. Such effects are later reflected in greater landings and were confirmed for several species, particularly for *Mullus* spp. and *Sparus aurata*. A particular problem related to aquaculture and sea warming is the case of *S. aurata* since higher abundance and biomass recorded in last years along the Mediterranean coast are probably result of multiple sources, but escape from aquaculture as a result of poor husbandry measures on fish farms or incidents caused by bad weather conditions (unexpected severe storms) are undouble contributing to this phenomenon.

Previous studies highlighted the winners and losers in fisheries under climate change scenario based on alterations in biomass, species composition and potential catches. Identifying changes in population processes of target species due to environmental effects is important in order to enable management strategy evaluations to elucidate the potential benefits and costs of changing management targets. Understanding how climate change influences the fisheries revenues is a crucial step towards the development of effective socio-economic policy and food sustainability strategies in adaptation efforts [90]. Particularly, fish prices and cross-oceans connections through distant water fishing operations may significantly modify the projected climate change impacts on fisheries revenues. Global fisheries might lose approximately $10 billion of their annual revenue by 2050 if climate change continues unchecked, and countries that are most dependent on fisheries for food will suffer the hardest [90]. Regionally, the projected increases in fish catch in high latitudes may not translate into increases in revenues because of the increasing dominance of low value fish, and the decrease in catches by these countries' vessels operating in more severely impacted distant waters. Therefore, Lam *et al.* [90] suggest the need to conduct full-fledged economic analyses of the potential economic effects of climate change on global marine fisheries. While aquaculture seems as a logical alternative solution to ease the financial burden of fishing losses and improve food security under climate change, Lam *et al.* [90] suggest that aquaculture may drive down the price of seafood, leading to further decreases in fisheries revenues.

Many coastal communities rely on living marine resources for livelihoods and food security. These resources are already under significant stress from overfishing, pollution, coastal development and habitat degradation. Climate change is an additional stressor impacting coastal systems and communities. However, it may also lead to opportunities for some species and the people they sustain [91]. In the future, we should concentrate our efforts to predicting the future of marine food resources in the light of climate change especially in sensitive, coastal areas. Basis for this should be better cooperation and transfer of data among the marine regions which are under the greatest impact of climate change. Although socio-political and resource dependency differences among countries will be the greatest challenge in tackling this issue, if prevailed, the benefits will largely surpass the potential costs of this effort. This is especially important in marine regions like the Mediterranean where a lot of countries share some common but not necessarily coordinated interests toward relatively small marine realm but one which already experiences large changes.

CONFLICT OF INTEREST

The author (editor) declares no conflict of interest, financial or otherwise.

ACKNOWLEDGEMENT

This work has been supported in part by Croatian Science Foundation under the project IP-2016-06-5251.

REFERENCES

[1] UNEP-MAP-RAC/SPA. 2014. Status and Conservation of Fisheries in the Adriatic Sea.

[2] Jardas I, Pallaoro A, Vrgoč N, Jukić-Peladić S, Dadić V. Crvena knjiga morskih riba Hrvatske Ministarstvo kulture, Državni zavod za zaštitu prirode. Republika Hrvatska 2008. (In Croatian).

[3] Zore-Armanda M, Grbec B, Morović M. Oceanographic properties of the Adriatic Sea-A point of view. Acta Adriat 1999; 40 (Suppl.): 39-54.

[4] Marini M, Campanelli A, Sanxhaku M, Kljajić Z, Grilli F. Late spring characterization of different coastal areas of the Adriatic Sea. Acta Adriat 2015; 56(1): 27-46.

[5] Orlić M, Gačić M, La Violette PE. The currents and circulation of the Adriatic Sea. Oceanol Acta 1992; 15(2): 109-24.

[6] Civitarese G, Gačić M, Lipizer M, Eusebi Borzelli GL. On the impact of the Bimodal Oscillating System (BiOS) on the biogeochemistry and biology of the Adriatic and Ionian Seas (eastern Mediterranean). Biogeosciences 2010; 7: 3987-97. [http://dx.doi.org/10.5194/bg-7-3987-2010]

[7] Batistić M, Garić R, Molinero JC. Interannual variations in Adriatic Sea zooplankton mirror shifts in circulation regimes in the Ionian Sea. Clim Res 2014; 61: 231-40. [http://dx.doi.org/10.3354/cr01248]

[8] Buljan M. Fluctuations of salinity in the Adriatic, Institut za Oceanografiju i Ribarstvo – Split (Croatia), Reports.1953; 2(2): 64.

[9] Vučetič T. Neuobičajena pojava meduze *Pelagia noctiluca* u Jadranu. III. - Utjecaj dinamike vodenih

masa na distribuciju meduze *Pelagia noctiluca* u Jadranu. Acta Adriat 1981; 23(1-2): 105-15. [In Croatian].

[10] Pallaoro A. On the possibilites of the occurrence of certain fish species in the Middle Adriatic connected with the Adriatic ingressions in years 1986/1987. Morsko Ribarstvo 1988; 3: 82-7. [In Croatian].

[11] Coll M, Piroddi C, Steenbeek J, *et al.* The biodiversity of the Mediterranean Sea: estimates, patterns, and threats. PLoS One 2010; 5(8): e11842. [http://dx.doi.org/10.1371/journal.pone.0011842] [PMID: 20689844]

[12] Zenetos A, Katsanevakis S, Poursanidis D, Crocetta F, Damalas D, Apostolopoulos G, *et al.* Marine alien species in Greek Seas: additions and amendments by 2010. Mediterr Mar Sci 2011; 12: 95-120. [http://dx.doi.org/10.12681/mms.55]

[13] Dulčić J, Grbec B. Climate change and Adriatic ichthyofauna. Fish Oceanogr 2000; 9: 187-91. [http://dx.doi.org/10.1046/j.1365-2419.2000.00128.x]

[14] Lipej L, Dulčić J. The current status of Adriatic fish diversity. Balkan Biodiversity Pattern and process in Europe's Biodiversity Hotspot. Dordrecht: Kluwer Academic Publishers 2004; pp. 291-306.

[15] Lipej L, Mavrić B, Orlando Bonaca M, Malej A. State of the Art of the Marine Non-Indigenous Flora and Fauna in Slovenia. Mediterr Mar Sci 2012; 13(2): 243-9. [http://dx.doi.org/10.12681/mms.304]

[16] Pećarević M, Mikuš J, Bratoš Cetinić A, Dulčić J, Čalić M. Introduced marine species in Croatian waters (Eastern Adriatic Sea). Mediterr Mar Sci 2013; 14: 224-37. [http://dx.doi.org/10.12681/mms.383]

[17] Monteiro P. Fisheries and climate change: inevitability or prophylaxis? Contributing to a necessary debate. Croatian Journal of Fisheries: Ribarstvo 2016; 74(3): 130-40.

[18] Azzurro E. The advance of thermophilic fishes in the Mediterranean Sea: overview and methodological questions. In: Briand F, Ed. Climate warming and related changes in Mediterranean marine biota. In: No:35. CIESM Workshop Monographs; 2008; pp. 39-46.

[19] Stephens JS, Hose JH, Love MS. Fish assemblages as indicators of environmental change in nearshore environments.Marine Organisms as Indicators. New York: Springer-Verlag 1988; pp. 91-105. [http://dx.doi.org/10.1007/978-1-4612-3752-5_5]

[20] Bettoso N, Dulčić J. First record of the oilfish *Ruvettus pretiosus* (Pisces: Gempylidae) in the northern Adriatic Sea. J Mar Biol Assoc U K 1999; 79: 1145-6. [http://dx.doi.org/10.1017/S0025315499001496]

[21] Eissa AE, Zaki MM. The impact of global climatic changes on the aquatic environment. Procedia Environ Sci 2011; 4: 251-9. [http://dx.doi.org/10.1016/j.proenv.2011.03.030]

[22] Pörtner HO, Peck MA. Climate change effects on fishes and fisheries: towards a cause-and-effect understanding. J Fish Biol 2010; 77(8): 1745-79. [http://dx.doi.org/10.1111/j.1095-8649.2010.02783.x] [PMID: 21078088]

[23] Karas J. Climate change and the Mediterranean Region 2000. http://www.greenpeace.org/ climate-change-and-the-medite r.pdf

[24] Ferrarese S, Cassardo C, Elmi A, *et al.* Air-sea interactions in the Adriatic basin: simulations of Bora and Sirocco wind events. Geofizika 2009; 26(2): 157-70.

[25] Shaltout M, Omstedt A. Recent sea surface temperature trends and future scenarios for the Mediterranean Sea. Oceanologia 2014; 56(3): 411-43. [http://dx.doi.org/10.5697/oc.56-3.411]

[26] Vilibić I, Šepić J, Proust N. Weakening thermohaline circulation in the Adriatic Sea. Clim Res 2013; 55: 217-25.

[http://dx.doi.org/10.3354/cr01128]

[27] Ben Rais Lasram F, Guilhaumon F, Albouy C, Somot S, Thuiller W, Mouillot D. The Mediterranean Sea as a 'cul-de-sac' for endemic fishes facing climate change. Glob Change Biol 2010; 16(12): 3233-45.
[http://dx.doi.org/10.1111/j.1365-2486.2010.02224.x]

[28] Perry AL, Low PJ, Ellis JR, Reynolds JD. Climate change and distribution shifts in marine fishes. Science 2005; 308(5730): 1912-5.
[http://dx.doi.org/10.1126/science.1111322] [PMID: 15890845]

[29] Riera F, Grau AM, Pastor E, Pou S. Faunistical and demographical observations in balearic ichthyofauna. Meridionalization or subtropicalization phenomena. Act Coll Sci OKEANOS La Mèditerranèe: variabilités climatiques, environment et bio-diversité Montpellier 1995; 213-20.

[30] Francour P, Boudouresque CF, Harmelin JG, Harmelin-Vivien ML, Quignard JP. Are the Mediterranean waters becoming warmer? Information from biological indicators. Mar Pollut Bull 1994; 28(9): 523-6.
[http://dx.doi.org/10.1016/0025-326X(94)90071-X]

[31] Dulčić J, Grbec B, Lipej L, Beg Paklar G, Supić N, Smirčić A. The effect of the hemispheric climatic oscillations on the Adriatic ichthyofauna. Fresenius Environ Bull 2004; 13(3b): 293-8.

[32] Brander K. Impacts of climate change on fisheries. J Mar Syst 2010; 79: 389-402.
[http://dx.doi.org/10.1016/j.jmarsys.2008.12.015]

[33] Pozdnyakov DV, Johannessen OM, Korosov AA, Pettersson LH, Grassl H, Miles MW. Satellite evidence of ecosystem changes in the White Sea: a semi-enclosed arctic marginal shelf sea. Geophys Res Lett 2007; 34: L08604.
[http://dx.doi.org/10.1029/2006GL028947]

[34] Grbec B, Dulčić J, Morović M. Long-term changes in landings of small pelagic fish in the eastern Adriatic - possible influence of climate oscillations over the Northern Hemisphere. Clim Res 2002; 20: 241-52.
[http://dx.doi.org/10.3354/cr020241]

[35] Azzurro E, Moschella P, Maynou F. Tracking signals of change in Mediterranean fish diversity based on local ecological knowledge. PLoS One 2011; 6(9): e24885.
[http://dx.doi.org/10.1371/journal.pone.0024885] [PMID: 21966376]

[36] Boughedir W, Rifi M, Shakman E, *et al.* Tracking the invasion of *Hemiramphus far* and *Saurida undosquamis* along the southern Mediterranean coasts: a local ecological knowledge study. Mediterr Mar Sci 2015; 2015(16): 628-35.
[http://dx.doi.org/10.12681/mms.1179]

[37] Sabates A, Martin P, Lloret J, Raya V. Sea warming and fish distribution: the case of the small pelagic fish, *Sardinella aurita*, in the western Mediterranean. Glob Change Biol 2006; 12: 1-11.
[http://dx.doi.org/10.1111/j.1365-2486.2006.01246.x]

[38] Dulčić J, Kraljević M, Pallaoro A, Glamuzina B. Unusual catch of bluefish *Pomatomus saltatrix* (Pomatomidae) in Tarska cove (northern Adriatic). Cybium 2005; 29: 207-8.

[39] Glamuzina B, Bartulović V, Skaramuca B. Impact of recent ichthyofauna changes on local fishery in the Neretva delta estuary. 43rd Estuarine and Coastal Sciences Association International symposium: Book of Abstracts. Lisbon: Faculty of Sciences of the University of Lisbon 2008; pp. 46-6.

[40] Jardas I. Jadranska ihtiofauna. 1996.

[41] Dulčić J. First record of larval *Brama brama* (Pisces: Bramidae) and *Coryphaena hippurus* (Pisces: Coryphaenidae) in the Adriatic Sea. J Plankton Res 1999; 21(6): 1171-4.
[http://dx.doi.org/10.1093/plankt/21.6.1171]

[42] Dragičević B, Dulčić J, Pallaoro A, Paladin A, Stagličić N. First record of dolphin-fish juveniles,

Coryphaena hippurus (Linnaeus, 1758), in the Eastern Adriatic. Annales. Series Historia Naturalis 2010; 20(2): 157-60.

[43] Dulčić J, Pallaoro A, Dragičević B. First record of the blue runner, *Caranx crysos* (Mitchill, 1815), in the Adriatic Sea. J Appl Ichthyology 2009; 25(4): 481-62. [http://dx.doi.org/10.1111/j.1439-0426.2009.01243.x]

[44] Kožul V, Antolović N. Occurrence of the false scad, *Caranx rhonchus* Geoffroy Saint-Hilaire, 1817 in the Adriatic Sea. J Appl Ichthyology 2013; 29(2): 449-50. [http://dx.doi.org/10.1111/jai.12086]

[45] Psomadakis PN, Bentivegna F, Guistino S, Tranvaglini A, Vacchi M. Northward spread of tropical affinity fishes: *Caranx crysos* (Teleostea: Carangidae), a case study from the Mediterranean Sea. Ital J Zool (Modena) 2011; 78(1): 113-23. [http://dx.doi.org/10.1080/11250001003680933]

[46] Glamuzina B, Tutman P, Kožul V, Glavić N, Skaramuca B. The first recorded occurrence of mottled grouper, *Mycteroperca rubra* (Serranidae), in the southeastern Adriatic Sea. Cybium 2002; 26: 156-8.

[47] Dulčić J, Tutman P, Ćaleta M. Northernmost occurrence of the white grouper, *Epinephelus aeneus* (Perciformes: Serranidae), in the Mediterranean area. Acta Ichthyol Piscat 2006; 36: 73-5. [http://dx.doi.org/10.3750/AIP2006.36.1.10]

[48] Mavruk S, Avsar D. Non-native fishes in the Mediterranean from the Red Sea, by way of the Suez Canal. Rev Fish Biol Fish 2008; 18: 251-62. [http://dx.doi.org/10.1007/s11160-007-9073-7]

[49] Parin NV. Hemiramphidae.Fishes of the north-eastern Atlantic and the Mediterranean. Paris: UNESCO 1986; Vol. 2: pp. 620-2.

[50] Rakaj N. Iktiofauna e Shqipërisë Shtëpia Boutese "Libri Universitar", Tiranë. 1995.

[51] Ungaro N, Marano G, Pastorelli AM, De Zio V, Rositani L. Nuove informazioni sulla fauna ittica del basso Adriatico. Biol Mar Mediterr 2004; 11(2): 648-51.

[52] Dulčić J, Antolović N, Kožul V, Dragičević B, Lipej L. First records of juveniles of two Lessepsian migrants, *Fistularia commersonii* Rüppell, 1838 and *Siganus luridus* (Rüppell, 1829), in the Adriatic Sea. J Appl Ichthyology 2013; 29: 661-2. [http://dx.doi.org/10.1111/jai.12129]

[53] Galil BS, Zenetos A. A sea change - Exotics in the Eastern Mediterranean.Invasive Aquatic Species of Europe Distribution, Impacts and Management. Dordrecht, The Netherlands: Kluwer Academic Publishers 2002; pp. 325-36. [http://dx.doi.org/10.1007/978-94-015-9956-6_33]

[54] Shigesada N, Kawasaki K. Biological invasions: Theory and practice. Oxford, UK: Oxford University Press 1997.

[55] Tzanatos E, Raitsos DE, Triantafyllou G, Somarakis S, Tsonis AA. Indications of a climate effect on Mediterranean fisheries. Clim Change 2014; 122: 41-54. [http://dx.doi.org/10.1007/s10584-013-0972-4]

[56] Gamito R, Costa MJ, Cabral HN. Fisheries in a warming ocean: trends in fish catches in the large marine ecosystems of the world. Reg Environ Change 2015; 15: 57-65. [http://dx.doi.org/10.1007/s10113-014-0615-y]

[57] Rijnsdorp AD, Peck MA, Engelhard GH, Möllmann C, Pinnegar JK. Resolving the effect of climate change on fish populations. ICES J Mar Sci 2009; 66: 1570-83. [http://dx.doi.org/10.1093/icesjms/fsp056]

[58] Cheung WW, Watson R, Pauly D. Signature of ocean warming in global fisheries catch. Nature 2013; 497(7449): 365-8. [http://dx.doi.org/10.1038/nature12156] [PMID: 23676754]

[59] Teixeira CM, Gamito R, Leitão F, Cabral HN, Erzini K, Costa MJ. Trends in landings of fish species potentially affected by climate change in Portuguese fisheries. Reg Environ Change 2014; 14: 657-69. [http://dx.doi.org/10.1007/s10113-013-0524-5]

[60] Haynie AC, Pfeiffer L. Why economics matters for understanding the effects of climate change on fisheries. ICES J Mar Sci 2012; 69(7): 1160-7. [http://dx.doi.org/10.1093/icesjms/fss021]

[61] Fortibuoni T, Aldighieri F, Giovanardi O, Pranovi F, Zucchetta M. Climate impact on Italian fisheries (Mediterranean Sea). Reg Environ Change 2015; 15(5): 931-7. [http://dx.doi.org/10.1007/s10113-015-0781-6]

[62] Tsikliras AC, Peristeraki P, Tserpes G, Stergiou KI. Mean temperature of the catch (MTC) in the Greek Seas based on landings and survey data. Front Mar Sci 2015; 2: 23.

[63] Hughes KM, Dransfeld L, Johnson MP. Climate and stock influences on the spread and locations of catches in the northeast Atlantic mackerel fishery. Fish Oceanogr 2015; 24: 540-52. [http://dx.doi.org/10.1111/fog.12128]

[64] Belhabib D, Lam VW, Cheung WW. Overview of West African fisheries under climate change: Impacts, vulnerabilities and adaptive responses of the artisanal and industrial sectors. Mar Policy 2016; 71: 15-28. [http://dx.doi.org/10.1016/j.marpol.2016.05.009]

[65] Miller K, Charles A, Barange M, *et al.* Climate change, uncertainty, and resilient fisheries: institutional responses through integrative science. Prog Oceanogr 2010; 87: 338-46. [http://dx.doi.org/10.1016/j.pocean.2010.09.014]

[66] Perry RI, Cury P, Brander K, Jenning S, Möllmann C, Planque B. Sensitivity of marine systems to climate and fishing: concepts, issues and management responses. J Mar Syst 2010; 79: 427-35. [http://dx.doi.org/10.1016/j.jmarsys.2008.12.017]

[67] Planque B, Fromentin JM, Cury P, *et al.* How does fishing alter marine populations and ecosystems sensitivity to climate? J Mar Syst 2010; 79: 403-17. [http://dx.doi.org/10.1016/j.jmarsys.2008.12.018]

[68] Colloca F, Cardinale M, Giannoulaki M, *et al.* Rebuilding Mediterranean fisheries: toward a new paradigm for ecological sustainability in single species population models. Fish Fish 2011; 14: 89-109. [http://dx.doi.org/10.1111/j.1467-2979.2011.00453.x]

[69] Pranovi F, Caccin A, Franzoi P, Malavasi S, Zucchetta M, Torricelli P. Vulnerability of artisanal fisheries to climate change in the Venice Lagoon. J Fish Biol 2013; 83(4): 847-64. [PMID: 24090551]

[70] Pranovi F, Colla S, Valeri P, Marco Anelli M. Present and future status of artisanal fisheries in the Adriatic Sea (western Mediterranean Sea). Ocean Coast Manage 2016; 122: 49-56. [http://dx.doi.org/10.1016/j.ocecoaman.2016.01.004]

[71] Johnson JE, Welch DJ. Climate change implications for Torres Strait fisheries: assessing vulnerability to inform adaptation. Clim Change 2016; 135: 611-24. [http://dx.doi.org/10.1007/s10584-015-1583-z]

[72] Hare JA, Morrison WE, Nelson MW, *et al.* A Vulnerability Assessment of Fish and Invertebrates to Climate Change on the Northeast U.S. Continental Shelf. PLoS One 2016; 11(2): e0146756. [http://dx.doi.org/10.1371/journal.pone.0146756] [PMID: 26839967]

[73] Valencia-Gasti JA, Baumgartner T, Durazo R. Effects of ocean climate on life cycles and distribution of small pelagic fishes in the California Current System off Baja California. Cienc Mar 2015; 41(4): 315-48. [http://dx.doi.org/10.7773/cm.v41i4.2571]

[74] Golani D, Orsi-Relini L, Massuti E, Quignard JP. Fishes.CIESM Atlas of Exotic Species in the

Mediterranean. Monaco: CIESM Publisher 2002; Vol. I: pp. 1-256.

[75] Nunes AL, Katsanevakis S, Zenetos A, Cardoso AC. Gateways to alien invasions in the European seas. Aquat Invasions 2014; 9: 133-44. [http://dx.doi.org/10.3391/ai.2014.9.2.02]

[76] Sala E, Kizilkaya Z, Yildirim D, Ballesteros E. Alien marine fishes deplete algal biomass in the Eastern Mediterranean. PLoS One 2011; 6(2): e17356. [http://dx.doi.org/10.1371/journal.pone.0017356] [PMID: 21364943]

[77] Bianchi CN. Biodiversity issues for the forthcoming tropical Mediterranean Sea. Hydrobiologia 2007; 580: 7-21. [http://dx.doi.org/10.1007/s10750-006-0469-5]

[78] Kallianiotis A, Lekkas V. First documented report on the Lessepsian migrant *Etrumeus teres* De Kay, 1842 (Pisces: Clupeidae) in the Greek Seas. J Biol Res (Thessalon) 2005; 4: 225-9.

[79] Giakoumi S. Distribution patterns of the invasive herbivore *Siganus luridus* (Ruppell, 1829) and its relation to native benthic communities in the central Aegean Sea, Northeastern Mediterranean. Mar Ecol (Berl) 2014; 35: 96-105. [http://dx.doi.org/10.1111/maec.12059]

[80] Tsikliras AC, Stergiou KI. The mean temperature of the catch increases quickly in the Mediterranean Sea. Mar Ecol Prog Ser 2014; 515: 281-4. [http://dx.doi.org/10.3354/meps11005]

[81] Galil BS. Lessepsian immigration: Human impact on Leventine Biogeography.The Biodiversity Crisis and Crustacea - Proceedings of the Fourth International Crustacean Congress Crustacean Issues. CRC Press 2000; pp. 50-1.

[82] Golani D, Ben-Tuvia A. Armantrout NB. Lessepsian migration and the Mediterranean fisheries of Israel.Conditions of the World's Aquatic Habitats Proc World Fish Congr Theme I. New Delhi: Oxford & IBH Pub. Co. Pvt. Ltd 1995; pp. 279-89.

[83] Dulčić J, Đođo Ž, Dragičević B, Ćukteraš M, Glamuzina B. New species in the Adriatic ichthyofauna and socio-economic consequences for Croatian marine fisheries. Ribarstvo 2012; 70: 111-23.

[84] Tsikliras AC. Climate-related geographic shift and sudden population increase of a small pelagic fish (*Sardinella aurita*) in the eastern Mediterranean Sea. Mar Biol Res 2008; 4: 477-81. [http://dx.doi.org/10.1080/17451000802291292]

[85] Mustać B, Sinovčić G. Eksploatacija i smrtnost srdele goleme Sardinella aurita, Valenciennes, 1847. u istočnom dijelu srednjeg Jadrana. Croatian Journal of Fisheries - Ribarstvo 2012; 70 (Suppl. 1): 1-13.

[86] Deidun A, Vella P, Sciberras A, Sammut R. New records of *Lobotes surinamensis* (Bloch, 1790) in Maltese coastal waters. Aquat Invasions 2010; 5 (Suppl. 1): 113-6. [http://dx.doi.org/10.3391/ai.2010.5.S1.023]

[87] Dulčić J, Dragičević B, Lipej L, Štifanić M. Range extension of tripletail, *Lobotes surinamensis* (Lobotidae) in the Adriatic Sea. A northernmost record in the Mediterranean. Cybium 2014; 58(2): 153-4.

[88] Grofit E. The trawl fishery in 1986. Fish Fishbreed Isr 1987; 20: 6-42. [in Hebrew with English Abstract].

[89] Ben-Yami M, Glaser T. The invasion of *Saurida undosquamis* (Richardson) into the Levant Basin – an example of biological effect of interoceanic canals. Fish Bull 1974; 72: 357-73.

[90] Lam VW, Cheung WW, Reygondeau G, Sumaila UR. Projected change in global fisheries revenues under climate change. Sci Rep 2016; 6: 32607. [http://dx.doi.org/10.1038/srep32607] [PMID: 27600330]

[91] Hobday AJ, Cochrane K, Downey-Breedt N, *et al.* Planning adaptation to climate change in fast-warming marine regions with seafood-dependent coastal communities. Rev Fish Biol Fish 2016; 26: 249-64.
[http://dx.doi.org/10.1007/s11160-016-9419-0]

CHAPTER 2

Biodiversity Status of and Threats to Freshwater Fish of Croatia and Eastern Adriatic Countries

Perica Mustafić* and **Milorad Mrakovčić**

University of Zagreb, Faculty of Science, Zagreb, Croatia

Abstract: Eastern Adriatic countries (ecoregion Dalmatia 419) have a unique array of endemic freshwater fishes. Our knowledge about the diversity of its freshwater fishes is still incomplete, many new species having been described only in recent years. It is widely held that fish biodiversity constitutes a natural resource that is valuable not only economically, but also in terms of culture and aesthetics, science and education. We have reviewed the characteristics of fish biodiversity in the eastern Adriatic countries (Slovenia, Croatia, Bosnia and Herzegovina as well as Montenegro), with a special emphasis on Croatia. Endemic species are found in the river Soča, Istria Mirna River, the region of Lika and in the Krka, Cetina, and Neretva River basins, in the Moraca and Skadar Lake. Usually, a single catchment supports between 6 and 10 species. Out of the 52 European threatened freshwater species, 28% occur in the eastern Adriatic, making this area a hotspot for threatened freshwater fish. Conservation management of these resources should be considered critical by all the governments involved. The area of occupancy of endangered freshwater fish is declining and if this does not change and if the loss of species continues at the current rate, the opportunity to conserve many of the remaining species will vanish. The speciation processes of endemic species started here about 5 million years ago, but the present-day distribution patterns essentially started to be established in the Pleistocene. A special mechanism for surviving drought has evolved in these populations and in general fish in the karstic area have become very hardy; however, negative anthropogenic effects have changed much of this evolution. An overview of endemic behaviour and some examples of patterns that undermine normal existing mitigation measures are provided. Human threats to global freshwater fish biodiversity fall into five categories: over-exploitation; water pollution; flow modification; destruction or degradation of habitat; and invasion by exotic species. Combined and interacting influences on fish biodiversity are now generally well known and have been exacerbated by climate change. We believe that knowledge of these threats is insufficiently incorporated into the development of water resources development and requires wider dissemination.

Keywords: Eastern adriatic coast, Ecoregions, Endemic, Freshwater fishes.

* **Corresponding author Perica Mustafić:** Faculty of Science, University of Zagreb, Rooseveltov trg 6, Zagreb, Croatia; Tel: +385-1-6189-717; E-mail: perica.mustafic@biol.pmf.hr

Panagiotis Berillis (Ed)

INTRODUCTION

Freshwater fish are an important European natural heritage and provide ecosystem services for tens of millions of people. Freshwater fish are also crucial to almost all food chains, especially in freshwater ecosystems [1, 2]. The current IUCN Red List publication [3] shows not only that there are declining populations in 17% of Europe's freshwater fish but also that in 76% of species population trends are still unknown [3]. Within the Mediterranean basin, eastern Adriatic countries have been identified as one of the world's biodiversity hotspots [4 - 6]. The region (Fig. **1**) supports a unique freshwater ichthyofauna comprising numerous endemic species with restricted distribution areas [4] and more than half of these are currently considered to be threatened [3, 5]. Defined by Freshwater Ecoregions of the World (FEOW) the coastal parts of eastern Adriatic countries, comprising Slovenia, Croatia, Bosnia and Herzegovina, and Montenegro, are defined as ecoregion Dalmatia 419 [7]. Eastern Adriatic countries, in this sense, include coastal drainages along the eastern Adriatic Sea from Slovenia to Montenegro, including some endorheic basins. These areas are bordered on the west by the Adriatic Sea and on the east by the Dinaric Alps. The northern and southern borders coincide with the Julian Alps and North Albanian Alps, respectively. Generally, eastern Adriatic countries have a rich and unique assemblage of endemic freshwater fish. Although speciation processes, especially of endemic species, started to take place here about 5 million years ago, the present-day distribution patterns are probably the result of events that occurred more recently, from the Pleistocene until historical times [8].

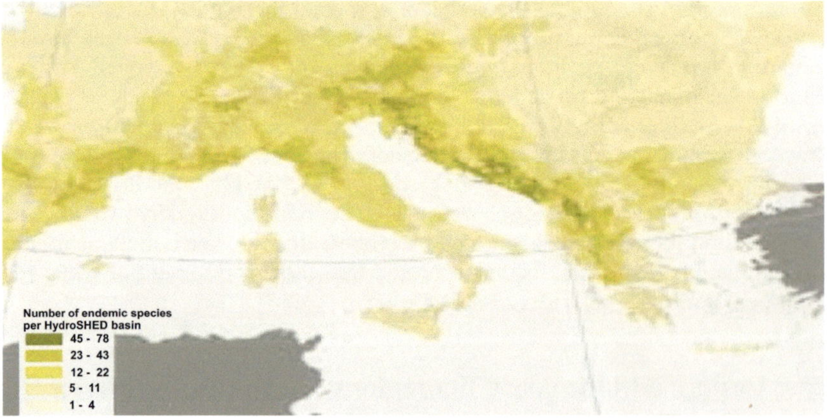

Fig. (1). Most important areas of endemic European freshwater fish species (enlarged to eastern Adriatic countries IUCN red list 2011).

One important question is whether we have already identified all the freshwater fish from this region. The question is particularly pertinent in view of the number of species that have been discovered only recently. The total number of species recorded in Europe in 1979 was between 170 and 213 [8, 9], rising to 358 in 1989 [10, 11], and finally to 571 by 2016 (Freyhof, personal communication). Extrapolating from the new discoveries of the last 16 years, it is reasonable to estimate that in the eastern Adriatic countries 3-5 fish will be discovered each year. Without detailed surveys and accurate taxonomy, it is impossible to evaluate the conservation value of fish or areas and it is very difficult to carry out fish protection. From around 52 European threatened freshwater species, 28% occur in an eastern Adriatic country, making the eastern Adriatic the most important hotspot for threatened freshwater fish (Fig. **2**).

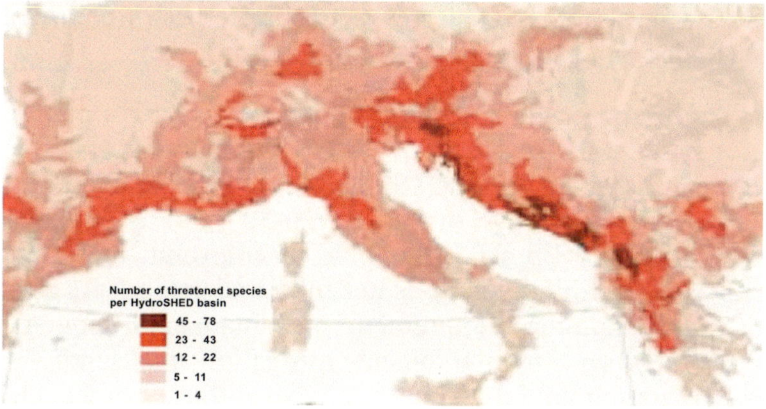

Fig. (2). Most important area of threatened European freshwater fish species (enlarged to eastern Adriatic countries IUCN red list 2011 [3, 4].

On the other hand, a proper appreciation of the value of freshwater fish biodiversity is essential for its preservation. Inland fish and their biodiversity constitute a natural resource that is important not only economically, but also in terms of culture and aesthetics, science and education. Conservation and management should be considered a matter of critical importance to the governments of eastern Adriatic countries. In spite of the goal of using freshwater fish as umbrella species for all freshwater biodiversity, awareness of the need to conserve freshwater fish seems very limited. Since they live in an aquatic habitat and yet are subject to the domination of a species from a different habitat, man, freshwater fish are experiencing declines in biodiversity; if these trends are unchanged, and species losses continue at current rates, the opportunity to conserve many of the remaining species will vanish in a relatively short time. We believe that knowledge of these threats is insufficiently incorporated into the management of water resources and thus requires much wider dissemination.

Problems of Conservation

Freshwater fish in eastern Adriatic countries from Slovenia to Montenegro have many common characteristics. Most importantly, the region has a great number of endemic species as a result of the glacial refugial history of this part of Europe. The region supports a unique freshwater ichthyofauna comprising numerous endemic species with restricted distribution areas [4, 5], and more than half of these are currently considered to be threatened [6].

The lack of knowledge of fish in this region was clearly shown by an investigation that took place in 2011 in one of the countries close to the Adriatic region, where for the first time 6 species of freshwater fish were recorded for this country's list (Radek Sandra, personal communication).

Comparative studies on eastern Adriatic rivers and their fish populations reveal commonalities but also differences in the co-evolution of rivers and societies. Countries in the area have a lower GDP than the European average. They are all post-communist states that have in the last 20-30 years experienced economic transition, in which only lip service is paid to environmental concerns and in which the importance of biodiversity is not fully understood. As a result of rapid development and the statutory requirement to evaluate anthropogenic impacts on freshwater diversity and to implement mitigation measures, a variety of organizations has sprung up, intending to achieve authority in this area. There are private agencies that engage no professional ichthyological support, the coastal Oceanographic Institutes and NGOs, all wanting to be involved. In spite of opinions to the contrary [12], the issue of the conservation of marine fish species is completely different than in fresh water habitats.

Some of the east Adriatic countries have institutes that have been working for many years but mainly deal with populations of economically important fish species (mainly as advisors to fish farmers), while others have completely restructured institutes which, for various reasons, have no experience in freshwater fish biodiversity.

The problem is complicated by a state administration that often blindly multiplies the importance of endangered freshwater species in Europe while neglecting the species that are in this region really endangered. Threatened freshwater fish provide ecosystem services, but their economic invisibility has thus far been a major cause of their undervaluation and mismanagement. The problem is further complicated by a government bureaucracy that is confused and can only copy species from the European Red List of Freshwater Fishes, underestimating some regional almost vanished fish species and emphasising instead those that are not endangered in the region, although they are in the rest of Europe. As a result of all these circumstances, the research of domestic ichthyologists, the taxonomy of species in the area and the proposed mitigation methods are unreliable.

Fig. (3). Illegal structure at a water source, theoretically the home of three endemic species of fish. Buildings of more than 50-meter-deep wells for drinking water can be seen in the background.

Threats to freshwater fish fall into five categories: flow modification; water pollution; destruction or degradation of habitat (Fig. **3**); overexploitation and invasion by exotic species. In the investigated area, freshwater fish protection is complicated by the omnipresent karst that greatly complicates the study of hydrogeomorphology and the general pattern of water flow. Karst formations with subterranean rivers, karstic springs, lakes, and swallow holes are distinct hydrological features of this ecoregion [13]. Prominent karst fields are very long: Livanjsko polje, Glamočko Polje, Grahovsko Polje, Drvarsko Polje, Duvanjsko Polje, Popovo Polje (Fig. **4**), Dabarsko Polje, Nevesinjsko Polje and Gatačko Polje in Bosnia and Herzegovina; Cerknica Polje in Slovenia; Nikšićko Polje in Montenegro; Ličko Polje Velika Kapela, Mala Kapela and Krbava Polje, Imotsko Polje and Cetinsko Polje in Croatia. Polja in Bosnia and Hercegovina, Croatia and some other areas have various endemic species of the genus *Sqalius, Chondrostoma, Knipowitschia, Delminichthys* and *Telestes,* the biology of which is still very poorly known.

Fig. (4). Due to underground water leakage in the karst, 64 km of the Trebisnica riverbed in Popovo Polje was paved.

Unfortunately, groundwater in karst areas is just as easily polluted as surface streams, where underground waters flow in unknown directions and sinkholes have often been used as community trash dumps. Lack of knowledge of underground water circulation systems and worse still damming them to create reservoirs for hydropower plants led to the disappearance of many populations of individual species and probably the disappearance of some species that had not even been discovered. The problem is huge because the knowledge of karst geology is not widespread in the world. In general, the construction of reservoirs in the karst should be prohibited because it destroys many unknown underground water paths (Fig. **4**).

The area including the closed, cascade karstic valleys is a special region in the inland part of the maritime domain that is usually full of endemic fish. It is connected by an underground network of rivers of various lengths and the surface streams of which some usually dry up in summer. It is regrettable that very often the immense flood waters in these valleys have been controlled, which has completely changed the ecological conditions of the waters. Intensive agriculture involving the use of fertilizers and pesticides has also worsened water quality in this region.

It is interesting that all the countries involved most often have two basins: the Adriatic, with a relatively large number of endemic species, often of a relatively small area, and the other, the Black Sea basin where in all the countries together 8-12 endemic species appear.

Some of the larger rivers in the Adriatic ecoregion include the Soča, Mirna,

Zrmanja, Krka, Cetina, Neretva, Trebižat, Trebišnjica and Morača rivers. There are also unknown numbers of karstic subterranean rivers. There are relatively few lakes; they include Vransko Lake, Lake Vrana, Lake Ričice, and the Baćina Lakes, some small lakes in the Neretva delta as well as Scadar Lake. There are also 8 artificial reservoirs, Butoniga, Peruća, Busko jezero, Brljan as well as numerous accumulations. From all of these translocated predatory fish intrude into the Adriatic watershed and threaten numerous endemic freshwater fish species in this area.

Comparative studies on eastern Adriatic rivers and their fish populations reveal commonalities but also differences in the co-evolution of rivers and societies.

Slovenia's freshwater streams flow into two large river basins, the smaller Adriatic and the larger Danube basins. The whole country is inhabited by 84 freshwater fish, of which 67 are indigenous and 12 are endemic. Slovenia has a relatively small number of endemic fish species and none of the endemics live exclusively in Slovenia. Around 37 species and subspecies of fish live in the Adriatic catchment of Slovenia. *Salmo marmoratus*, *Lethenteron zanandreai* and *Squalius janae*, *Chondrostoma soeta* and *Protochondrostoma genei* are threatened. The major threats are habitat destruction, water pollution, and water extraction (Fig. **3**). The introduced *Chondrostoma nasus* is a major current threat and was responsible for the near extirpation of *C. soeta* and *P. genei* from Slovenia 20 years ago. The South European Nase was declared extinct as a result of the introduction of *Chondrostoma nasus,* although later on it was found to occur in other locations. Modern conservation methods have been used in the area of the Soča River where *Salmo marmmoratus* seemed to have disappeared due to the intensive stocking of brown trout (*Salmo trutta*). According to some authors, marble trout was the only trout of the Soča river basin until the introduction of the brown trout in 1906 [14].

In Bosnia & Herzegovina, the former wetland Buško Blato is today the biggest reservoir in Europe (55.8 km^2). This water basin, an extension of Livanjsko polje, is theoretically a source metapopulation for the threatened *Aulopyge huegeli* (Fig. **5**), *Chondrostoma, Phoxinus* and *Squalius tenellus.*

These three species, like most endemics, are highly vulnerable to alien species invasion due to their long evolution in the absence of competitors and predators. Unfortunately, Danube species, mostly predators translocated from the Black Sea basin for angling (pike, European catfish) and small, hardy, accidentally introduced, actually translocated fish species are in competition with the endemics. In the last 10 years, it has become evident that the reservoir is experiencing numerous changes. The older generation of people says that the rare

Aulopyge was previously more abundant. In present days it can be found only at a certain time of year in the supply channels. On the other hand, in the reservoir itself, alien species such as the prussian carp, pumpkinseed, European catfish, pike and stone moroko flourish.

Fig. (5). The threatened *Aulopyge huegeli.*

In general, the problems of freshwater fish conservation in Croatia and Bosnia and Herzegovina are very similar. Their faunas are extremely diverse because of the large quantity of endemic fish species and subspecies that live there. With more than 44 species endemic to the area, we believe that those rivers form the most interesting ichthyo-faunistic region in Europe. Almost every river in this region has a few endemic species. In recent years these have been investigated by Croatian and other scientists [15 - 21]. Although all experts agree that this region is very interesting [22, 23], relatively little research has been done on the biology, ecology, and interspecific relations of the species.

Fig. (6). Dinaric minnow/*Phoxinellus alepidotus,* **IUCN***(EN)* in Croatia data deficient.

From 44 endemic species and subspecies of fish, many are linked exclusively to the Adriatic habitats in Bosnia and Herzegovina and Croatia. Those are 4 species of the family *Delminichthys, 3 of Phoxinellus* (Figs. **6**, **7**), *11 of Telestes, 2 species of Chondrostoma*, 9 species of *Sqalius (S. svallize,* S. *zrmanjae, S. microlepis, S. illiricus, S. janae, S. tennelus, S. miloradi, S.turskyi* and *S. ukliva) Aulopyge hugeli, Knipowitschia croatica, Orsinigobius croaticus*, 3 subspecies of *Salmo obtusirostris* and 4 species of *Cobitis.*

Fig. (7). Dalmatian minnow/*Phoxinellus dalmaticus,* a stenoendemic of the Krka Basin, endemic of the Adriatic Basin IUCN (EN).

Overall, our knowledge about the diversity of eastern Adriatic freshwater fish is still incomplete, as can be seen from the number of new species that have been described in recent years. Well known hotspots of endemism include the poljas and streams of the Dinaric karst in Croatia and Bosnia and Herzegovina. The first detailed lists of Adriatic endemic species were made in the year 1858 [24, 25]. The oldest data on salmonids are given by Munda [26], Karaman and others [27 - 30]. A comprehensive review of the fish fauna vas given by Vukovic [31]. He named 28 endemic taxa for the Adriatic watershed. Some important localities for endemic fish in the Adriatic catchment were given [32, 33]. Endemic taxa are also mentioned in many papers by European authors [34, 35], usually just as short descriptions of species without data on their distribution. Generally, the distribution of rare species is often very poorly described and is not systematic. Individual water bodies are frequently analyzed for different purposes [36, 37]. Very often studies made for the purposes of fisheries were concerned with common taxa. During a scientifically prolific time in Bosnia and Herzegovina, much research was carried out on the chromosomes of endemic species [38 - 40]. Lately, revisions have been made of the genera *Chondrostoma* [41], *Aulopyge* [34], *Petromyzon* [42] and *Salmotymus* [43]. For various reasons, many authors have investigated the growth and condition of different species [44, 45].

Growth characteristics of the genus *Chondrostoma* were described [46]. In recent years attention has been given to endangered species and their protection [47 - 58]. There is very few data on new introductions and translocations [59 - 61] even though alien species can be found in all the rivers of the Adriatic basin. On the ecology of spawning, maximum temperature survival, and general ecology, there has been very little work published. Most of the literature involves the recording of where a species lives, the general community structure and descriptions of new species. We are a long way from being able to define the ecological requirements of certain species and through the given parameters enable their survival.

The rivers of Montenegro also belong to two basins – the Black Sea, in which some 30 fish species have been recorded, and the Adriatic basin with 60 freshwater fish species [62]. Of the thirty-one endemic species, eight are exclusively endemic only to this area. Three species have been described lately (*Pomatoschistus montenegrensis, Knipowitschia montenegrina* and *C. Scodrensis).* Within the region, Lake Skadar is a particularly interesting location. It is actually the largest lake in all these countries, situated in the south-eastern part, in the Zeta-Skadar valley, and in the karstic terrain of the southeast Dinaric Alps in Albania. The Moraca River provides 62 percent of the lake's water with the rest coming from underground springs. A comprehensive list of native and introduced fish of Lake Skadar and its catchment area are given [62]. In Lake Skadar, 34 native fish species, 9 euryhaline and 13 introduced fish species were determined. Seven fish species are endemics. Among the fish fauna of Lake Ohrid and Lake Skadar 14 native identical fish species were determined. The endemic indigenous fishes from Lake Skadar and its catchment area are *Alosa sp.*, *B. zetensis*, *C. Scodrensis*, *G. skadrensis*, *Pomatoschistus montenegrensis* and *S. zetensis*. In 1962 *Chondrostoma scodrense* was classified as extinct by the IUCN Red List of Threatened Species. This species was described from nine specimens caught 100 years ago in Skadar Lake. Since then, in spite of intensive investigations of its only known previous range, no specimens have been recorded [64].

More recent changes that have influenced natural hydrological functioning are pollution, river regulation and the construction of hydroelectric and water supply reservoirs. In Croatia, in the Adriatic region, there are 12 relatively new big hydroelectric plants and more than 30 are in the planning stage (Fig. **8**), which if built will completely change the basic physico-chemical characteristics of the water such as temperature, water level, the quantity of dissolved oxygen, current speed and discharge. In general, reservoirs made for hydropower block the ecological route of fishes and transform the river into a series of lakes. The great hydroelectric potential of Adriatic rivers is being planned and partially exploited without any prior investigation of the fish fauna or environmental impact assessments. In 1990 the subspecies *Salmo* (*Salmothymus*) *oxyrhynchus krkensis*

(Fig. **9**) disappeared when the power plant Krčić of only 1 GWh was built within Topoljski buk waterfall on the River Krka (Croatia). According to the ten endemic rare species found, the River Krka belongs to one of the richest ichthyologic areas in Dalmatia. Fourteen water basins for water use are planned on this river.

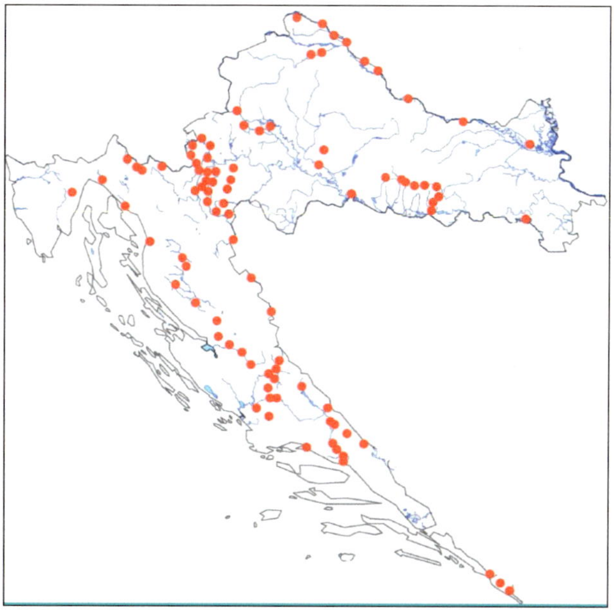

Fig. (8). More than 30 planned locations for hydropower plant construction in Croatia.

Fig. (9). *Salmo (Salmothymus) oxyrhynchus* krkensis.

The various habitats that are particularly specific to the specialized endemic species in this region have been drastically changed in recent years by human activities. Despite the hardiness and resilience of some species, it is questionable how long they will last here if special protective measures are not taken [4].

According to government fisheries policies, up until 1980, fish passes were not required by law on eastern Adriatic rivers. This policy was implemented in all eastern Adriatic countries but today things are changing. Unfortunately, the present fish passes are not miracles. Experience shows that, regardless of fish passes, if several dams are built in a row, the negative cumulative effect of these dams on fish is often catastrophic [4]. Furthermore, there are many threatened species in the Adriatic rivers that depend on intact wetlands along lowland rivers or that are restricted to small rivers or even to a few springs in the Dalmatian karst. Recently, it was discovered that some endemic fish retreat in the fall to underground streams far from their summer habitats (Fig. **10**). For these species, alterations of river ecosystems into artificial lakes will totally degrade and change their habitats and presumably lead to the disappearance of the species.

Fig. (10). A small stream where *Aulopyge huegelii* and *Telestes turskyi* from the river enter underground winter habitats in huge numbers (Courtesy of Tanja Mihinac).

Engineers are for various reasons led by a desire to enrich the waters of the Adriatic rivers at the time of year when water is not abundant. Often, for more hydropower, water is diverted from the Black Sea basin into the Adriatic watershed, without any understanding of how much damage this causes the indigenous fish. An idea currently underway in Montenegro is to create a connection between the Morača drainage area (with many endemic and threatened species) and Skadar Lake in order to increase the supply of water for hydropower. This is likely to change communities of native species of the Morača-Lake Skadar basin. A similar situation is in the Neretva basin in Bosnia & Herzegovina, where the water transferred among basins and water from the Bergava subbasin is redirected to the Trebisnica subbasin through a tunnel in Dabarsko Polje. Several new similar connections are planned, which will lead to the introduction of alien species, threatening the often highly endemic biodiversity. Often bypasses are built on the river, leaving part of the riverbed without water, or sometimes even shortening the river by 10 km (river Krka).

It is certain that the majority of endemic species will not survive changes arising in consequence of the strong development in the region. We believe that cold-water and particularly hypogean species living for most of the year underground will suffer the most [56]. The destruction of this natural heritage simply cannot be justified. These fish have survived many geomorphological changes, and although reduced in numbers they have managed to survive since the Pleistocene. What is more, freshwater fish are only one part of the ecoregion's biodiversity, which also includes many other stenoendemic and relict cave animals.

SOME CHARACTERISTICS OF THE CROATIAN ADRIATIC RIVER SYSTEM

The great abundance of fish species in Croatia (140) is a result of the existence of two large river systems within the country. In comparison with the rivers of the Danube basin, in the Adriatic basin, they are short, cold and isolated. Frequently, rivers emerge at one end of a polje (closed valley) and disappear at the other, sometimes in the dry summer season on the surface they disappear completely. The Adriatic basin with 60 freshwater fish species can be divided into three regions: the Istrian Peninsula, the Croatian maritime region and the karstic region with valleys and underground streams in the inland Adriatic basin.

On the Istrian Peninsula, about 38 freshwater fish with 12 endemic species have occurred. There are two rivers, the Mirna and Rasa with their tributaries Boljnčica, Butoniga, Pazinčica, Raša, Račica, Podmerišće and the water basin Butoniga. The majority of existing data on the freshwater fish fauna of this region date from the 19th and the start of the 20th century [59]. Strong changes and the apparent disappearance of many fish are the consequence of areas such as Cepic polje Lake having been drained.

Dalmatia with 40 endemic freshwater species itself stands out as a special ichthyological subregion of the western Balkan Peninsula. The endemic ichthyofauna includes species of the genera of trout, chubs, rudds, barbelgudgeons, nases, barbels, daces, minnows, loaches, roaches, gobies and more (genera: *Salmo*, *Aulopyge*, *Chondrostoma*, *Telestes*, *Phoxinellus*, *Delminichthys*, *Squalius*, *Rutilus*, *Scardinius*, *Cobitis*, *Barbus and Knipowitschia*).

The fish fauna of the River Krka, Cetina and Neretva drastically changed since the construction of the water reservoirs for hydropower. Many alien species have been introduced. From the Cetina river *Salmo dentex* and *Chondrostoma kneri* have disappeared, and the previously common species *Leuciscus ukliva* has severely decreased in the last 20 years. Of the five main rivers in the maritime region, the longest is the River Neretva (218 km), which runs through Croatia for only 20 km before its delta entrance to the sea. The rest of the river is in Bosnia and Herzegovina. It has around *38* species of fish, but the river has never been thoroughly investigated. The species found here are: *Salmo trutta, Thymallus thymallus, Salmo dentex, Oncorhynchus mykiss, Salmo marmoratus, Salmo obtusirostris oxyrhynchus, Rutilus rubilio, Gambusia affinis, Gasterosteus aculeatus, Cyprinus carpio, Scardinius pottiza, Leuciscus cephalus albus, Leuciscus svallize, Chondrostoma kneri, Alburnus albidus, Gobio gobio, Anguilla anguilla, Phoxinus sp. Rhodeus sereus amarus* and *Alosa fallax nilotica* [62]. On the River Neretva, a series of four hydroelectric power plants were built, and the fifth is located on the tributary Rama. Therefore, in this area, we cannot anymore speak about the pristine lotic ecosystem. The river is full of translocated intruders from the Danubian watershed [53, 63].

The karstic closed valleys area has in recent time been recognised as a very special region in the inland part of maritime Croatia. It is connected by an underground network of rivers of various lengths, the surface streams which usually dry out in summer, with the streams continuing to exist underground. These valleys include Velika Kapela, Mala Kapela, Krbavsko Polje, Licko Polje, Sinjsko Polje, Imotsko Polje, Cetinsko Polje, and Hrvatacko Polje, while the Koljansko and Ribaricko polje were flooded when the biggest Croatian reservoir Peruca was built. Various endemic species occur in this region. It is regrettable that the flood waters in these valleys have been controlled, completely changing the hydrological conditions. Also, extensive agriculture has also worsened water quality in this region.

CONCLUSION

It would be wise to list all habitats with rare and interesting species from all the eastern Adriatic countries. Boundaries are big problems as each side of the border has a completely different idea of conservation. Some of the endemically important rivers (in Croatia the River Krka, a National Park) ought to be left in their pristine natural conditions, but this is an almost impossible task. Small streams, sources, and wells in the inland karstic region and their fish fauna should be evaluated and, if necessary, protected. We are sure that a lot of interesting endemic species remain to be discovered in these areas.

Among the most endangered species of the Adriatic catchment in Croatia and Bosnia and Herzegovina are all the species of the genera *Delminichtis Phoxinellus Orsinigobius Knipowitschia* and *Salmo* (*Salmothymus*), plus the species *Telestes polylepis, T. carsticus, P. pseudalepidotus* and *Aulopyge huegelii.* Action plans for some species have been given to the government by ichthyologists but have very rarely been implemented. It is important to describe and protect Freshwater Key Biodiversity Areas in the entire area of the eco-region of Dalmatia 419. If politically possible, these areas should be integrated into the Natura 2000 network (4). Whenever necessary, development and implementation of species recovery plans should be made for all threatened freshwater fish species. In our opinion, the government administration, in its eager development of action plans for charismatic flag terrestrial species (bear, wolf, lynx, otter and others) is committing an injustice, for they are quite safe in comparison to some invisible freshwater fish.

The Following Measures are Proposed

Assessment and regulation of the extraction of waters in areas with threatened freshwater fish species and improvement of the enforcement of existing legislation on water abstraction.

Stringent prohibition of the creation of water reservoirs in zones with many endemic species. Such artificial bodies of water are dissemination zones for alien species and the introduction by fishers of fish species into lakes and river basins where they are not native.

Reduction of the loss of lotic river habitats to ensure the future of freshwater fish in the eastern Adriatic region of Europe. This is especially important where the construction of new dams is planned.

Imposition of stricter requirements for environmental impact assessment for river perturbation by the construction of canals, dams and hydropower plants. The assessments should be totally objective and take an account of the often very seriously neglected threats that such infrastructure is posing to freshwater fish biodiversity.

Support for the development and maintenance of *ex-situ* facilities, to halt the extinction of the most threatened species.

Conduct of research on threatened fish, monitoring of threatened fishes regularly by a competent ichthyological body and action where necessary if new data become available.

CONFLICT OF INTEREST

The author (editor) declares no conflict of interest, financial or otherwise.

ACKNOWLEDGEMENT

Declared none.

REFERENCES

[1] Holmlund CM, Hammer M. Ecosystem services generated by fish populations. Ecol Econ 1999; 29: 253-68. [http://dx.doi.org/10.1016/S0921-8009(99)00015-4]

[2] Olden JD, Kennard MJ, Leprieur F, Tedesco PA, Winemiller KO, García□Berthou E. Conservation biogeography of freshwater fishes: recent progress and future challenges. Divers Distrib 2010; 16: 496-513. [http://dx.doi.org/10.1111/j.1472-4642.2010.00655.x]

[3] Freyhof J, Brooks E. European Red List of Freshwater Fishes. Luxembourg: Publications Office of the European Union 2011.

[4] Freyhof J. Threatened freshwater fish and mollusc of the Balkan, Potential impact of hydropower project. Unpublished report ECA Watch Austria & EuroNatur 2012; 81.

[5] Povž M. Andre Gregory i Mary GregoryJ 2016, Slatkovodne ribe in piškuriji v Slovenija. 2011; p. 61.

[6] Smith KG, Darwall WR. (Compilers) The Status and Distribution of Freshwater Fish Endemic to the Mediterranean Basin Gland (Switzerland) and Cambridge. UK: IUCN 2006.

[7] Abell R, Thieme ML, Revenga C, *et al.* Freshwater Ecoregions of the World: A New Map of Biogeographic Units for Freshwater Biodiversity Conservation. Bioscience 2008; 58: 403-14. [http://dx.doi.org/10.1641/B580507]

[8] Bianco PG. Potential role of the palaeohistory of the Mediterranean and Paratethys basins of the early dispersal of Euro-Mediterranean freshwater fishes. Ichthyol Explor Freshwat 1990; 167-84.

[9] Blanc M, Banarescu P, Gaudet J-L, Hureau J-C. European inland water fish A multilingual catalogue. London: Fishing News 1977; pp. 22-131.

[10] Maitland PS. Hamlyn guide to freshwater fishes. London: Hamlyn 1976; pp. 1-256.

[11] Kottelat M. Systematics, species concepts and the conservation of freshwater fish diversity in Europe. Ital J Zool (Modena) 1998; 65(S1): 65-72. [http://dx.doi.org/10.1080/11250009809386798]

[12] Mc Reid G, Contreras MB, Csatádi K. Fish conservation in freshwater and marine realms: status, threats and management. Biol Conserv 2016; 72: 281-95.

[13] Tvrtković N, Veen P, Eds. The Dinaric Alps: Rare habitats and species Conservation Project in Croatia Part A – Hrvatski prirodoslovni Muzej and Royal Dutch Society for Nature Conservation. KNNV 2006.

[14] Crivelli A, Poizat G, Berrebi P, Jesensek D, Rubin JF. Conservation Biology Aplied to fish: The example of a project for rehabilitating the marble trout (*Salmo marmoratus* in Slovenia). Cybium 2000; 24: 211-30.

[15] Bless R, Riehl R. Biology and egg morphology of the Dalmatian barbell gudgeon *Aulopyge huegeli*, an endangered endemic species in Croatia. Environ Biol Fishes 2002; 63: 451-6. [http://dx.doi.org/10.1023/A:1014992626062]

[16] Jelić D, Špelić I, Žutinić P. Introduced species community over-dominated endemic Ichthyofauna of high Lika plateau (Central Croatia) over a 100 year period. Acta Zool Acad Sci Hung 2016; 62: 191-216.
[http://dx.doi.org/10.17109/AZH.62.2.191.2016]

[17] Buj I, Vukić J, Šanda R, *et al.* Morphological comparison of bleaks (Alburnus, Cyprinidae) from the Adriatic Basin with the description of a new species. Folia Zool (Brno) 2009; 59: 129-41.

[18] Bogutskaya NG, Zupančič P, Bogut I, Naseka AM. Two new freshwater fish species of the genus *Telestes* (Actinopterygii, Cyprinidae) from karst poljes in Eastern Herzegovina and Dubrovnik littoral (Bosnia and Herzegovina and Croatia). ZooKeys 2012; 180(180): 53-80. [Top of FormBottom of Form].
[http://dx.doi.org/10.3897/zookeys.180.2127] [PMID: 22539906]

[19] Popović J. Idiobiology and taxonomy of *Leuciscus iliricus*. Sarajevo University 1989. PhD thesis.

[20] Freyhof J, Lieckfeldt D, Bogutskaya NG, Pitra C, Ludwig A. Phylogenetic position of the Dalmatian genus *Phoxinellus* and description of the newly proposed genus *Delminichthys* (Teleostei: Cyprinidae). Mol Phylogenet Evol 2006; 38(2): 416-25.
[http://dx.doi.org/10.1016/j.ympev.2005.07.024] [PMID: 16214378]

[21] Ahnelt H, Tutman P, Dulčić J, Glamuzina B. A new species of freshwater goby of the genus Knipowitschia (Teleostei: Gobiidae) from Hutovo Blato wetland (Neretva River basin), Bosnia and Herzegovina. 13th European congress of ichthyology, Book of Abstracts, Klaipeda, Lithuania, 6–11 September 2009.

[22] IL. Holcik J. The freshwater fishes of Europe. General introduction to fishes. Acipenseriformes.Wiesbaden: Aula-Verlag 1989; I: pp. 18-147.

[23] Šanda R, Kovačić M. Freshwater gobies in the Adriatic drainage basin of the Western Balkans. Ann Ser Hist Nat 2009; 19: 1-10.

[24] Heckel J, Kner R. Die Susswasserfische der Ostreichischen Monarchie. Leipzig: Verlag von Wilhelm Engelmann 1858.

[25] Steindachner F. Die Flussfische des südlichen Spaniens und Portugals. Anzeiger der Kaiderlichen Akademie der Wissenschaften, Mathematisch-Naturwissenschaftlichen Classe 1866; p. 3.

[26] Munda A. Ribe v slovenskih vodah [The fishes in the waters in Slovenia]. Ljubljana: Slovensko ribarsko društvo 1926; p. 65.

[27] Karaman S. Salmonidi Balkana. Almanah Vode i ribe Jugoslavije. Zagreb. 1937.

[28] Curcic V. Neretva i njezine pastrve (Salmonidae). Sarajevo: Štamparija Risto V. Savić 1938; pp. 1-89.

[29] Taler Z. Rasprostranjenje i popis slatkovodnih riba Jugoslavije. GI Pir Muz Srp Zem Ser B 1954; 5–6: 425-55.

[30] Sommani E. Il *Salmo marmoratus* Cuv.: sua origine e distribuzione nell Italia setentrionale. Boll Pesca Piscic Idrobiol 1960; 15: 40-7.

[31] Vuković T. Ribe Bosne i Hercegovine. Sarajevo: Svjetlost 1977.

[32] Zanella D, Mihaljević Z, Mrakovčić M, Ćaleta M. Ecology and diet of the endemic *Telestes ukliva* (Cyprinidae) in the Cetina River system, Croatia. Cybium 2009; 33: 97-105.

[33] Zupančič P, Bogutskaya NG. Description of a new species, *Phoxinellus dalmaticus* (Cyprinidae: Leuciscinae), from the Cikola river in the Krka river system, Adriatic basin (Croatia). Natura Croat 2000; 9: 67-81.

[34] Howes GJ. The phylogenetic position of the Yugoslavian cyprinid fish genus *Aulopyge* Heckel 1841, with appraisal of the genus *Barbus* Cuvier & Cloquet, 1816 and the subfamily Cyprinidae. Bull Brit Mus (Nat Hist). Zool 1987; 52: 165-96.

[35] Perea S, Böhme M, Zupančič P, *et al.* Phylogenetic relationships and biogeographical patterns in Circum-Mediterranean subfamily Leuciscinae (Teleostei, Cyprinidae) inferred from both mitochondrial and nuclear data. BMC Evol Biol 2010; 10: 265. [http://dx.doi.org/10.1186/1471-2148-10-265] [PMID: 20807419]

[36] Habeković D, Pažur K, Popović J. Ihtiofauna nekih ličkih tekućica. Ribarstvo 1992; 47: 23-33.

[37] Mrakovčić M, Schneider D, Kerovec M, Mustafić P, Bukvić I. Change of native fish community after closure of the water reservoir on river Butoniga (Istria, Croatia).Freshwater Fish Conservation: Options for the Future. Lisboa: Faculdade de Ciencias de Lisboa 2000; p. 67.

[38] Berberović Lj, Hadžiselimović R, Pavlović B, Sofradžija A. Sistematski polozaj vrste Aulopyge hugeli Heckel 1841 u svjetlu osnovnih podataka o njenoj hromo¬somskoj garnituri I Jugoslovenski simpozijum iz sistem¬atike, Sarajevo. 1971; p. 31.

[39] Berberović Lj, Hadžiselimović R, Pavlović B, Sofradžija A. Uporedni pregled osnovnih podataka hromosomskih garnitura vrsta *Chondrostoma phoxinus* Heckel i *Chondrostoma kneri* Heckel. Ichthyologia 1971; 2: 25-30.

[40] Tsigenopoulos CS, Berrebi P. Molecular phylogeny of north mediterranean freshwater barbs (genus Barbus: cyprinidae) inferred from cytochrome b sequences: biogeographic and systematic implications. Mol Phylogenet Evol 2000; 14(2): 165-79. [http://dx.doi.org/10.1006/mpev.1999.0702] [PMID: 10679153]

[41] Elvira B. Taxonomic revision of the genus *Chondrostoma* Agassiz, *1835* (Pisces, Cyprinidae). Cybium 1987; 11: 111-40.

[42] Šorić V. New data on the taxonomy and distribution of lampreys (Petromyzontidae) in southern Europe: revision of data on the lampreys in the waters of Serbia and Montenegro. Ichthyologia 2002; 34: 23-34.

[43] Snoj A, Melkič E, Sušnik S, Muhamedagić S, Dovč P. DNA phylogeny supports revised classification of *Salmothymus obtusirostris*. Biol J Linn Soc Lond 2002; 77: 399-411. [http://dx.doi.org/10.1046/j.1095-8312.2002.00130.x]

[44] Treer T, Šprem N, Torcu-Koc H, Sun Y, Piria M. Length–weight relationships of freshwater fishes of Croatia. J Appl Ichthyology 2008; 24: 626-8. [http://dx.doi.org/10.1111/j.1439-0426.2008.01084.x]

[45] Treer T, Matulić D, Bogdanović G, *et al.* The condition of allochtonous fishes in the Mediterranean Vransko Lake. J Appl Ichthyology 2010; 27: 965-7. [http://dx.doi.org/10.1111/j.1439-0426.2010.01610.x]

[46] Mikavica D, Kosorić Đ. Uzrasna i polna struktura, dužinsko i težinsko rastenje podbile (*Chondrostoma phoxinus* Heck., 1843) iz Buškog jezera. God Biol In Un 1986; 39: 71-80.

[47] Kosorić Đ. Neki podaci o rasprostranjenosti i migracijama neretvanske glavatice (Salmo marmoratus Cuv.) u slivu Neretve. Glasnik Zemaljskog muzeja Bosne i Hercegovine u Sarajevu. 1969; 8: 79-82.

[48] Zupančič P. Rijetke i ugrožene ribe jadranskog slijeva Hrvatske, Slovenije i Bosne i Hercegovine [Rare and endangered freshwater fishes of Croatia, Slovenia and Bosnia and Herzegovina - Adriatic basin]. Dolsko: Atlas u boji. AZV d.o.o. 2008; p. 79.

[49] Ćaleta M, Mrakovčić M, Buj I, Mustafić P, Zanella D, Marčić Z. Threatened fishes of the world: Chondrostoma phoxinus Heckel, 1843 (Cyprinidae). Environ Biol Fishes 2009; 84: 229-30. [http://dx.doi.org/10.1007/s10641-008-9413-z]

[50] Mrakovcic M, Misetic S. Status, distribution and conservation of the salmonid, *Salmothymus obtusirostris* (Heckel) and the cyprinid, *Aulopyge hugeli* (Heckel) in Yugoslavia. J Fish Biol 1990; 37: 241-2. [http://dx.doi.org/10.1111/j.1095-8649.1990.tb05052.x]

[51] Mrakovčić M, Mišetić S, Povž M. Status of freshwater fish in Croatian Adriatic river systems. Biol

Conserv 1995; 72: 179-85.
[http://dx.doi.org/10.1016/0006-3207(94)00080-A]

[52] Leiner S, Popovic J. Rod *Leuciscus* (Cyprinidae, Pisces) u vodama jadranskog sliva s osvrtom na nalaz *L. svallize* (Heckel, Kner 1858) i *L. souffia* Risso 1826 u Cetini. Ichthyologia 1984; 16: 111-20.

[53] Škrijelj R. Populacija riba Neretvanskih Jezera Ihtiološka monografija. Prirodno - Matematički Fakultet, Universitet u Sarajevu 2002.

[54] Cowx G, Ed. Freshwater Fish Conservation: options for the Future Fishing News Book. 55-64.

[55] Povž M, Leiner S, Mrakovčić M, Popović J. Rare and endangered fishes from Yugoslavian Adriatic rivers. J Fish Biol 1990; 37: 247-9.
[http://dx.doi.org/10.1111/j.1095-8649.1990.tb05055.x]

[56] Ćaleta M, Buj I, Mrakovčić M, *et al.* Hrvatske endemske ribe. Zagreb: Agencija za zaštitu okoliša. 2015.

[57] Bianco PG, Knežević B. The *Leuciscus cephalus* complex (Pisces, Cyprinidae) in the western Balcanic area. In: Kullander SO, Fernholm B, Eds. Congress of European Ichthyologists. Stockholm. 1987; pp. 49-55.

[58] Mustafić P, Marčić Z, Duplić A, *et al.* A new loach species of the genus Cobitis in Croatia. Folia Zool (Brno) 2008; 57: 4-9.

[59] Leiner S, Povž M, Mrakovčić M. Freshwater fish in Istrian Peninsula. Anales. 7:215-223.Kosorić D, Vuković T, Kapetanović N, Guzina N. Ihtiofauna rijeke Neretve i njene promjene posljedicno izgradnji hidroelektrana. Zbornik radova - Savjetovanje o ribarstvu na hidroakumulacijama. Mostar 1995; 1989: 133-8.

[60] Treer T, Matulić D, Bogdanović G, *et al.* The condition of allochtonous fishes in the Mediterranean Vransko Lake. J Appl Ichthyology 2010; 27: 965-7.
[http://dx.doi.org/10.1111/j.1439-0426.2010.01610.x]

[61] Kosorić D, Vuković T, Kapetanović N, Guzina N. Ihtiofauna rijeke Neretve i njene promjene posljedicno izgradnji hidroelektrana Zbornik radova - Savjetovanje o ribarstvu na hidroakumulacijama. Mostar 1989; pp. 133-8.

[62] Talevski T, Milosevic D, Maric D, Petrovic D, Talevska M, Talevska A. Biodiversity of ichthyofauna from Lake Prespa, Lake Ohrid and Lake Skadar. Biotechnol Biotechnol Equip 2009; 23: 400-4.
[http://dx.doi.org/10.1080/13102818.2009.10818449]

[63] Brigic A, Mrakovčić M, Mustafić P, Buj I, Ćaleta M, Zanella D. Freshwater Ichthyofauna Of The Neretva River Delta And Its Tributaries Estonian Marine Institute Report Series No 12 Toomas Saat (ur). Tallinn, Estonia: Estonian Marine Institute, University of Tartu 2004; p. 146.

[64] Mrdak D, Nikolić V, Joksimović V, Simonović P. Data collecting for sustainable fishery program of Skadar Lake. Podgorica, Belgrade 2001.

CHAPTER 3

Challenges Facing Marine Aquaculture in the EU-Mediterranean

Ivan Katavic*

Institute of oceanography and Fisheries, Split, Croatia

Abstract: In addressing the issue of Mediterranean aquaculture, and having the fact that the aquaculture sector is partly associated with the relevant European Common Fisheries Policy, the study provide both global and European circumstances relevant for current state, growth target and future challenges. Marine aquaculture, in the Medterranean has shown a rapid growth in 1990s, and by the time became an important source of seafood, employment and income. However, it has not progressed significantly over the last decade, due to several factors that influence performance of the main fish species production. Replacement with ingredients from plant origin has influenced FCR, presence of mycotoxins in feed and micro ingredient deficiency Many breeding, health and welfare issues still remain unresolved. Lack of spatial planning on the integrated principles, access to quality fish feed, eggs and fry, has restricted aquaculture in some developing countries. Complicated administrative procedures related to national and regional environmental regulations were constraining aquaculture development in many EU-Mediterrenan states. Fragmented approach to marketing of mariculture products requires a cooperation of different actors across the Mediterranean. Vision for 2030 has projected doubling finfish production. However, for such a development a set of practical solutions are needed in production, governance, cooperation, policy management, market development, and spatial planning. If aquaculture tends to meet its goals, it must improve production models and environmental performances. It is expected that future production economy may take advantage of diversifying species and combining land based recirculating aquaculture systems (RASs) with more open sea farming by employment of more robust cage installations. Integration with non-fed organisms is a promising option in achieving better economic and environmental benefits.

Keywords: Challenges, EU-Mediterranean context, Fish nutrition, Future direction, Global aquaculture, Health management, Public concerns, Spatial planning, Standards.

* **Corresponding author Ivan Katavic:** Institute of Oceanography and Fisheries, Split, Croatia; Tel: +385 2140844; E-mail: katavic@izor.hr

Panagiotis Berillis (Ed)

INTRODUCTION

The UN publishes its population projections every two years, demonstrating that over the past decades the population has grown more quickly than expected. The total population of 7.3 billion has increased 7 fold in only 200 years. United Nations Department of Economic and Social Affairs in its 2015 Revision report [1], (in an average variant projected 8.5 billion inhabitants by 2030, 9.7 billion by the middle of this century), and then is expected to level off at around 11.2 billion by the end of it.

Such a world population growth over the past two centuries is largely the result of advances in modern food biotechnology, medicines, and improvements in living standards. Global population has a relatively large proportion of older persons, and it will certainly have global implications for sustainable development [2]. The world food system is expected to experience high pressures, creating demand for new food biotechnology to produce more varied, high-quality, nutritious, and safe diet. More food will be produced through the implementation of existing knowledge and biotechnology, and by investment into a wide range of sciences and innovation [2]. In meeting the challenges of the global food security and safety critical contributions are required from the scientific community [3]. Being the fastest-growing animal production sector when measured by annual percentage growth rate, aquaculture is expected to continue its contribution to food security and poverty reduction [4].

Faced with the world challenges, of critical importance for people and the planet is to include food security for the growing human population. In September 2015 the UN Member States adopted the 2030 Agenda for Sustainable Development. The 2030 Agenda among others sets aims for the contribution of fisheries and aquaculture towards food security and nutrition in the use of natural resources as to ensure sustainability in economic, social and environmental development [5].

Case studies of EU-Mediterranean aquaculture are considered in a context of world fish supply and demand at the globalized seafood market. The chapter then described state of art of EU fisheries and aquaculture sectors by identifying barriers that have prevented past improvement in productivity and environmental performance in EU aquaculture including Mediterranean member states. Finally, the paper distils key factors relevant to future EU-Mediterranean aquaculture development and offers recommendations for how to address the barriers and ensure a projected future of this seafood industry.

Global Fish Supply and Demand Indicators

Global food systems must supply enough calories and protein for a growing human population and provide important micronutrients. Through centuries, captured fisheries have important roles in meeting ever-rising demand for fish and other products such as fishmeal and fish oil and enhancing incomes for fishermen and quality seafood for consumers [6]. World fish supply has been growing faster than- the world population growth over the past six decades, with an annual rate of 3.2 percent compared with a 1.6 percent increase in population growth [7]. In fact, population growth doubled while fish supply *per capita* tripled, from 6 kg/ year in 1950, 9.9 kg in the 1960s to 14.4 kg in the 1990s and 19.7 kg in 2013, with preliminary estimates for 2014 and 2015 pointing towards further growth beyond 20 kg [8]. Estimated global fish consumption is expected to increase to 22 kg/capita/year in 2024, with increases in all regions [9, 10].

The world's marine fisheries expanded continuously to a production peak of 86.4 million tonnes (mt) in 1996 but have since exhibited a general declining trend [4]. In total current catch of 93.4 mt, marine waters contributing 81.5 mt, and 11.9 mt is obtained from inland waters. Most of the stocks are fished within biologically sustainable levels (68.5 percent), and just short-lived pelagic stocks have certain potential for increases in production [8].

It is not protein that provides nutritional needs of growing human populations. The benefits of fish and fish-like products are associated in part with high concentrations of unsaturated long-chain omega-3 fatty acids, vitamins and bioavailable minerals such as iron, zinc, all essential to nutrition and human health [11, 12].

Based on the data of dietary nutrition, and fish catch, it is predicted that more than 10% of the global population could face micronutrient and fatty acid deficiencies driven by fish declines over the coming decades, especially in the developing nations [13]. It is has been proven that fish consumption reduces mortality risk from heart disease by 36% [14]. Deficiencies of vitamin A and zinc increase the risks of child deaths; deficiencies of iodine and iron can contribute to children not reaching their developmental potential. Combined effect may result in the risks of perinatal and maternal mortality, growth retardation, and reduced immune function [15].

While the fisheries sector is contributing directly and indirectly to the global food security and nutrition, locally, it is an important source of income and livelihood for millions of people around the world [16, 17]. At present, global fisheries provide more than 4.5 billion consumers with quality seafood [18]. According to the latest FAO [8] estimates more than 75 percent of fish consumption occurs in developing countries. Fisheries sector support the livelihoods of 12 percent of the world's population and has already accounted for 17% of the global population's intake of animal protein. More than 90 percent of all employees in seafood sectors live in developing countries.

What is critical for the future of capture fisheries is an estimate of 31.5 percent of fish stocks being classified as overfished [8].This may cause negative ecological consequences, losses in biodiversity, and reduced fish production that all leads to negative social and economic impacts. The situation in capture fisheries, where many fish stocks are fully or over-exploited require temporarily reducing harvest to allow the stock to recover. Some progress has been made in reducing fishing pressure and restoring overfished stocks through effective management actions in some areas [4].

Could Global Increases in Fish Farming Meet the Nutritional Shortfalls?

In spite of the decline in global capture fisheries and doubled human population within a single generation, global fish supply is progressing thanks to rapidly growing aquaculture which is one of the world's fastest-growing animal food producing sectors [4, 19]. The World Bank described aquaculture as an expanding sector which will continue providing high quality and nutritious seafood with an estimate of a 30 million tons increase by 2030 [20].

According to the State of World Fisheries and Aquaculture (SOFIA) farming now provides every second fish for human consumption. SOFIA has been published every 2 years, aiming to provide a comprehensive and global view of the seafood industry including associated policy issues. Summarizing the World review of fish supply from SOFIA 2016 some key facts relevant for aquaculture are highlighted [21].

World Aquaculture Production by Species Groups and by Volume

- 580 species are reared in global aquaculture. These species include 362 finfishes (including hybrids), 104 mollusks, 62 crustaceans, 6 frogs and reptiles, 9 aquatic invertebrates, and 37 aquatic plants.
- Global production of aquatic animals amounted to 73.8 mt with an estimated value of US$160.2 bn. This consisted of 49.8 mt of finfish, 16.1 mt of molluscs, 6.9 mt of crustaceans, and 7.3 mt of other aquatic animals.
- Combined production of aquatic animals and plants reached 101 mt in 2014, with an estimated value of US$165.8 bn. From 2005–2014, global aquaculture production grew at 5.8% annually, compared to from 7.2% achieved in the previous decade (1995–2004). In 2014, 63.8% (47.1 mt) of fish farmed for human consumption came from inland aquaculture; 36.2% (26.7 mt) came from marine and coastal aquaculture.
- Aquaculture supply of fish for human consumption surpassed that of wild-

caught fish for the first time in 2014. Farmed aquatic animals constitute around ¾'s of the total aquaculture production by volume, and farmed aquatic plants 1/4. By volume, 50 percent of world aquaculture production was produced without feeding; non-feed aquaculture products include seaweeds and microalgae (27%) and filter-feeding animal species (22.5%). The most important non-fed aquatic animal species are two finfish species, silver carp and bighead carp produced in inland freshwater. Shellfish, mussels, oysters, clams, *etc.* and other filter feeding animals are from marine and coastal waters.

Aquaculture Production by Region/Countries

- Aquaculture is globally a highly concentrated seafood industry. In 2014, 25 countries produced 96.3% of farmed fish and 99.3% of farmed aquatic plants in the world. The top 5 countries produced 82.8% of the world's total aquaculture production.
- In the past two decades, Asia has accounted for about 89% of world aquaculture production of fish for human consumption. China remains the major producer with a share of 62% in the world fish supply from aquaculture.

Aquaculture by Production Systems

Inland finfish aquaculture is the most common aquaculture operation in the World. Earthen pond culture of finfish is by far the largest contributor from aquaculture to food security and nutrition in the developing world. Improvements in management practices and production technology are leading to decreases in water pollution from aquaculture [22].

Sustainability in Feed Ingredients

A declining proportion of world fisheries production is processed into fishmeal (FM) and fish oil (FO). Owing to high prices of marine sourced fish feed ingredients, a growing share of fishmeal is being produced from fish by-products, which previously were often discarded. An intensive research is focused on finding replacements for FM and FO in aquaculture feeds, saving these strategic feed ingredients for the hatchery, brood stock and finishing diets.

International Trading of Aquaculture Products

Expansion in aquaculture production has contributed significantly to increased consumption and international fish trade. In spite of the recent improvement in trade classifications, international trade statistics do not distinguish between wild and farmed origin of the products. Between 20-25% of relatively high-value aquaculture products is traded by quantities on the international markets, but 1/3 in value terms. Because of declined capture fisheries, the fish feed is widely regarded as a major constraint to projected aquaculture growth. Measured in value terms, salmonids are now the largest single commodity (16.6%), the share of world trade for shrimps and prawns are 15.3%, and 3% for bivalves.

Outlook

It is expected that future growth in fish production and related fish consumption will mainly originate from aquaculture. Based on observed regional trends in seafood production and consumption, the multi-market model projects a predicted fish supply increase for an additional 32 million tonnes in 2030, with aquaculture entirely responsible for the increase [23]. To meet projected demand in 2050 aquaculture production will need to double by mid-century to roughly 140 mt [24].

MATERIAL AND METHODS

This review examines available published materials dealing with the current state of global fish food supply in the context of capture fisheries and aquaculture, and current status and future development of the EU aquaculture sector focusing on the Mediterranean environment. The first source of the statistical data has been taken from the recent Food and Agriculture Organization of the United Nations (FAO) statistics. Additional data sources include the European Market Observatory for Fisheries and Aquaculture Products (EUMOFA) and the European Commission Scientific, Technical and Economic Committee for Fisheries (STECF) that has been providing among others an overview of the latest available information on the structure, social, economic and competitive performance of the aquaculture sectors at national and EU level (https://stecf.jrc. ec.europa.eu/web/stecf/ewg1410). Otherwise, data on the economic performance of the European aquaculture sector is collected under the Data Collection Framework (DCF).

Further information sources were fishery statistics collected by Eurostat for the members of the European Economic Area (EEA) and Turkey. In the fisheries context this means that EEA apply to the EU-28, plus Iceland and Norway Since 2014, Eurostat publishes the annual Statistical book on agriculture, forestry and fishery statistics. Some useful aquaculture figures were also found in the annually updated information on aquaculture that is provided by the EEA Member States and published in the Eurostat Yearbook Statistics Explained (http://ec.europa.eu/ eurostat/statistics-explained/index.php/ Fishery_statistics_in_detail). A useful set of aquaculture data re-organized by species/country/culture system type were found in the European Aquaculture Technology and Innovation Platform (EATiP) and Federation of European Aquaculture (FEAP) reports. An evaluation report from the Commission to the European Parliament and the Council was examined.

The last report was published in 2015. The European Commission recently published a Summary of the 27 Multiannual National Aquaculture Plans analyzing the main objectives and challenges of the aquaculture sector identified by 27 Member States (Luxembourg not included) in their plans within the overall context of the CFP [25].

In addition to the sources mentioned above, complementary sources of information such as the national reports to the GFCM- CAQ, Workshops, questionnaires and other published information in the scientific journals were used and cited as to provide the most updated figures relevant to the aquaculture sector in the Mediterranean.

Methods and Techniques in Mediterranean Aquaculture

Nowadays, the Mediterranean aquaculture industry consists of various segments, depending on the species produced. There is a long-established and a rather traditional culture of shellfish (mussels and oysters), the freshwater pond culture of carp and rainbow trout in raceways. Modern marine water finfish production was made possible only after technical difficulties in reproduction and fry rearing were solved. On growing phase was moved to the coastal areas following already developed farming technology of salmon and miscellaneous species in Norway and Japan.

Aquaculture in the Mediterranean has been traditionally practicing a variety of methods and breeding techniques in shellfish (mussels and oysters). The EC website on maritime affairs and fisheries recognized a variety of rearing systems and techniques (https://ec.europa.eu/fisheries/sitemapen) including extensive or intensive systems, in natural settings or tanks, in fresh water or sea water, in flow-through or recirculation systems, traditional or modern, classic or organic, sheltered or exposed, and so on. A technology-based classification primarily considers the nature of the containment system, and a primary distinction is made between land-based and water-based systems.

Land-based aquaculture may include as follows:

i. fresh water (FW) and marine water (MW) pond aquaculture;
ii. flow-through in tanks and raceways;
iii. land based recirculated aquaculture system (RAS), both in FW and MW for hatchery, nursery, and ongrowing, respectively.

Water-based systems are as follows:

i. small cage system in FW and inshore MW;

ii. large cage system in open sea environment by employing offshore technology;
iii. marine bottom culture for non-fed animals and plants;
iv. suspended culture from floating rafts or longlines for shellfish and plant culture.

Furthermore, food production through aquaculture may be classified according to various primary characteristics including:

- the intensity of the system (extensive, semi-intensive, intensive);
- feeding behavior of cultured species (carnivorous, herbivorous, omnivorous, filter-feeders);
- the matrix of the environment (salinity - marine, freshwater, brackish water, or temperature -warm water, temperate, cold water);
- species or species group (monoculture, polyculture, integrated multitrophic aquaculture – IMTA).

Technical Classification of Aquaculture System

Following FAO aquaculture production statistics [4] there are 11 most dominant ongrowing technologies in EU aquaculture, including the hatchery [26]. These characterizations of aquaculture systems were based on the miscellaneous zootechnical characteristics including environmental conditions, intensity and the type of culturing system, its design and implementation [27]. Characterisation of prevailing aquaculture systems in the Mediterranean has adjusted accordingly.

Extensive and semi-intensive freshwater aquaculture in ponds maintained in such a way as to promote the development of aquatic fauna at a yield greater than that found in the natural ecosystem. Density is low and fish feed naturally. Certain producers may provide additional nutrients and/or additional feed to increase energy flow within the trophic food chain. These ponds play an important and positive role in the landscape, water management, and biodiversity. Common carp, in mixed farming with other species (pike, catfish, *etc.*) is widely introduced in freshwater ponds. It is the third most frequently farmed species worldwide [28]. In some European countries the majority of freshwater fish production (*i.e.* Poland) comes from common carp reared in ponds [29].

Intensive freshwater aquaculture in tanks/raceways where fishes are farmed until they reach marketable size. There are two main techniques: open flow (river water enters tanks upstream and leaves downstream) and a recirculating one (the water remains in a closed circuit and is recycled and 'recirculated' in the tanks). Recirculation systems (RASs) are more costly (energy), but offer better control of breeding conditions (temperature and oxygen) and water quality (rainbow trout, eel, catfish, sturgeon, tilapia, *etc.*) [30].

Extensive brackish water aquaculture in lagoons like coastal environment in which aquatic organisms are often brought by the marine flow. These organisms are kept in lagoons that are developed for this purpose (*i.e.* Italian valliculture). The semi-intensive nature of this production is reinforced by introducing hatchery fry and providing additional feed. This type of aquaculture plays an important role in the conservation of the natural coastal heritage [31].

Marine fish in floating net cage is a kind of intensively designed production for both FW in the rivers or lakes and the MW, neither nearshore nor offshore. The fish are kept in cages anchored to the seabed and maintained on the surface by means of a floating plastic framework. This form of breeding is practiced mainly in sheltered zones near shore, but more sophisticated mooring techniques with automatic feeding and remote monitoring or submersible cages *etc.* may make it possible to move further to the open sea – offshore. The economic interest of offshore mariculture is today primarily related to finfish [32, 33], but from a technological point of view, seaweed and mollusks production have good opportunities for offshore farming.

Marine fish in shore-based installations (particularly flatfishes) can also be bred in artificial shore-based tanks supplied with seawater. Recirculation of the water creates a closed and controlled environment (RAS) that is necessary for optimal production in hatcheries and nurseries for marine species (turbot, common sole, Senegalese sole, sea perch, gilt-head sea bream, European sea bass…). Ongoing developments show significant development in the environmental sustainability of RAS [34]. Further, enhancement towards economic sustainability of fish production in RAS is needed for expansion by country and a competitive production of species in an ongrowing system.

Shellfish farming is based on the collection of wild or hatchery spat, which feeds on natural nutrients found in the environment (filter-feeding animals). Such a production on the floating longlines with suspended ropes below rafts (hanging culture) is a typical extensive production model [35].

Algae farming - Seaweed is cultivated on big ropes or nets in the coastal area, protected from the winds and strong currents where they can be constantly immersed underwater.

Integrated multi-trophic aquaculture (IMTA) includes organisms from different trophic levels of an ecosystem (*e.g.* fish, shellfish, algae) so that the byproducts of one become the inputs of another [36]. As a combination, the cultivation of fed finfish with aquatic plants or seaweeds as inorganic extractive species and organic extractive invertebrates (oysters, mussels…). Such a system aims to increase the productivity and reduce the environmental impacts of aquaculture. Its development in the Mediterranean, including the Adriatic Sea, is still in the early stages [37]. It is clear that in an oligotrophic Mediterranean environment, even in the vicinity of the fish farm, there may not be enough nutrients possible to grow seaweeds in an open water IMTA system. The effects of IMTA systems on bivalves have yielded contradictory results. Some studies have demonstrated the benefits, both economic and environmental of IMTA systems [38 - 41]. Other studies reported no significant effects on growth and CI of mussels cultivated close to the fish farm [42 - 44].

EUROPEAN UNION SEAFOOD SUPPLY

The EU attributed itself as a major seafood consumption market in the world with a consumption that has continuously increased over the past decades. Estimated seafood consumption per capita reached 25.5 kg where more than half is imported. Consumption is dominated by capture fisheries where 80% of the EU-28's total fishery production relates to catches. The live weight of catches amounted to 5.1 mt in 2015, that is 21% or 1.4 mt less than in 2000 [45].

The live weight of catches for the EU-28 was 5.1 million tonnes in 2015, 5.0% less than in 2014. However, an overall decline of about 21% or 1.4 million tonnes of live weight since 2000 and the growing demand offers an opportunity for aquaculture production to contribute much more to fish food market supply in the EU [46].

The aquaculture sector in EU-28 can be divided into three main subsectors: Shellfish, Marine and Freshwater production. The production of crustaceans and seaweeds was rather minor. The EU28 production of fish and shellfish in 2015 was 1.3 million tonnes, a total value of around EUR 4 billion, that is slightly over 1% of the global aquaculture production [46].

Despite a large number of species produced only a few species and species groups make up the bulk of aquaculture production. The most important species in the EU aquaculture are mussels, oysters, salmon, trout, carp, seabass and seabream. Some other species like bluefin tuna turbot, catfish and clams are also produced in lesser volume. Marine water fish are reared in floating cages placed in the protected coastal areas (salmon, sea bass gilthead seabream, bluefin tuna), while the freshwater fish species like carp and trout are reared in semi-intensive ponds, and intensive raceways or recirculation systems. France is the main producer of oysters, Spain is the main producer of Mediterranean mussels, whereas the main producers of clams is Italy. The UK is the leader in salmon production in marine waters, and Greece is the main producing country in gilthead sea bream and sea bass farming. Carp is mostly produced in Eastern Europe, where the main producer is Poland.

EU fish production accounted for 674,336 tonnes that is slightly more than half of total EU-28 production quantity in 2015 [47]. Norway's fish production amounted to *1.38 million tonnes* of live weight that is double to the estimated value for the entire EU-28 (Table **1**).

Table 1. Fish farming production (tons/year) in Europe (adjusted from FEAP Production Report, October, 2016).

Country 2008 2015	2008	2015	Δ%
NORWAY	904,623	1.382,800	+35%
TURKEY	149,589	234,000	+36%
UTD.KINGDOM	144,031	201,578	+29%
GREECE	148,509	112,159	-32%
FARO ISLANDS	45,506	66,090	+31%
SPAIN	65,835	64,186	-3%
ITALY	64,073	55,480	-13%
FRANCE	47,110	44,595	-5%
DENMARK	39,831	38,829	-3%
POLAND	34,370	38,590	+12%
CZECH REPUBLIC	19,765	19,113	-3%
GERMANY	34,964	15,341	-128%
HUNGARY	15,860	16,124	+2%
FINLAND	12,000	12,500	+4%
IRELAND	12,020	12,000	-1%
SWEDEN 6,703 11,144 +66%	6,703	11,144	+66%
CROATIA	7,635	12,093	+58%
ICELAND	5,014	8,249	+65%
NETHERLANDS	9,340	6,155	-52%
PORTUGAL	4,024	5,919	+48%
CYPRUS	2,452	5,409	+121%
AUSTRIA	2,206	3,121	+41%
Grand total EU-28	670,728	674,336	0.9%
Non-EU Total	1.099,718	1.884,477	+71%

Fish farming in EU-28 is mainly concentrated in 5 countries: Spain, United Kingdom, France, Italy, and Greece, together accounted for 71 percent of total EU-28 fish production in 2015 [48]. European finfish production excluding non-EU countries showed no appreciable increase, less than <1%. UK has the most important increase in volume adding +29% to EU-28 in the observed period (Table **1**). There were some other success stories (Cyprus, Sweden, Croatia) but they are relatively limited in production scale. However, the overall picture of a sector compared to the competitors in the region and the rest of the world is not rosy. A dramatic declining trend in Germany, Netherlands, Greece and Italy, less so in France and Denmark was recorded in the last decade (Table **1**).

While the EU aquaculture production is facing a period of stagnation, other non-EU member states experience strong growth. Norway, Turkey, Iceland, and the Faroe Islands, all have shown positive overall trends with a production that has increased by 71 percent in the observed time (from 1.099,718 in 2008 to 1.884,477 in 2015).

Table 2. Production data for fish species reared in European aquaculture 2008 and 2015 (adjusted from FEAP Production Report, January 2016).

Species/Years	2008	2015	Change: 2015 over 2014
Salmon	917,188	1.568,118	+71%
Trout[1]	372,527	386,859	+4%
Sea bass	127,399	158,479	+24%
Sea bream	165,871	147,649	-12%
Carp sp[2].	70,821	62,290	-14%
Turbot	9,122	11,270	+23%
African Catfish	6,911	6,285	-9%
European eel	7,938	6,266	-27%
Artic char	4,297	6,032	+40%
Meagre	1,821	4,156	+128%
Sturgeon sp.	2,454	2,559	+4%
Halibut	1,626	1,500	-8%
Sole	87	1,457	+ 1,575%
European Catfish	859	1,110	+33%
Cod	19,554	74	-99.6%

[1] portion size trout represents 63%, [2]92% belong to Common carp *Cyprinus carpio*

Atlantic salmon has the highest economic value. It is cultivated in only a few EU countries and mostly in Norway and the United Kingdom representing 61% of the total European. From 2008 to 2015 production increased to 71% in volume (Table **2**). Otherwise, about 87% of Salmon is produced in non - EU-28 countries. At the same time, an appreciable increase (23 to 40 percent) was an increase in Arctic char, catfish, sea bass, and turbot. Increasing trends in meagre and in sole production, although in small volume are indicative in terms of diversification.

Why did the EU Aquaculture Sector Fail ?

Environmental regulations continue to be the most important areas to be addressed to increase production growth in the EU aquaculture sector. The majority of experts found administrative barriers and regulatory restriction highly problematic as well as difficulties in the licensing process due to multilevel governance and competition for space both on the land and the coastal environment.

The EATiP (European Aquaculture Technology and Innovation Platform) completed its Vision for 2030 and Strategic Research and Innovation Agenda (SRIA) in 2012, examining the future of European aquaculture and how research and innovation could contribute to sustainable aquaculture in Europe [54]. This is achieved through the European project 'Aquainnova'. The Vision for EU aquaculture is:

- EU aquaculture will be sustainable and globally competitive
- Aquaculture production will grow and diversify by providing a market and consumer range of new of high quality and nutritious products
- EU Aquaculture production will adapt to climatic and geographic circumstances
- Aquaculture will be in harmony with nature and society

All of the above will be achieved by enhancing husbandry, welfare, technology and knowledge management ensuring the sustainability of aquaculture and its global role in technological leadership.

Along with Vision 2030, the European Union intends to boost the aquaculture sector through the Common Fisheries Policy reform and in 2013 published Strategic Guidelines for the Sustainable Development of EU aquaculture [49] presenting common priorities and general objectives at EU level. For future development of the sector four priority areas were highlighted: i) simplification of administrative procedures, ii) ensure access to space through co-ordinated spatial planning, iii) increase competitiveness and, iv) exploiting advantages due to high quality, health, and environmental standards. EU countries have been asked to set up multiannual plans to increase the sector's production and competitiveness. In their Multiannual National Strategic Plan for the development of aquaculture activities (Article 34 of Regulation (EU) No 1380/2013) Member States have defined measures to address these issues at national and regional scales. The EC has analyzed the main objectives and challenges of the aquaculture sector identified by 27 Member States (all EU countries, except Luxembourg) in their plans within the overall context of the CFP and Environmental Policy [50].

According to the summary of the multiannual national aquaculture plans the EU demand for fish is met by EU aquaculture only by 10% percent and EU fisheries by 30%, while the remaining 60% of wild and farmed fish consumed are imported from third countries [25]. According to the figures presented in Member States' plans, the estimated projection for aquaculture production in 2024 will increase 25% to a total of more than 1.5 million tonnes.

MEDITERRANEAN AQUACULTURE: CURRENT STATE, CHALLENGES AND PERSPECTIVES

Current State of Mediterranean Aquaculture

In addressing issues of aquaculture development in the Mediterranean one must acknowledge that this sector is partly associated with the relevant European Common Fisheries Policy. While demand for seafood in the region continued to rise, the assessment of Mediterranean fisheries has shown declining trends that characterized global fisheries. Demersal stocks experience high fishing mortality rates close to the target, while small pelagic stocks are fished at a biologically unsustainable level [8].

To the dramatic decline of capture fisheries in the Mediterranean is added downward trends in the European aquaculture including aquaculture in most of the EU-Member states. After rapid and dynamic marine aquaculture development in the 1990s, in the last decade, there was a downward trend in European aquaculture including mariculture in EU-Mediterranean member states.

The EU shellfish production reached almost 550,000 tonnes. The main species produced included mussels, oysters and other bivalve species including clams, cockle, carpet shell, abalone, and scallops. Shellfish aquaculture relies on extensive techniques that employ suspended installations and spat mostly collected at rearing areas. Mussels production is about 450,000 tonnes. Spain, with almost 50% of total production is followed by Italy and France, with 20% each. Oyster production is 95,000 tonnes, which is produced 90% in France. Production of 30,000 tonnes of clams is mainly from Italy. Production efficiency and environmental conditions with potential algal blooms and the spread of biotoxins with links to human and shellfish health also have a strong effect on production growth [25].

Aquaculture finfish production of the Mediterranean region is mainly concentrated in six countries (Egypt, Greece, Italy, Spain, France and Turkey) that account for nearly 95 percent of marine and freshwater aquaculture in the region [8]. Present production of the two main species in the Mediterranean mariculture is 146,249 tonnes of sea bream and 157,479 tonnes of sea bass, reaching 99% of the total European production of these species [48] (Table **3**).

Table 3. Production of the principal marine finfish species in the EU-Mediterranean countries 2008 *vs* 2015 (calculated from FEAP Production Report 2016).

PRODUCTION (Tons) YEAR SPECIES	SPECIES	YEAR 2008	2015	Δ%
European sea bass (*Dicentrarchus labrax*)	GREECE	50,000	45,000	-10%
	SPAIN	9,840	21,324	+117%
	ITALY	9,800	6,450	-34%
	CROATIA	2,700	4,500	+67%
	FRANCE	3,968	1,980	-50%
	CYPRUS	752	1,725	+129%
	Total	77,060	80,979	+5%
Gilthead seabream (*Sparus aurata*)	GREECE	94,000	65,000	-31%
	SPAIN	23,930	16,231	-31%
	ITALY	9,600	7,360	-23%
	CROATIA	1,800	4,500	+150%
	CYPRUS	1,600	3,656	+128%
	FRANCE	1,636	1,502	-8%
	Total	132,566	98,249	-26%
Meagre (*Argirosomus regius*)	SPAIN	1,300	1,642	+26%
	FRANCE	235	226	-4%
	ITALY	300	190	-37%
	CROATIA	-	70	+
	CYPRUS	-	28	+
	Total	1,835	2,156	+17%
Turbot (*Scophtalmus maximus*)	SPAIN	7,870	7,915	
	FRANCE	850	303	
	CROATIA	-	8	
	Total	8,720	8,126	-7%
Sole (*Solea solea*)	SPAIN	55	736	+1,280%

(Table 3) contd.....

PRODUCTION (Tons) YEAR SPECIES	SPECIES	YEAR 2008	2015	Δ%
	FRANCE	-	256	+
	ITALY	19	-	-
	Total	74	992	+1,240%
GRAND TOTAL		**220,255**	**190,502**	**-14%**

The main producer of gilthead seabream and seabass was Greece (42.8 percent), but also several other EU Member States including Spain, Italy, Croatia, France, and Cyprus. In the last decade, seabass production stagnated at around 80,000 tonnes. Total gilthead seabream production was 98,000 that is 26% less compared to 2015 production statistics [51].

Iversification in Mediterranean mariculture is still focussed on meagre and two flatfish (sole, turbot) with no clear indicators for the sustainable future of these fish. Globally, meagre as a perspective candidate for future Mediterranean aquaculture showed a 17% increase in volume thanks to Spain, while both France and Italy indicated a negative trend.

In Southern EU countries, and in particular in the Mediterranean Sea, the total EU production is close to 150,000 tonnes at a value of EUR 800 million. The main EU producers are Greece with 62% of EU production, Spain (20%) and Italy (7%). These two species are also grown in several other EU Member States including Croatia, Cyprus, France, Malta, and Portugal but at modest levels (less than 4,000 tonnes annual production). Portugal and Spain are the main producers of turbot (10,000 tonnes). Italy and Spain are the two main Member States pioneering production of other marine finfish species.

Challenges - The main challenge to enhancing production is the lack of available space in inshore sheltered areas. In response, the United Kingdom aims to develop offshore aquaculture while Greece, Italy, Ireland and Spain are focusing on reorganizing production sites to optimize the use of available space. Competitiveness of products compared to similar production in third countries is another important challenge identified by the Member States. Cyprus, France, Greece, Ireland, Italy and Spain aim to address this issue through the development of more cost-effective production techniques incorporating research and development outputs, improved marketing of products and better linkages between economic operators in the industry (*i.e.* feed manufacturers, fry producers). All Member States listed also intend to simplify administrative procedures related to national and regional environmental laws which have an effect on aquaculture activities in marine inshore areas. Objectives - The objective for EU marine finfish aquaculture is to increase production to 480,000 tonnes by

2020, a 60% increase compared to current production levels.

Turkey showed a steady increase in fish production quantity resulting in an output of 234 thousand tonnes in 2015 compared to 149 thousand tonnes in 2008. This was relatively larger than the fish production quantity increase in any of the EU Member States. Marine water fish production increased from 50% to 100% with a remarkable growth in gilthead sea bream production growth (82%) and 100% in large trout. (Tables **4**, **5**).

Table 4. Changes in 2008 sea bass and gilthead seabream production over 2015 in Turkey (calculated from FEAP Production Report 2016).

Species/Year	2008	2015	Δ%
European sea bass (*Dicentrarchus labrax*)	31,670	48,000	+52%
Gilthead seabream (*Sparus aurata*)	42,270	77,000	+82%
Sea bass & Seabream Total	73,940	125,000	+69%

Lack of suitable space creates significant restrictions on the development of the freshwater aquaculture in the Mediterranean. Labor costs, land and other inputs in many Mediterranean states imposes uncompetitivness against cheap imports from overseas countries. Carp farming in ponds faces some additional specific damage by protected predators, especially cormorants. Seasonality in market demand is another problem which must be addressed by processing and product diversification, and marketing campaigns.

Turkey production growth in portion size trout was 54% from 2008 to 2015 while EU countries have shown a 15% decrease. EU- Mediterranean seabass and seabream hardly compete with expanded Turkey export on the European market. According to FEAP [48], Northern European markets are mainly supplied by Turkey which export has increased 236% from 2010 (2964 tonnes) to 2015 (9965 tonnes).

Table 5. Changes in 2008 over 2015 production in portion size of trout and large trout (>1200 gr) in Turkey (calculated from FEAP Production Report 2016).

Species/Year	2008	2015	Δ%
Trout – portion size (*Oncorhynchus mykiss*)	65,928	101,550	+54%
Large Trout	2,721	5,450	+100%
Trout Total			+56%

The International Commission regulates the bluefin tuna fishery for the Conservation of Atlantic Tunas (ICCAT), which in 2006 adopted a 15-year recovery plan for bluefin tuna in the East Atlantic and Mediterranean. Thanks to good management and sacrifices a recovery of the (eastern Atlantic Bluefin tuna) stock were recovered. Consequently, from 2013 there has been about a 20% increase of the catch quota resulting in 13,500 tonnes of total allowable catch for the EU in 2017.

The eight Member States actively involved in catching bluefin tuna - Spain, France, Italy, Croatia, Greece, Portugal, Malta and Cyprus, share the EU quota. The EU quota is shared by the eight Member States actively involved in catching bluefin tuna - Spain, France, Italy, Croatia, Greece, Portugal, Malta and Cyprus. Spain and France have the largest shares those vessels, which catch the fish alive for farming purposes, share 61% of the total EU quota. Having 15% increase in biomas by fattening an estimate of 9.500 tonnes will be harvested from cages. This does not include Croatian tuna farming that last from 18 to 30 months, and in 2017 is expected to amount 3 000 tonnes export to Japanese market.

Spain and France have the largest shares of those vessels, which catch the fish alive for farming purposes, share 61% of the total EU quota. Having 15%, increase in biomass by fattening an estimate of 9,500 tonnes will be harvested from cages in 2017. This does not include Croatian tuna farming that lasts from 18 to 30 months [52] and in 2017 is expected to amount to a 3, 000 tonne export to the Japanese market.

Egypt is the main producer of gray mullet (*Mugil cephalus*) with 97.2 percent of the total. A drastic reduction from 2006 with 250,000 tonnes declined to 120,000 tonnes in 2010. Such a decrease is linked to more restrictive regulations in place for the collection of wild juveniles [53].

Crustacean aquaculture is still very limited for the GFCM member countries and in 2010 about 840 tonnes produced mainly by Spain (in the Atlantic area) and by Egypt (in the brackish water zones along the coast). The dominant species was the kuruma prawn (*Penaeus japonicus*) which accounted for 91 percent of total crustaceans' production [53].

Outlook

Overall, there has been little or no improvement in the observed period, in technical and marketing performance that are both crucial for long-term sustainability of the sector. Many known factors influence the performance of the Mediterranean aquaculture sector in the observed period that corresponded to the great economic crises.

The Mediterranean marine aquaculture excluding Turkey has not progressed significantly over the last decade (8 to 10 years). There has been no significant improvement, neither in volume nor in its technical performance. Actually, the quality of juveniles and quality of feed ingredients affecting growth performance as well as susceptibility to infection and survival may have gotten worse! The feed conversion ratios for sea bass and gilthead seabream are still too high (FCR 2 to 2.3) compared to salmon (FCR 1.2) [48]. The lack of available space in the inshore sheltered area prevented many countries from optimizing production and to be competitive with a similar production in third countries. A further challenge is the development of more cost-effective production techniques bringing in line research and development outputs, improved marketing of products and better linkages between government responsibities and aquaculture industry. Complicated administrative procedures related to national and regional environmental laws which have an effect on aquaculture activities must be simplified [25].

Furthermore, lack of a multidisciplinary approach in aquaculture to develop economically effective and environmentally sustainable production technologies as well as technologies and systems to reduce the incidence of disease and parasite infestations has paid its price.

Mediterranean Aquaculture Projected by Production Technology

At present, the EU Mediterranean member states aquaculture development will follow the Common Fishery Policy (CFP) supported by European Maritime and Fisheries Fund (EMFF). Following detailed EATiP Vision document with an estimation of the aquaculture development trends for 2030 [54], the long-term economic and ecologic impact of larger sustainable aquaculture was prepared for the European Parliament [26]. Further inputs were provided with an assessment of the economic contribution of EU aquaculture production and the influence of EU policies on its sustainable development [27].

Having our categorization of the sector by production technology and species groups, the EATIP projected a vision for Mediterranean aquaculture productions for 2030 [54], statistical information from EUMOFA, 2015 [47], Eurostat 2016 [45], and the Food and Agriculture Organization of the United Nations 2016 [8], the following trends are projected.

1. ***Small cage-based aquaculture*** in both freshwater lakes and rivers, as well as nearshore in protected and isolated coastal bays and estuaries, has been attackeds due to concerns over environmental impacts.
2. ***Large cage system in open or semi-open sea environment*** by employing (semi)offshore technology may increase both economic and environmental

performances of Mediterranean warm water farming. This will mitigate aquaculture space conflict with other activities and improve public perception of the aquaculture sector and its product.

3. ***Land based recirculated aquaculture system(RAS)***, both in FW and MW, for hatchery and nursery production will increase. RAS increased use may go two ways. Firstly, as to meet requirements for fingerlings while doubling production of warm water fish as was estimated by EATiP for 2030. Secondly, production economy may take advantage of combined and prolonged land-based RAS employment for ongrowing robust "seed", and offshore technology for market. It is not likely that that RAS will be employed for sea bass and sea bream ongrowing because of the more competitive cage production. However, RAS may play a role in production of some high-value warm water species such as turbot, sole, sturgeon, pike-perch, eels, some shrimps *etc.* [27].
4. ***Coastal pond aquaculture*** in EU Mediterranean countries will certainly decline or stagnate because-of lower yields and competition for space [27]. However, that may not be the case with some of South Mediterranean countries. Semi-intensive aquaculture in earthen ponds is the most dominant fish farming system in Egypt representing approximately 85 percent of the total production, with Nile tilapia and mullets as the major culture species [55].
5. ***Freshwater pond production*** will remain stable with a slight increase (1.5 percent/year) amounting to a 40% increase for 2030. Integrating RAS will lead to better production efficiency focusing on local economies. Diversification, and improved market understanding and environmental services may raise productivity of traditional farms.
6. ***Intensive flow-through systems*** for freshwater will probably decline. The range of this decline will be dependent on a combination of several factors such as market demand, water availability and diversification towards specialized/niche markets (*e.g.* organic label) where lower intensity is demanded [27].
7. ***Shellfish production*** is projected by EATiP [54] to increase by 30% and reach 859 tonnes of which 70 percent are mussels on longlines suspended cultivation system. An adjustment of rearing technology for production in more deeper and clean water, diversfied species spectrum, increased hatchery supply of spat, and genetic improvement for disease resistance and higher productivity is forecasted. A much less increase is expected from other bivalves such as European flat oysters, carpet shells, cockles, scallops, or clams, and abalone (gastropods).

Challenges Facing Mediterranean Aquaculture

Spatial Planning

Early development of marine finfish caging industry that started by the early

1990's occupied shallow and protected locations with not much interest for environmental issues. UNEP/MAP/PAP-RAC initiative "Approaches for zoning of coastal areas with reference to Mediterranean aquaculture" [56] has not been widely applied while planning and placing cage farms in the majority of Mediterranean countries, including leading ones.

In the meantime, the UNEP MAP Protocol on Integrated Coastal Zone Management (ICM) in the Mediterranean has been signed (January 2008), representing a unique regional regulatory tool for the ICM process as a standard approach to solving the issues in coastal areas. The Protocol includes the principles of the ICM, institutional measures, recommendations for new tools and measures regarding international collaboration. It provides guidelines for defining coastal areas, principles and elements of ICM, collaboration, public participation and other relevant issues [57]. The General Fisheries Commission for the Mediterranean (GFCM) adopted a resolution on Allocated Zones for Aquaculture (AZAs) in 2012, in which marine spatial planning, Environmental Impact Assessment (EIA) and a monitoring program are considered essential for the responsible development and management of aquaculture activities in the Mediterranean and Black Sea [58]. Further to this resolution, the Committee on Fisheries of FAO [28] emphasized the need for spatial planning to ensure the allocation of adequate space for aquaculture. It highlighted the benefits of spatial planning with multiple outcomes.

Spatial planning now takes benefits from advances in satellite technology, digital mapping technology, ecological modeling, information systems that enable use and exchange of environmental data at a regional and global level. A large amount of data can be collected and validated, and indicators could be produced and presented to decision-makers and the public in the form of text, graphic, or visualized in GIS Google maps. A huge progress has also been made in setting up indicator based information systems and developing websites. Integrating all these technologies could help responsible authorities to improve spatial planning, help the industry to properly plan and demonstrate environmental and economic sustainability of aquaculture operations.

In most of the Mediterranean countries, the main obstacles for marine aquaculture development are spatial positioning [59]. Such an environmental pressure can make the site selection and site management for aquaculture quite tough because of its high dependency on ecosystem good health. No doubt, the most important issues for making a good CZMP are knowledge, co-operation, and compromise: knowledge about the different interests, their needs in the coastal zone, and knowledge about the coastal zone itself; co-operation between different interests, and co-operation across borders of municipalities and counties; a compromise among different coastal users.

Environmental goals and preserving of the wildlife in the Mediterranean [60] (Natura 2000) is another important interest in the coastal zone. To protect wildlife and protected environment, a number of rules are established. Potential competitors to use the coastal areas are numerous, and their competition for space has resulted in negative interactions with traditional and new coastal users [61].

Here traditional fishermen, local passengers, and different kinds of industry are located in the coastal zone and occupy the areas. Tourism, which in the Mediterranean is a big industry, is another competitor to the coastal zones, especially when also counting local recreation possibilities. In the Mediterranean, it is common to have summerhouses, weekend cottages to use on holidays. Naturally, many of the places used for recreation are found along the coast. As the number of fish farming locations increased, many "weekend guests" find the fish farming plants problematic. The term "visual pollution" is often used as an argument against mariculture from local people or organizations with interests in tourism or related to local recreation. This trend is expected to continue because, along with existing inshore and new offshore aquaculture operations, other human activities in coastal and open sea waters will increase significantly over the next 20 years, including offshore energy generation [62].

Therefore, when planning in the coastal zone, one must be aware of all existing and new planned activities in the area concerned. Activities like fish farming plants and archaeologically important places, the traditional fishing grounds, the best beach for the midsummer night, and so on are obviously easy to plot in the plan. The big challenge is to make room for as many interests as possible in a limited area, or in other words: optimize the yield from the area – of course within environmental sustainability.

When it comes to aquaculture in the coastal zone the challenge is to make local planners give aquaculture equal rights and responsibilities [56]. To do this, a framework for the industry must be established that makes sure aquaculture also benefits the local community, through generating jobs, taxes or other. Generated yield from aquaculture to the local community are a result of established framework for the industry, and this is important to keep in mind for everybody that works for developing of the aquaculture industry.

To make an optimal plan for all interests in the coastal zone, knowledge is the key. This relates to knowledge about the existing industry, knowledge about fishing grounds, about recreation areas, and knowledge about guidelines and rules from different authorities. When it comes to aquaculture itself, knowledge about aquaculture zoo technology will be best suited to prevailing environmental conditions at the specific site. What the farmed species requires in a location relates to knowledge about the quality of waters to plan for. There's not much use of a plan that reserves an area for fish farming where the water is polluted and there is no fish, and particularly shellfish can be cultivated. Therefore, knowledge must be established about the different species that can be farmed in the area – what is required of temperature, how strong the currents should be, and how much exposure to waves can the farm take? Similarly: what are the actual conditions in the different parts of the coastal zone related to the same: conditions regarding temperature, current, and exposure of waves? As the number of species that are likely to be farmed in the area increases, the need for knowledge increases as well.

Croatian Case

Marine aquaculture in Croatia has started from being the local entrepreneurs to these days' participants in a globally competitive industry. Fish farmers started with the small units, and the farming was based on the Norwegian and Japanese experience in cage farming, more than on research. Bluefin tuna farmed in Croatian cages destined Japan, whilst finfish is competing on the EU seafood market. After the painful process of the transitional economy and big losses due to privatization in almost every big farm, and then problems in the most important markets connected to the global economic crises in 2008, finfish farming in Croatia has gone through a big restructuring. Generally speaking, the industry is now characterized by fearless owners, and a leading company is organized as an integrated enterprise. Professionals run the companies and we see ownership across borders of both municipalities and counties, and even across borders of nations, which is the case with several Bluefin tuna farms. In addition: research starts to be support for almost every detail, from hatching to sales. After 35 years of industry, the processing plants have started by adding new value to the product.

Given the fact that economic and social patterns of community life could be positively changed by income and employment generated by aquaculture investment and operation brought new ways of development, especially for the heavily depopulated islands and depressed rural areas. The existing inhabitants participated in these new opportunities that probably attracted the interest of ex-islanders who have left for seeking jobs in the nearby urban areas. It has positively changed the employment pattern in touristic areas in the sense that new activities provide all-year-around jobs and in that sense, overturn the negative demographic and employment trends (*i.e.* Croatian islands Ugljan and Brač).

A decade or so ago we came to the point of responsibility for making CZMPs. It has finally been decided to be at a county level. The reason why CZMP is suggested to be implemented on the county level is to make co-ordination more efficient, and deliver a cooking book for the municipalities to use. The aim is to ensure a better process and good substance in coastal zone planning as has been demonstrated in the Croatian case.

On a county level, we were acting for a development that furthers the interest of the entire regions, overlooking municipality borders. For the municipalities, however, the goal will often be to get as many jobs as possible and thus generating income and socio-economic stability to the municipality. Through projects, meetings and direct information municipalities benefits by building knowledge, and by having the opportunity to make their plans based on CZMP achievement.

From the planning of bigger farm's point of view, the focus must be on a regional level. For instance; if three finfish farms in three different municipalities can generate the need of a processing plant in a fourth municipality, this is a good thing for the region, based on the many jobs and the expected derived activity generated by the processing plant to properly manage with this new opportunity. The challenge is to make acceptance for regarding it at a regional level rather than only the municipal level.

On the other hand, if municipalities are requested to accommodate fish farms only, the result of this entire situation, are that they are less interested in reserving "their" areas for aquaculture if not receiving something in return, preferably more than a few jobs. Namely, the fish farming itself does not require many employees, and the bigger companies are not necessarily located in the economically depressed areas. Experiences with the lesser yield from the aquaculture in terms of taxes, jobs, and derived activity, make the local authorities give priority to other interests that provide something in return – preferably more than a few jobs (*i.e.* tourism and its related activities). Then you get the situation of "what's in it for me?"

Having the majority of Mediterranean marine aquaculture inshore, and strong competition with other coastal users for space [59], numerous environmental problems appeared, and a negative public perception of both the sector itself and the aquaculture product as it is. A global assessment of the potential for offshore mariculture development in spatial terms indicates that the potential is large for inshore and offshore mariculture [33]. The authors suggested using a GIS-based Decision Supporting System relevant for the assessment of aquaculture potential. For these countries not practicing coastal zone planning yet, an integration of spatial data is suggested with administrative, social, economic and ecological criteria. This may facilitate mariculture development but also be used in the context of wider development and management.

Outlook

The most important issues for making a good Coastal Zone Management Plan (CZMP) for aquaculture is an integrated approach, both a horizontal (across sectors) and vertical one connection for administrative authorities (from top to down). To proceed with such a planning for aquaculture that is based on integrated CZM principles, knowledge, co-operation, and compromise are requested: knowledge about the different user's interests in the coastal zones and knowledge about the coastal zone itself; co-operation between different branches and industries, and co-operation across borders of municipalities and counties; compromise among different coastal users and public interests. Proper planning and management of aquaculture operations will mitigate conflicts among users, strengthen complementarity while respecting surrounding ecosystem's holding capacity.

Fish Nutrition

Feeding farmed fish with fishmeal and fish oil is considered one of the most controversial environmental issues in the predicted future expansion of mariculture [63, 64]. Inadequate feed ingredients and feeding management are considered leading contributors to the global burden of disease [65]. Fish feed is usually the largest operational cost, often accounting for 50 percent or more of all production costs [66].

The aquaculture sector has been the largest consumer of fishmeal and fish oil for over a decade [63, 67], consuming 68% of the total global fishmeal and 74% of the total global fish oil production [68]. By increasing worldwide aquaculture production while capture fisheries stagnating over three decades,

FM and FO used as fish feed ingredients are in a much higher demand than supply [7]. Use off fishmeal and fish oil produced from wild population to feed carnivorous farmed fish is not sustainable.

The current supply cannot meet the long-term ambitions of the seafood industry and there needs to be improvements in fisheries management based on a wider ecosystem approach [69]. Small pelagic fish as the main source for fishmeal and fish oil are short-lived species highly dependent on the environmental condition [70] demonstrated that the overall trends for fisheries over the last few decades is one of decline in global catch contributions.

In any case, the future with fishmeal and fish oil is not promising. It will not be the only problem of higher uncertainty changes but also feed prices that make the aquaculture industry unsustainable. Little information is available about the social and economic impacts of the use of wild fish to feed carnivorous fish in aquaculture [71].

The demand for fishmeal and fish oil will likely become stronger, given the fast expansion of the global aquaculture and stagnation of the global capture fisheries that supply their ingredients the prices of these commodities will rise. Therefore, the growth in aquaculture production, while not increasing the total amount of fishmeal used, is coming through the partial replacement of fishmeal in the diets of almost all species, A key question is how effective replacement will be with alternative proteins from plants, waste products from fish and terrestrial animals [72].

Fishmeal replacement is influenced not only by an increase in the price of fishmeal but also by growing demand and concern about the fluctuating supply. Nonetheless, with significant improvements anticipated in the efficiency of feed and management practices, the expansion of aquaculture continued [73]. This replacement success will be even more important as the proportion of small pelagic fish used for human consumption is expected to increase in the years to come [74].

There has been a further effort to reduce the use of both fishmeal and oil in fish feed after the possibility of partial replacement of fishmeal in the diets of many species were tested. The dependency of fish oil demand can be more limiting than the dependency of fishmeal for the expected development of marine fish aquaculture [72].

Discard proportion of total catch returned to the sea (in most case dead, dying or badly damaged), represent a significant part of the world's marine catches and is generally considered a wasteful misuse of marine resources. Mortality among discarded fish is not adequately known, but has, for some species, been estimated up to 80%, and even perhaps higher [75, 76].

Discards from marine fisheries are considered the single largest proportion of any food source produced or harvested from the wild. The first global assessment was published in 1994 and it identified a total discard of 27 million tonnes [77]. More recent estimates are similar and go up to 30 percent compared to total landings [78]. In any case, discarded fish from marine fisheries is the single largest proportion loss of any food source produced or harvested from nature. The proportion is particularly high for trawl fisheries [79].

The additional potential source of animal protein are a variety of processed animal by-products as an underutilized protein source with high digestibility (meat meal, blood meal *etc.*). These products can provide the amino acid balance in fish diets and meet their nutritional requirements for good growth and health status [80, 81]. The total replacement of fishmeal with poultry by-product showed a low growth rate and mortality that could be due to limiting essential nutrients like fatty acids [82].

Replacement of Fishmeal and Fish Oil with Crop-Based Ingredients

Over the past decade, the scarcity of fishmeal led to increased use of cereal grains, especially wheat, as a source of protein in commercial aquaculture. This replacement of fish meal and fish oil has never been of such importance. Plant nutrients already represent the major dietary protein source for lower trophic level fish species (carp, tilapia, catfish) and fishmeal and fish oil replacement for high trophic level fish species [72]

Some positive results were recorded in growth performances, survival and protein efficiency in rainbow trout (*Oncorhynchus mykiss)* with partial fishmeal substitution by poultry by-product meal [83]. Most of the substitution of fishmeal with alternative protein sources has shown unbalance in aminoacid balance [67]). The effect of dietary fatty acids on gut microbiota in fish are mostly focussed on salmonids. It has been demonstrated that plant-based protein ingredients in feedstuffs for carnivorous fish often presents anti-nutritional factors that may interfere with fish health and can negatively affect growth and feed efficiency [84]. Recent studies have demonstrated that plant protein sources influence gut microbiota in European sea bass juveniles [85]. A decrease in gut microbiota diversity and richness when sea bass juveniles are fed with soybean meal and rapeseed meal were also noticed [86]. In this context, the efficient digestive use of plant dietary ingredients is probably limited by fish digestion physiology and capacity.

Mycotoxins are toxic metabolites produced by a diverse group of fungi that contaminate agricultural crops prior to harvest, and their effects in aquaculture species are recognized threats since 1960, when an outbreak of aflatoxicosis in hatchery-reared rainbow throats was observed (*Onchorhynchus mykiss*) [87]. It is general knowledge that an increase of plant-based ingredients in fish feed is a potential for mycotoxin poisoning in fish with an increase in mortality and significant economic losses. Feed components such as cereal grains (wheat, corn, and barley) frequently contain mycotoxin, produced by *Aspergillus, Fusarium,* and *Penicillium* genera [88]. It is manifested as tissue damage or through immunosuppression, while feed contamination by mycotoxins is a potential risk to both human and animal health [89].

Consumer Confidence

Modern aquaculture production integrates new techniques emerging from combining recent research with empirical knowledge. The challenge is to ensure that food biotechnology is developed and applied safely, without damaging the environment or threatening human health.

A number of evidence have confirmed that eating fish provides a health benefit. The availability of fish is much facilitated by supermarkets and retail channel enabling the potential consumer to reach fish all year around.

The intensification of production systems and increases in fish food supply also carry microbiological and toxicological risks, as well as problems related to animal welfare. The majority of new diseases affecting humans caused by pathogens originate from an animal or from products of animal origin [90]. A World Health Organization (WHO) consultation session on "future trends in veterinary public health" stated that both human and animal health are closely linked to the surrounding environment [91].

Public concerns include contaminants (pesticides, heavy metals, organic pollutants *etc.*) which could be harmful to consumers' health [92]. Dioxin-like compounds and polychlorinated biphenyls (PCBs) are introduced via the feed of farmed fish. Bacteria found in wild fish have been found to be resistant to several antibiotics [93].

Other contaminants could be higher in farmed than in wild fish, and consuming such fish may pose a health risk [94]. It would seem that the feed of farmed salmon is a source of organic contaminants to these fish. Chemicals have been applied to combat sea lice, while orally administered pesticides and antifouling paints applied to salmon cages are used to prevent fouling [95].

Further concerns include the use of large quantities of antibiotics in the fish feed that may result in the presence of residual antibiotics in fish products and the aquatic environment [65, 96]. Antibiotic-resistant determinants in the marine environment will, in turn, pass their antibiotic resistance genes to other bacteria, thus affecting the entire ecosystem [97]. A major cause of concern with the use of antibiotics in farmed fish is that many of these antibiotics are also used to treat human diseases. Allowing disease microbes to become resistant to antibiotic treatments makes the treatments of human diseases more difficult.

The research has progressed and the knowledge and technology have been globally shared over the last decades enabling commercial aquaculture expansion. Opposite to that seems that public opinion about aquaculture and its food product fluctuate from bad to good.

The quality of farmed-raised fish and food safety are issues that should not be ignored. Consumers might be confused about production methods, fish nutrition, zootechnics, benefits of eating fish and the risks from many unknown relations. The majority of people along the coast have an opinion that eating fish and fish-like the product is globally good for their health as it may prevent coronary disease, cancer and stroke but they have also lessened media about the risks associated with their popular seafood product. When it comes to evaluation of native wild seafood product compared to the farmed one, again the latter one is the looser. Here we come to imported feed and ingredients badly connected with GMO issues, the toxicity of heavy metals and pregnancy, antibiotics and resistance.

Sector image may have a significant impact on the success of the seafood business. Some studies clearly demonstrated that a positive image influencing consumer's perceptions of price level for goods persuading clients to purchase a certain product [98].

Outlook

Although research in fisheries sciences and biotechnology has progressed acceptance of seafood product become more a question of public opinion than science based knowledge. Theoe who oppose aquaculture may not deal with real information. This confirmed the globally known proverb that selling product is not always the evidence supported by real science. It is more a matter of trust that is much easier to loose than to keep. Opponents have always an advantage as they have knowledge and experience on how to manipulate public opinion. Poorly informed public create a solid starting point for misunderstandings and manipulations by opponents. A proactive approach to the public is needed, especially to young people interested in the food sector. A joint effort of both producers and professionals might be included into the process of education. Good examples are Croatian companies presenting all year around educational documentary on the ferry operating Island to a nearby city and back, aiming to better explain specifies of the fish farming process and products obtained. Aiming to further improve social relations they started to be open for visitors, schoolers, and potential customers who are given the opportunity to follow aquaculture process from "cage to plate", and thus providing more realistic knowledge for a wider audience. Providing jobs for locals, quality seafood supply for the market, accurate data obtained from farm monitoring, and the best available science, all this has to be confronted with "mental environmentalist".

Fish Health Management

One of the main challenges in any farming activities is efficient health management, which is crucial for maintaining and further developing the food industry. As happens with terrestrial animals, fish raised in crowded conditions enable spreading of diseases. Fish are susceptible to pathogens such as bacteria, viruses, and parasites, which may also be transmitted to wild fish [99]. High stocking density may cause stress altering some physiological and biochemical parameters in overcrowded marine fish and produce stress-related disease [100].

In animal health management there is a common understanding that prevention is better than treatment, this is also so in aquaculture. Fish diseases affect the survival and growth rates of fish that may result in huge economic losses. To lower risks of such losses, it is crucial to take precautions to prevent fish diseases and reduce pathogen levels in water bodies, and to strengthen the innate immunity of cultured fish stock. To minimize the necessity of usage of drugs against pathogens causing fish diseases, a good farm management has no alternative. As once Ilan Paperna said – “we, fish ichthyopathologists are needed to bad farm managers at most”.

The groups of experts recently discussed and evaluated the most important sea bream and seabass health status in the Mediterranean concluding that the most important problems are almost the same in the whole area [101].

Viral infections are particularly dangerous in this regard since fishes who survive their infections can be carriers of the viruses and transmit them to non-infected animals, even if they present no symptoms [102]. Viral Encephalopathy and Retionpathy – Viral Nervous Necrosis [103, 104] was concluded as the most important disease in the Mediterranean aquaculture associated with a high mortality rate and growth reduction resulting in an unpredictable economic loss. Groups have concluded that no control methods are available, and the only preventive measure is checking each individual batch of juvenile fish before seeding [101].

Vibriosis is the second most important disease in farmed marine fish in the Mediterranean. To prevent a clinical appearance of disease a quite different approach is performed. In some countries, only dip vaccination is applied while in others a combination of dip and one or two IP injections to achieve long term and effective specific immunity. It was reported that in some cases antibiotic resistance has started to develop [101]. The same authors concluded that an efficient management of the bacterial pathogens remains a priority in spite of vaccines and antibiotic treatments concerning bacterial disease Vibriosis caused by *Vibrio anguillarum* and Photobacteriosis/Pasteurellosis due to *Photobacterium damsalae* subsp. *piscicida.* Photobacteriosis remains a greater problem because vaccines were found not to be a fully effective treatment for already affected fish. Parasitic infestations ("gill flukes") are considered the third most important cause of mortalities mainly due to difficulties in the parasitic treatment of the affected batches. Widespread in the area, isopod infestation poses a serious threat to production cost due to high mortality. It was pointed out that control of this pathogen with medications used for salmon sea lice could be effective.

Developing a resistance to specific antibiotics, including genetic mutation are causing concerns for humans and environmental health. This resistance has diminished the effectiveness of antibiotics in treating human and animal disease. Resistant bacteria isolated from agriculture and aquaculture agriculture share the same resistance mechanisms, indicating a combined effect to the resistance already reached by terrestrial agriculture. The most alarming findings were, that out of 51 antibiotics commonly used in aquaculture and agriculture 39 (or 76%) are also of importance in human medicine [97].

Vaccines against specific fish pathogens have proved a good management tool to control particular diseases and to reduce mortality and the use of chemotherapeutics [105]. It is the administration of antigenicmaterial that stimulates the immune system to develop adaptiveimmunityto a pathogen (immunization). Development of vaccines and improvement of biosecurity (control and containment of diseases) has greatly reduced the need for antibiotics [106]. Benbroock [107] stated that this decline is partly attributed to the use of more effective antibiotics. Let us not forget that the cost of antibiotics used for treatment of disease by increased demand became rather high, and from the desire to search for effective preventive measures further efforst have been made to test new compounds to prevent fish diseases. Last but not less important was growing public concern over antibiotic resistances and consumer perception of seafood quality [106].

Vaccination has proved highly effective in controlling bacterial diseases in the aquaculture while more variable efficacy is apparent for viral vaccines. Live virus vaccines in high doses were demonstrated to have the ability to induce protection in fish, but it is considered to be environmentally unsafe for field use [108]. No vaccines against fish parasites are commercially available [109].

In order to strengthen an inate immunity of cultured fish stock, the aquaculture industry has developed the concept of functional phytobiotics additives in fish diet. The first application of immunostimulants in aquaculture was made in 1986, following a great success by the use of probiotics in human medicine and poultry disease prevention [110]. Glucans from yeast and plants showed an excellent effect when injected or orally administered to fish.

The use of phytobiotic additive as an immunostimulant appears to be a promising and useful tool for the prophylactic treatment of farmed fish [111]. A promising protection of the fish that are released into environments where the nature of the pathogen is unknown and immunization by the specific vaccine may be futile [112]. They are safer than chemotherapeutics, can be easily obtained, are not expensive, and their range of efficacy cover a broad spectrum of pathogens (Viruses, Bacteria, Fungi, and Parasites). However, these compounds will not replace vaccines, proper nutrition or good management techniques [113].

After decades of traditional immunostimulants, a new generation of immunostimulants, such as prebiotics (polysaccharides, oligosaccharides), herbs, microorganisms and miscellaneous compounds offered a great potential in supporting fish growth, health and production time. Several immunostimulants have been evaluated in fin fishes. Propolis enriched diet was found effective in preventing adverse effects of temperature-induced stress and promoting the growth of European sea bass juveniles [114]. Studies of microbial ecology in aquaculture systems, correlating microbial communities with fish growth and its relationship to the water quality is highly recommended [115]. The highest efficacy of immunostimulation is achieved after administration of immunostimulants to fish by injection. The efficacy of oral and immersion methods decreases with long-term administration. In some cases, overdoses of immunostimulants might induce immunosuppression in fish [116]. When evaluating the usefulness of the potential probiotic in aquaculture it was recommended to take into consideration a panel of biomarkers that represent pro-/anti-inflammatory, pro-/anti-apoptotic and anti-bacteria/viral responses of the fish [117].

Standards and Procedures

In the site of appreciable growth marine aquaculture in the Mediterranean is still a relatively new food industry with poor public understanding and a rigid consumer acceptance [118]. In the years to come food safety, environmental standards, allochthone species, chemicals and pharmaceutical products, genetic engineering will all play its role in creating misunderstandings and ignorance leading towards uncertainty and consumer's confusion and generally ending with a bad image of such a product. To improve public awareness of aquaculture, the industry needs standards and a more open, broader dialogue that will increase trust and transparency in the sector [119].

Product quality starts from good aquaculture practices during grow out, and continue over harvesting, handling, processing and storage procedures. However, the fish product is extremely susceptible to physical, chemical and microbiological deterioration during and after harvest. Better management practices (BMPs) for aquaculture aiming to standardize on-farm practices and reduce the economic risk associated with environmental degradation [120]. In order to maintain health benefits while replacing fishmeal and fish oil in fish feed with vegetable sources, standards on the minimum levels of essential omega-3 fatty acids in aquaculture product have to be established [121].

After harvesting, product quality may be ensured by reducing or eliminating microbial activity and tissue degradation. Although spoilage bacteria associated with aquaculture product are generally harmless, but do cause changes in color, texture, organoleptic characteristics, and reduce the shelf-life of aquaculture products [8]. Therefore, quick freezing and stable low storage temperature will slow biochemical activities associated with enzymatic changes during frozen storage. Many of these food safety procedures need standards to contribute to the quality, longevity, and safety of products.

Considerable irreversible post-mortem changes occur not only on the fish surface but also in fish muscle. These changes depend on the physical condition at death, primarily on temperature. The level of deterioration after harvest varies from species to species, may differ from wild to cultured one, and can be slowed down if proper handling standards and storage procedures are followed. Implementing such practices for fish on-growing, handling and post-harvesting can enhance the quality and safety of the fish product [122].

Developing effluent aquaculture standards will help the improvement of environmental standards and aquaculture to operate in harmony with nature [123]. This also implies Standard Methods for the Examination of Water and Wastewater [124].Water quality criteria and standards relevant to fish health could minimize the needs for pharmaceuticals [124]. Other procedures and standards may cover technologies ranging from the production of seed material to on-growing structures and other farming support facilities [125].

Such a program should include improvement and verification of aquatic animal health with site-specific procedures as to assist for early disease detection, surveillance, reporting, and response to prevent pathogen dissemination via movement and trade of aquatic animals. Such a program of standards and procedures may improve health management, protect and expand aquaculture business opportunities, promote and facilitate trade as well as improve natural resources protection and environmental sustainability. The program may also be incorporated into existing standards and guidelines for aquaculture management in order to conduct regulatory activities such as certification, and thus to improve trust among producers and consumers [126].

Having supply chains more vertically integrated, private standards became an important mechanism that facilitates traceability and transparency of product processes [126]. Combined with effluent standards this may provide producers and retailers with insurance against attacks from a variety of environmental groups and negative consumer perception.

Serious food safety concerns are linked to the many pathogens, unsafe residues, and the potential to contribute to antimicrobial resistance. When food safety problems appear, there will be a range of negative impact for producers, processors, buyers, and importers. The increase in the negative perception of aquaculture may disrupt the market. Minimizing the risk of market disruption due to the food safety concerns, and choosing an approach to obtain customer confidence in aquaculture is extremely important [127].

Standards, procedures, and certifications are seen as key issues for sitting and monitoring the aquaculture sector, as well as for improving understanding and acceptance of aquaculture products. They are also expected to establish the conditions for the proper aquaculture development as to increase consumer confidence into the concept of a new food biotechnology and to protect fish farmer's investment in a technologically advanced seafood production system. The same standards and procedures should also ensure environmental sustainability through appropriate regulations while achieving economic efficiency [118]. Development of a specific code of conduct, based on accepted international standards among producers, is another way to introduce best practices [25]. The benefits of having standards can be in having a response to the concerns expressed by many NGOs, scientists and other stakeholders regarding environmental and sustainability issues including criteria for site selection and monitoring, benthic community indicators, quality of sediments, quality of feed ingredients *etc.* [123]. Then we come back to the standardized criteria for site selection and management procedures of cage operation, which are essential for both economic and environmental success.

CONCLUSION

In addressing issues of aquaculture development in the Mediterranean one must acknowledge that this sector is partly associated with the relevant European Common Fisheries Policy. The present study provide both global and European seafood environment for Mediterranean aquaculture development focussing on current state, predicted visions, and challenges.

The marine aquaculture plays an important role in the Mediterranean providing income, food and jobs in many handicapped areas including islands, that otherwise would not have much opportunity to develop any important industry. It is relatively a new industry that learned by painful mistakes in the recent history. Some recent tendency towards positive changes were recognized in fish nutrition, broodstock management, hatchery reproduction, genetics, health status, and farming system design.

We conclude that a positive health image of the Mediterranean diet could offer good social acceptance and market opportunity for Mediterranean aquaculture development. There is a huge potential in new biotechnology to tackle major problems if the industry aiming to become economically sustainable and environmentally-friendly aquaculture. Introduction of new species offer a wider range of products to the market but also to select species whose growth rates and endoskeleton will allow for processing and new range of product presentation. Adding new value to farmed fish trough processing is a new strategic advantage in achieving the high quality and freshnes of farmed product. Better communication and promotion of diversify quality products, branding and labelling may help in obtaining consumer confidences and creating new market niches. Obviously, aquaculturist need to envisaged market before producing fish if a profitability of investment will be secured.

A critical issue for further expansion will be in finding new spaces and better use of existing ones. Future production economy may take advantage of more robust, and fast growing species, combined with prolonged land based RAS for pre-growing "seeds", and offshore technology for market production. An improvement of fish nutrition, feed and feeding management should follow both positive practice and the best scientific knowledge. Main challenges are to improve feed efficiency, optimize production costs and thus improve competitiveness. In parallel, there will be further need to search alternatives to limited and highly costly marine source ingredients in fish feed.

Simplify administrative procedures related to national and regional environmental regulations, lesser organic enrichment, chemicals and pharmaceutical products, solution for genetic effect from escapees, improved health management are challenges for a competitive, economically stabile, environmentally sustainable and socially acceptable aquaculture. By encouraging integration of farmed finfish with other non-fed organisms, both economy and environment may benefit. Finally yet importantly, developing standards and procedures are expected to play an important role in improving fish farming practices as well as increase consumer confidences. These procedures and standards should be proposed and applied everywhere as the essential tool to guide policy makers, administrators, aquaculture producers, and other stakeholders. Pragmatic and practical solution are needed in production, governance, cooperation, policy and planning. An understanding of processes and responsible management are key issue for success. More efforts and resources should be directed towards a positive recognition around aquaculture.

CONFLICT OF INTEREST

The author (editor) declares no conflict of interest, financial or otherwise.

ACKNOWLEDGEMENT

Declared none.

REFERENCES

[1] UN, Department of Economic and Social Affairs, Population Division. 2015.

[2] Foresight The Future of Food and Farming Final Project Report. London: The Government Office for Science 2011.

[3] Beddington JR, Asaduzzaman M, Clark ME. The role for scientist in tackling food insecurity and climate change 2012.http://www.agricultureandfoodsecurity.com/content/1/1/10 [http://dx.doi.org/10.1186/2048-7010-1-10]

[4] FAO. The State of the World Fisheries and Aquaculture. Rome: FAO 2014.

[5] UN (United Nations) General Assembly. 2009.

[6] Béné C, Hersoug B, Allison EH. Not by rent alone: analysing the pro-poor functions of small-scale fisheries in developing countries. Dev Policy Rev 2010; 28: 325-58. [http://dx.doi.org/10.1111/j.1467-7679.2010.00486.x]

[7] FAO. The state of world fisheries and aquaculture 2012. Rome: FAO 2012.

[8] FAO. The State of the World Fisheries and Aquaculture 2016 Contributing to food security and nutrition for all. Rome: FAO 2016.

[9] OECD-FAO 2013 Agricultural outlook 2013-2022. Paris:Organization for Economic Co-operation and Development (OECD) and the Food and Agricultural Organization (FAO) of the United Nations.

[10] FAO. Post-harvest changes in fish 2015.http://www.fao.org/fishery/topic/12320/en

[11] Beveridge MC, Thilsted SH, Phillips MJ, Metian M, Troell M, Hall SJ. Meeting the food and nutrition needs of the poor: the role of fish and the opportunities and challenges emerging from the rise of aquaculture. J Fish Biol 2013; 83(4): 1067-84. [PMID: 24090563]

[12] Wheal MS, DeCourcy-Ireland E, Bogard JR, Thilsted SH, Stangoulis JC. Measurement of haem and total iron in fish, shrimp and prawn using ICP-MS: Implications for dietary iron intake calculations. Food Chem 2016; 201: 222-9. [http://dx.doi.org/10.1016/j.foodchem.2016.01.080] [PMID: 26868569]

[13] Golden CD, Allison EH, Cheung WWL, *et al.* Fall in fish catch threatens human health 2016. [http://dx.doi.org/10.1038/534317a]

[14] Mozaffarian D, Rimm EB. Fish intake, contaminants, and human health: evaluating the risks and the benefits. JAMA 2006; 296(15): 1885-99. [http://dx.doi.org/10.1001/jama.296.15.1885] [PMID: 17047219]

[15] Black RE, Victora CG, Walker SP, *et al.* Maternal and child undernutrition and overweight in low-

income and middle-income countries. Lancet 2013; 382(9890): 427-51.
[http://dx.doi.org/10.1016/S0140-6736(13)60937-X] [PMID: 23746772]

[16] World Bank/FAO/WorldFish. Hidden harvest: The global contribution of capture fisheries. World Bank Report, Washington 2012; 66469-GLB.

[17] Wijkström UN. The use of wild fish as aquaculture feed and its effects on income and food for the poor and the undernourished 2009.

[18] Béné C, Barange M, Subasinghe R, *et al.* Feeding 9 billion by 2050 – Putting fish back on the menu. Food Secur 2015; 7: 261.
[http://dx.doi.org/10.1007/s12571-015-0427-z]

[19] FAO. FAOSTAT: Production. 2014. Rome.FAO.

[20] Fish to 2030, Prospects for Fisheries and Aquaculture, World Bank Report, December, 2013, No.

[21] Fisheries and Aquaculture topics. The State if World Fisheries and Aquaculture (SOFIA). Topics Fact Sheets. Text by Jean-Francois Pulvenis In: FAO Fisheries and Aquaculture Department [online] Rome Updated 7 July 2016 [Cyted 28 January 2017] http://www.fao.org/fishery/sofia/en.

[22] Bouwman AF, Beusen AH, Overbeek CC. Hindcasts and Future Projections of Global Inland and Coastal Nitrogen and Phosphorus Loads Due to Finfish Aquaculture. Rev Fish Sci 2013; 21(2): 112-56.
[http://dx.doi.org/10.1080/10641262.2013.790340]

[23] Kobayashi M, Msangi S, Batka M, *et al.* Fish to 2030: The Role and Opportunity for Aquaculture. Aquac Econ Manag 2015; 19(3): 282-300.
[http://dx.doi.org/10.1080/13657305.2015.994240]

[24] Waite R, Beviredge M, Brummett R, *et al.* Improving Productivity and Environmental Performance of Aquaculture Working Paper, Installment 5 of Creating a Sustainable Food Future. Washington, DC: World Resources Institute 2014.http://www.worldresourcesreport.org

[25] EC (European Commission). Summary of the 27 multiannual national aquaculture plans. 2016.

[26] Lane A, Hough C, Bostock J. 2014.The long-term economic and ecologic impact of larger sustainable aquaculture http://www.europarl.europa.eu/RegData/etudes/STUD/2014/529084/IPOL_STU(2014)529084_EN.pdf

[27] Bostock J, Lane A, Hough C, *et al.* An assessment of the economic contribution of EU aquaculture production and the influence of policies for its sustainable development. Aquacult Int 2016; 24: 699.
[http://dx.doi.org/10.1007/s10499-016-9992-1]

[28] Statistical Yearbook FA. World food and aqriculture. Rome: FAO 2013.

[29] Woynarovich A, Moth-Poulsen T, Peteri A. Carp polyculture in Central and Eastern Europe, the Caucasus and Central Asia: a manual. Rome: FAO 2010.

[30] Funge-Smith S, Phillips MJ. Aquaculture systems and species.

[31] Cataudella S, Crosetti D, Massa F. Mediterranean coastallagoons: sustainable management and interactions amongaquaculture, capture fisheries and the environmentGeneral Fisheries Commission for the MediterraneanStudies and Reviews, no 95. Rome: FAO 2015.

[32] Lovatelli A, Aguilar-Manjarrez J, Soto D. 2013.http://www.fao.org/docrep/018/i3092e/i3092e00.htm

[33] Kapetsky JM, Aquilar-Manjarrez J, Jenness J. A global assessment of potential for offshore mariculture development from a spatial perspective 2013.

[34] Martins CI, Eding EH, Verdegem MC, Heinsbroek LT, Schneider O, *et al.* New developments in recirculating aquaculture systems in Europe: a perspective on environmental sustainability. Aquacult Eng 2010; 43(3): 83-93.
[http://dx.doi.org/10.1016/j.aquaeng.2010.09.002]

[35] Ravagnan G. Coastal aquaculture systems for fish and crustacea in the Mediterranean. Expert Consultation on Aquaculture Development in the Mediterranean Region convened by the Government of Greece in cooperation with FAO(GFCM) and UNEP. Athens, Rome. 1978.

[36] Chopin T, Cooper JA, Reid G, *et al.* Open-water integrated multi-trophic aquaculture: environmental biomitigation and economic diversification of fed aquaculture by extractive aquaculture. Rev Aquacult 2012; 4(4): 209-20. [http://dx.doi.org/10.1111/j.1753-5131.2012.01074.x]

[37] Župan I, Peharda M, Bavčević L. Potential for development of integrated multi-trophic aquaculture (IMTA) in the Adriatic Sea. Ribarstvo 2012; 70(1): 125-37. [full text: Croatian].

[38] Peharda M, Župan I, Bavčević L, *et al.* Growth and condition index of mussel Mytilus galloprovincialis in experimental integrated aquaculture. Aquacult Res 2007; 38: 1714-20.

[39] Troell MA, Joyce T, Chopin A, *et al.* Ecological engineering in aquaculture - Potential for integrated multi-trophic aquaculture (IMTA) in marine offshore systems. Aquaculture 2009; 297: 1-9. [http://dx.doi.org/10.1016/j.aquaculture.2009.09.010]

[40] Zühlke Lj, Engel M, Reid S, *et al.* Letter to the editor. Clin Epidemiol 2011; 3: 171-2. [http://dx.doi.org/10.2147/CLEP.S20223] [PMID: 21750626]

[41] Lander TR, Robinson MC, MacDonald BA, Martin JD. Enhanced growth rates and condition index of blue mussels (Mytilus edulis) held at integrated multitrophic aquaculture sites in the Bay of Fundy. J Shellfish Res 2012; 31: 997-1007. [http://dx.doi.org/10.2983/035.031.0412]

[42] Parsons GJ, Shumway SE, Kuenstner S, Gryska A. Polyculture of sea scallops (*Placopecten magellanicus*) suspended from salmon cages. Aquacult Int 2002; 10: 65-77. [http://dx.doi.org/10.1023/A:1021324610930]

[43] Navarrete-Mier F, Sanz-Lázaro C, Marín A. Does bivalve mollusc polyculture reduce marine finfish farming environmental impact? Aquaculture 2010; 306: 101-7. [http://dx.doi.org/10.1016/j.aquaculture.2010.06.013]

[44] Irisarri J, Cubillo AM, Fernández-Reiriz MJ, *et al.* Growth variations within a farm of mussel (*Mytilus galloprovincialis*) held near fish cages: importance for the implementation of integrated aquaculture. Aquacult Res 2015; 46: 1988-2002. [http://dx.doi.org/10.1111/are.12356]

[45] Eurostat-Statistics Explained. Fishery statistics in detail 2016.http://ec.europa.eu/eurostat/statistics explained/index.php/Fishery_statistics_in_detail

[46] The EU in the World 2016.http://europa.eu/eurostat

[47] EUMOFA (European Market Observatory for Fisheries and Aquaculture Product). 2015.http://www.eumofa.eu

[48] FEAP (Federation of European Aquaculture Producers). 2016.

[49] EC (European Commission) Strategic Guidelines for the sustainable development of EU aquaculture. Communication to the European Parliament, the Council, the European Economic and Social Committee and the Committee of the Regions. COM/2013/229- (29/04/2013) final. http://ec.europa.eu/fisheries/efp/aquaculture/officialdocuments/com_2013_229_en.pdf

[50] Science for Environment Policy. Sustainable Aquaculture 2015.http://ec.europa.eu/science-environment-policy

[51] Annual Report 2015.www. feap.info

[52] Katavic I, Tičina V, Grubišić L, *et al.* Comparison of the recent and historical information on the size composition of Bluefin tuna (*Thunnus thynnus*) in the Adriatic obtained by purse seine fisheries. Collectice Volume of Scientific Papers – International Commission for the Conservation of Atlantic

Tunas 2013; 69(2): 699-703.

[53] GFCM (General Fisheries Commission for the Mediterranean). 2013.

[54] The future of European aquaculture-Our Vision: a strategic agenda for research and inovation 2012.http://www.eatip.eu/default.asp?SHORTCUT=92

[55] El-Sajed AF. Analysis of feeds and fertilizers for sustainable aquaculture development in Egypt.Study and Analysis of Feeds and Fertilizers for Sustainable Aquaculture Development FAO Fisheries Technical Paper. Rome: FAO 2007; p. 497.

[56] Dosdat A, Héral M, Katavic I. Approaches for zoning of coastal areas with reference to Mediterranean aquaculture. 1996.

[57] UNEP/MAP. Protocol on integrated coastal zone management for Mediterrenean 2008.

[58] GFCM. 2012.http:// bit. ly/

[59] Sanchez-Jerez P, Karakassis I, Massa F, *et al.* Aquaculture's strugle for space: the need for coastal spatial planning and the potential benefits of Alocated Zones for Aquaculture (AZAs) to avoid conflict and promote sustainability. Aquacult Environ Interact 2016; 8: 41-54. [http://dx.doi.org/10.3354/aei00161]

[60] 1992.http://ec.europa.eu/environment/nature/natura2000/

[61] Dempster T, Sanchez-Jerez P. Aquaculture and coastal space management in Europe:an ecological perspective. 2008.

[62] Douvere F, Ehler CN. New perspectives on sea use management: initial findings from European experience with marine spatial planning. J Environ Manage 2009; 90(1): 77-88. [http://dx.doi.org/10.1016/j.jenvman.2008.07.004] [PMID: 18786758]

[63] Naylor RL, Hardy RW, Bureau DP, *et al.* Feeding aquaculture in an era of finite resources. Proc Natl Acad Sci USA 2009; 106(36): 15103-10. [http://dx.doi.org/10.1073/pnas.0905235106] [PMID: 19805247]

[64] Tacon AG, Metian M. Fishing for Aquaculture: Non-Food Use of Small Pelagic Forage Fish—A Global Perspective. Rev Fish Sci 2009; 17(3): 305-17. [http://dx.doi.org/10.1080/10641260802677074]

[65] Sapkota A, Sapkota AR, Kucharski M, *et al.* Aquaculture practices and potential human health risks: current knowledge and future priorities. Environ Int 2008; 34(8): 1215-26. [http://dx.doi.org/10.1016/j.envint.2008.04.009] [PMID: 18565584]

[66] White R, Beveridge M, Brummet R, *et al.* Improving productivity and environmental performances of Aquaculture Working Paper, Installment 5 of Creating Sustainable Food Future. Washington, DC: World Resources Institute 2014.http://www.worldresourcesreport.org

[67] Tacon AG, Metian M. Global overview on the use of fish meal and fish oil in industrially compounded aqua feeds: Trends and future prospects. Aquaculture 2008; 285: 146-58. [http://dx.doi.org/10.1016/j.aquaculture.2008.08.015]

[68] Tacon AG, Metian M. Feed Matters: Satisfying the Feed Demand of Aquaculture. Rev Fish Sci Aquacult 2015; 23(1): 1-10. [http://dx.doi.org/10.1080/23308249.2014.987209]

[69] Viegas I, Jarak I, Rito J, *et al.* www.jlr.org2017.

[70] Pauly D, Zeller D. Comments on FAOs State of World Fisheries and Aquaculture (SOFIA 2016). Mar Policy 2017; 77: 176-81. [http://dx.doi.org/10.1016/j.marpol.2017.01.006]

[71] Wijkström UN. The use of wild fish as aquaculture feed and its effects on income and food for the poor and the undernourished. 2009.

[72] Tacon AG, Hasan MR, Metian M. Demand and supply of feed ingredients for farmed fish and crustaceans: trends and prospects 2011.

[73] The World Bank. 2013. Fish to 2030. World Bank Report. November 83177-GLB, 80pp.

[74] World Bank. Fish to 2030 Prospects for Fisheries and Aquaculture. World Bank Report, 2014; No. 83177-GLB. Washington, DC: World Bank.

[75] Bettoli PW, Scholten GD. Baycatch rates and initial mortality of paddle fish in a commercial gillnet fishery. Fish Res 2006; 77: 343-7. [http://dx.doi.org/10.1016/j.fishres.2005.11.008]

[76] Broadhurst MK, Suuronen P, Hulme A. Estimating collateral mortality from towed fishing gear. Fish Fish 2006; 7: 180-2. [http://dx.doi.org/10.1111/j.1467-2979.2006.00213.x]

[77] Alverson DL, Freeberg MH, Murawaski SA, *et al.* A global assessment of fisheries bycatch and discards. Fisheries Technical Paper Rome FAO 1994; p. 339.

[78] Nellemann C, MacDevette M, Manders T. 2009.www.grida.no

[79] Global food losses and food waste – Extent, causes and prevention. Rome: FAO 2011.

[80] Bureau DP. Rendered products in fish aquaculture feeds.Essential rendering. Alexandria, Virginia, USA: National Renderers Association 2006; pp. 179-84.

[81] Shapawi RW, Mustafa S. Replacement of fishmeal with poultry by-product meal in diets formulated for the humpback grouper, Cromilepte saltivelis. Aquaculture 2007; 273: 118-26. [http://dx.doi.org/10.1016/j.aquaculture.2007.09.014]

[82] Barreto-Curiel F, Parés-Sierra G, Correa-Reyes G, Durazo-Beltrán E, Viana M-T. Total and partial fishmeal substitution by poultry by-product meal (petfood grade) and enrichment with acid fish silage in aquafeeds for juveniles of rainbow trout *Oncorhynchus mykiss.* Lat Am J Aquat Res 2016; 44(2): 327-35. [http://dx.doi.org/10.3856/vol44-issue2-fulltext-13]

[83] Burr GS, William RW, Frederic TB, Ronald WH. Replacing fish meal with blends of alternative proteins on growth performance of rainbow trout (*Oncorhynchus myk*iss), and early or late stage juvenile Atlantic salmon (*Salmo salar*). Aquaculture 2012; 334: 110-6. [http://dx.doi.org/10.1016/j.aquaculture.2011.12.044]

[84] Krogdahl A, Penn M, Thorsen J. Important antinutrients in plant feedstuffs for aquaculture: an update on recent findings regarding responses in salmonids. Aquacult Res 2010; 41: 333-44. [http://dx.doi.org/10.1111/j.1365-2109.2009.02426.x]

[85] Gatesoupe FJ, Huelvan C, Le Bayon N, *et al.* The effects of dietary carbohydrate sources and forms on metabolic response and intestinal microbiota in sea bass juveniles, *Dicentrarchus labrax.* Aquaculture 2014; 422: 47-53. [http://dx.doi.org/10.1016/j.aquaculture.2013.11.011]

[86] Serra GR, Gil F, Oliva-Teles A, Enes P, Tavares F. Gut mictobiota dinamics in European sea bass (Dicentrarchus labrax) juveniles feed plant-based diets. 2016.

[87] Kumar V, Roy S, Barman D, *et al.* Importance of mycotoxins in aquaculture feeds 2013.

[88] Matejova I, Modra H, Blahova J, *et al.* The effect of mycotoxin deoxynivalenol on haematological and biochemical indicators and histopathological changes in rainbow trout (*Oncorhynchus mykiss*). BioMed Res Int 2014; 2014: 310680. [http://dx.doi.org/10.1155/2014/310680] [PMID: 24729967]

[89] Wegulo SN. Factors influencing deoxynivalenol accumulation in small grain cereals. Toxins (Basel) 2012; 4(11): 1157-80. [http://dx.doi.org/10.3390/toxins4111157] [PMID: 23202310]

[90] Morse SS. Factors in the emergence of infectious diseases. Emerg Infect Dis 1995; 1(1): 7-15. [http://dx.doi.org/10.3201/eid0101.950102] [PMID: 8903148]

[91] 1999.

[92] Mozaffarian D, Rimm EB. Fish intake, contaminants, and human health: evaluating the risks and the benefits. JAMA 2006; 296(15): 1885-99. [http://dx.doi.org/10.1001/jama.296.15.1885] [PMID: 17047219]

[93] Barnes AC, Hastings TS, Amyes SG. Amoxycillin resistance in Scottish isolates of Aeromonas salmonicida. J Fish Dis 1994; 17: 357-63. [http://dx.doi.org/10.1111/j.1365-2761.1994.tb00231.x]

[94] Hites RA, Foran JA, Carpenter DO, Hamilton MC, Knuth BA, Schwager SJ. Global assessment of organic contaminants in farmed salmon. Science 2004; 303(5655): 226-9. [http://dx.doi.org/10.1126/science.1091447] [PMID: 14716013]

[95] Les , *et al.* Aquaculture 2010; 306: 7-23. [http://dx.doi.org/10.1016/j.aquaculture.2010.05.020]

[96] Cabello FC. Heavy use of prophylactic antibiotics in aquaculture: a growing problem for human and animal health and for the environment. Environ Microbiol 2006; 8(7): 1137-44. [http://dx.doi.org/10.1111/j.1462-2920.2006.01054.x] [PMID: 16817922]

[97] Done HY, Venkatesan AK, Halden RU. Does the Recent Growth of Aquaculture Create Antibiotic Resistance Threats Different from those Associated with Land Animal Production in Agriculture? AAPS J 2015; 17(3): 513-24. [http://dx.doi.org/10.1208/s12248-015-9722-z] [PMID: 25700799]

[98] Standifird SS. Reputation e-commerce:eBay auctions and the asymmetrical impact of positive and negative ratings. J Manage 2001; 27(3): 279-95. [http://dx.doi.org/10.1177/014920630102700304]

[99] Johansen LH, Jensen I, Mikkelsen H, *et al.* Disease interaction and pathogens exchange between wild and farmed fish populations with special reference to Norway. Aquaculture 2011; 315: 167-86. [http://dx.doi.org/10.1016/j.aquaculture.2011.02.014]

[100] Montero D, Izquierdo MS, Tort L, *et al.* High stocking density produces crowding stress altering some physiological and biochemical parameters in gilthead seabream,Sparus auratus, juveniles. Fish Physiol Biochem 1999; 20: 53-60. [http://dx.doi.org/10.1023/A:1007719928905]

[101] Vendramin N, Zrncic S, Padros F. Fish health in Mediterranean aquaculture, past mistakes and future challenges. BullEurAss Fish Pathol 2016; 36(1)

[102] Zrncic S, Vendramin N, Boutrup TS, *et al. Piscirickettsia salmonis* infection in european sea bass - an emerging desease in mediterranean mariculture. 17th International Conference on Diseases of Fish And Shellfish: Abstract book. 153-3.

[103] Munday BL, Kwang J, Moody N. Betanodavirus infections of teleost fish: a review. J Fish Dis 2002; 25: 127-42. [http://dx.doi.org/10.1046/j.1365-2761.2002.00350.x]

[104] Bovo G, Nishizawa T, Maltese C, *et al.* Viral encephalopathy and retinopathy of farmed marine fish species in Italy. Virus Res 1999; 63(1-2): 143-6. [http://dx.doi.org/10.1016/S0168-1702(99)00068-4] [PMID: 10509726]

[105] Ellis AE. Fish vaccination. London: Academic Press 1988; p. 255.

[106] Engelstad M. Vaccination and consumer perception of seafood quality. Dev Biol (Basel) 2005; 121: 245-54. [PMID: 15962487]

[107] Benbrook CM. Antibiotics drug use in U.S. aquaculture. IATP Report, February 2012. 18pp.

[108] Biering E, Villoing S, Sommerset I, Christie KE. Update on viral vaccines for fish. 2005.

[109] Sommerset I, Krossøy B, Biering E, Frost P. Vaccines for fish in aquaculture. Expert Rev Vaccines 2005; 4(1): 89-101.
[http://dx.doi.org/10.1586/14760584.4.1.89] [PMID: 15757476]

[110] Kozasa M. Toyocerin (*Bacillus toyoi*) as growth promotor for animal feeding. Microbiologie Aliments Nutrition 1986; 4(2): 121-35.

[111] Deivasigamani B, Subramanian V. Applications of Immunostimulants in Aquaculture: A Review Int J Curr Microbiol App Sci 2016.

[112] Shankar Murthy K, Kiran BR. Review on usage of medicinal plants in fish disease. Int J Pharma Bio Sci 2013; 4(3): 975-86.

[113] Barman D, Nen P, Mandal SC, Kumar V. Immunostimulants for Aquaculture Health Management. J Mar Sci Res Dev 2013; 3: 134.
[http://dx.doi.org/10.4172/2155-9910.1000134]

[114] Segvic-Bubić T, Boban J, Grubišić L, *et al.* Effects of propolis enriched diet on growth performance and plasma biochemical parameters of juvenile European sea bass (Dicentrarchus labrax L.) under acute low temperature stress. Aquacult Nutr 2013; 19(6): 877-85.
[http://dx.doi.org/10.1111/anu.12032]

[115] Martínez Cruz P, Ibáñez AL, Monroy Hermosillo OA, Ramírez Saad HC. Use of probiotics in aquaculture. ISRN Microbiol 2012; 2012: 916845.
[http://dx.doi.org/10.5402/2012/916845] [PMID: 23762761]

[116] Ganguly S, Paul I, Mukhopadhayay SK. Application and Effectiveness of Immunostimulants, Probiotics, and Prebiotics in Aquaculture: A Review. The Israeli J Aquaculture – Bamidgeh 2010; 62(3): 130-8.

[117] Mladineo I, Bušelić I, Hrabar J, *et al.* Autochthonous Bacterial Isolates Successfully Stimulate In vitro Peripheral Blood Leukocytes of the European Sea Bass (Dicentrarchus labrax). Front Microbiol 2016; 7: 1244.
[http://dx.doi.org/10.3389/fmicb.2016.01244] [PMID: 27551281]

[118] Bacher K. Perception and misconceptions of aquaculture: a global overview. Globefish Research Programme, Rome. FAO 2015; 120: 35.

[119] Freeman S, Vigoda-Gadot E, Sterr H, *et al.* Public attitudes towards marine aquaculture: a comparative analysis of Germany and Israel. Environ Sci Policy 2012; 22: 60-72.
[http://dx.doi.org/10.1016/j.envsci.2012.05.004]

[120] Kusumawati R, Bush SR. Co-producing Better Management Practice standards for shrimp aquaculture in Indonesia. Marit Stud 2015; 14-21.

[121] Sprague M, Dick JR, Tocher DR. Impact of sustainable feeds on omega-3 long-chain fatty acid levels in farmed Atlantic salmon, 2006-2015. Sci Rep 2016; 6: 21892.
[http://dx.doi.org/10.1038/srep21892] [PMID: 26899924]

[122] Borderías AJ, Sánchez-Alonso I. First processing steps and the quality of wild and farmed fish. J Food Sci 2011; 76(1): R1-5.
[http://dx.doi.org/10.1111/j.1750-3841.2010.01900.x] [PMID: 21535702]

[123] Kreeger K. Down on the Fish Farm: Developing Effluent Standards for Aquaculture. Bioscience 2000; 50(11): 949-53.
[http://dx.doi.org/10.1641/0006-3568(2000)050[0949:DOTFFD]2.0.CO;2]

[124] American Public Health Association. Standard Methods for the Examination of Water and Wastewater. 20th ed., 1998.

[125] Tolon MT. Sustaining consumer confidence in Middle East aquaculture secured by traceability systems. Aquaculture Engineering and Fisheries Research 2017; 3(1): 44-50. [http://dx.doi.org/10.3153/JAEFR17006]

[126] Washington S, Ababouch L. Private standards and certification in fisheries and aquaculture: current practice and emerging issues 2011.

[127] Whitmarsh D, Palmieri MG. Social acceptability of marine aquaculture: The use of survey-based methods for eliciting public and stakeholder preferences. Mar Policy 2009; 33: 452-7. [http://dx.doi.org/10.1016/j.marpol.2008.10.003]

Trends in Fisheries and Aquatic Animal Health, **2017, 85-101**

CHAPTER 4

Mediterranean Fisheries in the Framework of a New Common Fisheries Policy (CFP): Challenges and Opportunities

Sanja Matić-Skoko*

Institute of oceanography and Fisheries, Split, Croatia

Abstract: A detailed picture of fisheries issue, within the context of the Mediterranean Sea, in order to investigate possible challenges and opportunities in the framework of the reformed Common Fishery Policy (CFP) is presented as invited speech. The Mediterranean Sea has lots of diverse benthic communities in the coastal area that imply high number of species and complex trophic relationships. Since ancient times fisheries were very important on the Mediterranean, and nowadays it's even more crucial due growing human population. A majoritie of fisheries in that coasta, shallow area is small-scale, characterized by multi-purpose and multi-gear fleet operating on seasonal basi, with multi-species catches and extremely heterogeneous landing places and marketing. Thus, scientific research of small-scaled fisheries is a particularly complex. Moreover, the facts regarding serious overfishing of most Mediterranean stocks demands urgent reforms of the management measures aiming to guarantee the sustainability of resources, particularly in comparison with the improvement observed in NW Atlantic seas. According to the new, reformed CFP, all European fish stocks should be brought to a state where they can produce at maximum sustainable yield (MSY) by 2020 at the latest. The sustainable exploitation should be achieved through multiannual plans (MAPs) reflecting regionality and the specificities of different fisheries as it is established as new CFP objective. Creation of MAPs should be based on scientific, technical and economic advices with included conservation measures for restoring and maintaining Mediterranean fish stocks above the determined MSYs. Since, in mixed Mediterranean fisheries, small-scale fishers landed more than a hundred commercial species and each one has specific MSYs, it is extremely difficult to regulate the fishing mortality independently for each species. Moreover, population dynamics of all Mediterranean species is also influenced by the recent environmental and oceanographic changes provoked by global climate change impacts. To really improve state of Mediterranean resources and maintain fisheries as economic activity, many different contrasting socioeconomic and ecological interests need to be confronted. Further on, recent demands for drastic reductions of fishing effort and consequently reduction of landings, based on new CFP objectives, requests for socially unacceptable management measures. For sure, actively involving scientists, fisherman and policy makers in fisheries management can minimize those tensions.

* **Corresponding author Sanja Matić-Skoko:** Institute of Oceanography and Fisheries, Split, Croatia; E-mail: sanja@izor.hr

Panagiotis Berillis (Ed)

Keywords: Fisheries, Mediterranean Sea, Reformed Common Fisheries Policy.

INTRODUCTION

Mediterranean Sea has lots of diverse benthic communities in the coastal area that imply high number of species and complex trophic relationships [1]. The Mediterranean Sea is very rich in term of total number of species with >17,000 reported ones, of which 3000 species are considered to be endemic [2]. Unfortunately, this sea is amongst the most impacted regions globally in term of biodiversity vulnerability [3, 4], due number of antrophogenic threats that impact biodiversity at all levels [5 - 7], climate change impacts [8], and biological invasions [9, 10]. On the other side, since ancient times fisheries are very important on the Mediterranean, and nowadays it's even more crucial due growing human population [11 - 14].

Due historic evidence of Mediterranean fisheries existence since ancient times, the current fisheries framework probably resulted from synergic human and nature interactions through such long history, then the consequence of a specific management policy. For most commercial species landed along the Mediterranean coast, a significant feature is that most of the landings is exactly consist of juvenile fish. The mean annual landings fluctuated around 1.5 million tons in the last decade [15]. Thus, Mediterranean fishing production represents just a limited share of the world fisheries landings [16]. But, the average price of fresh fish from that fisheries on the Mediterranean markets, are almost five to ten times higher than in other world's regions. However, these landings represent nearly 20% by weight and 35% by the value of the Community's fishery production, so this Mediterranean fishery is far from being marginal in the European context. Also, in a socio-economic sense, Mediterranean fisheries employ thousands of fishers and persons indirectly involved in this activity. However, socioeconomic situations in all Mediterranean countries are quite different and thus Mediterranean fisheries are highly diverse and vary geographically, both in term of marine organisms and human resources [17]. Concerning the catch in recent years, countries that mostly mostly contribute to the total landing in 2014 for the Mediterranean Sea are Italy (177,019 t), Spain (78,467 t), Croatia (78,928 t), Greece (no data), France (15,076 t), Cypar (1,249 t), Malta (2,403 t) and Slovenia (247 t), all EU member states while just small proportion fall on African Mediterranean countries.

The Mediterranean fishery is consisted of three main categories: small-scale fisheries, trawling and seining fisheries [18]. The term "small-scale fisheries", trying to integrate terms of the "coastal" and "artisanal" fisheries to avoid the inconsistencies and differences of numerous definitions, is virtually absent in the official terminology of the most Mediterranean countries. Much of the fishery in the Mediterranean coastal zone is SSF (80%). According to official statistics in the nineties, in the EU countries operated 41900 SSF vessels, of which 46%, 39%, 8% and 7% are registered to Greece, Italy, Spain and France respectively [19]. Beside above-mentioned countries, one of those with large SSF fleet is also Croatia but it was not considering before 2013 when Croatia entered in EU. That multi-gear (the same species is fished by a number of different fishing gears), multi-species (catch of specific fishing gear is composed of a lots of different species) fisheries comprise numerous fleets, landing relatively low catches per fishing effort. Further on, SSF catches are mostly intended for direct supply to restaurants and market and thus often goes unrecorded in national official statistics [17, 20, 21]. Because of the small-scale nature of these fisheries they are often considered to have less ecological impact on the marine environment than large-scale fisheries [22]. Furthermore, coastal SSF are often characterized by their contribution to coastal community development and rural livelihoods, in the most Mediterranean counties, for their socio-cultural value [23, 24], strongly linked to national and religious identity. Trawls, both bottom and pelagic, are widely used in the Mediterranean and they are considered as semi-industrial or industrial vessels. Purse seine nets are one of the main types of fishing gear used in the Mediterranean. The purse-seiners are distinguished as those operating during the day or those operating at night. There are different by equipment and vessel construction since they employ a different fishing methodology and they are focus on different species. In the term of numbers, around 4300 "semi-industrial" fishing vessels, mostly trawlers and purse-seiners, operating in the EU ports, of which 45%, 32%, 17% and 6% are registered to Italy, Spain, Greece and France respectively [19].

Many human impacts besides fishing, like pollution or climate change, may considerably affect Mediterranean marine communities [25]. It's reported that fish communities and/or their populations may respond to disturbances on water column and sea bottoms (sewage discharges, thermal pollution, alterations to benthic substrates). Pollution affects biodiversity, abundance, mortality, fish fecundity and increase the susceptibility to infections and parasite infestation [26]. Contrary, climate change acting on a wider scale, affects species distribution, abundance and consequently the structure of ichthyocommunities [27, 28].

Scientific and expert community related to marine environment issue agree that fisheries mismanagement can produce the most far-reaching consequences [29] encompassing both direct (abundance decrease, changes in size and species composition, modifications of population parameters) and indirect effects (trophic shift, bottom disturbance) which act on short and long-term temporal scales [30, 31]. Also, it is very important to know if fisheries affected specific species or communities either as target species or through bycatch and discards.

Thus, recent research indicates a dramatic decrease in the number of peak predators, long-lived, carnivorius and piscivorius fish species. It seems that in the last decades, worldwide fisheries landed much more low trophic level, short-lived invertebrates and fishes than long-lived, high trophic ones [30] and in the Mediterranean Sea. These species are therefore not only important for fisheries viability, but also potential ecological indicators of environmental state [*e.g.* 24]. There is increasing evidence that many coastal marine species are under threat of become over-fished or in collapse due direct or indirect effects of selective fishing practice [32].

At present, 30% of the stocks in the European waters for which information exists are fished outside safe biological limits (SBL). Of the assessed commercial stocks in the NE Atlantic, 21% (Baltic Sea) to 62% (Iberian Peninsula waters) are outside SBL. Generally, between 30% and 54% of fish stocks are outside safe biological limits in the NE Atlantic. In the Mediterranean, reports on status of fish stocks are rare, sporadic and irregularly updated, but from conducted on until 2005, it seems that from 44% to 78% stocks are outside SBL with alarming situation in the Adriatic Sea. Nowadays, we still have the situation of assessing more stocks in the north than in the south of the EU. Thus, in the NE Atlantic, maximum of 30% non-assessed stocks exists. In the Mediterranean region, this percentage is higher, ranging from 23% (Adriatic Sea) to 70% (tuna and tuna like species for the entire Mediterranean). In the Black Sea assessment is just staring to be internationally coordinated [33] and evaluated.

In general, conducted stock assessments in the Mediterranean Sea suggest that pelagic stocks are in better condition than demersal ones. Moreover, demersal stocks remain outside safe biological limits for years, particularly hake, *Merluccius merluccius* and red mullet, *Mullus barbatus*. Small pelagic stocks have yearly different large-scale fluctuations, but are generally not fully exploited, except for anchovy and pilchard in the Southern Alboran and Cretan Seas. The latest swordfish assessment by the International Commission for the Conservation of Atlantic Tunas (ICCAT) suggests that this species is now considered overexploited. Strong concern still remains about the over-exploitation or even risk of stock collapse of bluefin tuna [33].

In general fisheries management is very conservative in comparison with that in northern European seas [18]. Quota systems are generally not applied, mesh-size regulations usually are set at low levels relative to scientific advice, and effort limitation is not usually applied, and mostly it's not related to results of some species assessment. However, in last years, management in terms of closed seasons and areas is more present and provides positive results [34 - 36]. The fisheries legislation of the different Mediterranean countries is generally divided in two major categories: the fishing effort control [37, 38] and the exploitation patterns framework [18]. The fishing effort control is generally based on the limited number or fishing capacity of the vessels, rather than on catch limits and control of discards and by-catches. Direct measures are conducted through formation of a licensing system in recent years, while indirect measures define upper limits of the fishing capacity of individual vessels, through engine power and tonnage limitations. Exploitation pattern or technical measures are based on legislative related to gear specification and deployment, rules of fishing practices or techniques, spatio-temporal limitations, and resource exploitation patterns [18]. For sure multiple control methods are more than welcome and necessary and large closed areas excluding all main gears should be used to buffer uncertainty [38] to be really effective in term of protection.

REFORMED COMMON FISHERIES POLICY (CFP)

A new, reformed Common Fisheries Policy (CFP) is effective from 1 January 2014. Europe's fisheries policy was in urgent need of reform because EU fleet still exhausting individual stocks and threatening the marine ecosystem. As stated above, a number of stocks is overfished in EU waters and the fishing industry report lower landings and is thus facing with an uncertain future. So, new CFP has set goals as targets that will make fishing in EU environmentally, economically and socially sustainable. If fish stocks can be restoring to sustainable levels, that will provide to EU citizens stable, secure and healthy food supply in a future on long-term basis. The new CFP have to bring new prosperity to the fishing sector through new opportunities for jobs and coastal areas development and rescind subsidies [39]. We have to establish rules (like Total Allowable Catches (TAC), fishing licenses, boat capacity management, reducing environmental impact, minimum fish and mesh sizes, design and use of gears, closed areas or seasons...) by which fisheries have to be regulated to safeguard fair access, sustainability and profitability to all stakeholders.

All these attempts can be listed as follows: ensuring maximising long-term catches of fish stocks; introducing multi-annual plans governed by ecosystem approach; simplifying rules and decentralised management; creating system of transferable fishing concessions; adopting measures beneficial to small-scale fisheries; introducing discards ban; creating new marketing standards and clearer labelling; establishing better framework for aquaculture; ensuring EU financial assistance to support sustainability objectives; holding up-to-date information on state of marine resources and reinforcing international responsibility [39].

According to new CFP, EU fisheries have to be managed by multi-annual plans (MAPs) based on the scientific advices resulted from ecosystem fishery approach (EAF) assessments and the precautionary principle reflecting the regional specificities of each fishery [*e.g.* 15, 40-42]. Without doubt, MAPs should be based on scientific, technical and economic advice and contain conservation measures to restore and maintain fish stocks above the maximum sustainable yield (MSY). These commitments of adopting an ecosystem approach are built in reformed CFP. Thus, new European fisheries management have to be designed taking into account biological, environmental, social, and economic aspects and how they are interconnected on local, regional and/or national scales [43]. However, there are numerous obstacles along that path of adopting an EAF in the current CFP particularly because there is no agreed guidance on the priority to be given to objectives when trade-offs have to be made. A decision-making system is still strongly influenced by short-term national interests, and low societal and political will to pay the high short-term costs to achieving sustainability [41, 44]. The rate of progress towards an operational EAF following CFP reform will be influenced by the requirement for a reformed management system to help achieve Good Environmental Status for ecosystem components and attributes, in support of European policy for the marine environment (principally the Marine Strategy Framework Directive) [45, 46]. So, the importance of addressing mismatches in environmental management is increasingly recognised [46].

Generally, up-to-date information about the state of marine resources has to be more reliable because the fishing industry needs to have a better and more stable basis for long-term planning and investment [37]. Reliability is essential for bringing sound management decisions and their effective implementation. But for reliable information accurate assessment of fishing effort, spatio-temporal distribution of fishing effort together with socio-economic issues of requested. From all the aforementioned it is evident that in-depth scientific research together with reliable official statistics would provide a comprehensive understanding of threats and the impact they have on fisheries resources. EU Member States have the obligation to collect, maintain and share scientific data about fish stocks and the impact of fishing at sea-basin level and thus national research programmes are established to coordinate this activity.

As part of the new CFP a landing obligation will be phased out until 2019. This wasteful practice of throwing unwanted fish overboard is estimated at 23% of total catches and it's substantially more in some kind of fisheries [47]. It explicitly requires that all the species captured by any fishery sector be subject to catch limits or minimum sizes although the discarding issue is more related to bottom fisheries [48]. Moreover, EU has recognized that the main challenges associated with implementing a discard ban will be in the SSF sector in the Mediterranean

[49]. Further on, it was suggested that effects of this measure in the long-term are unpredictable, but in the short term, a landing obligation is likely to bring more negative social, economic and ecological impacts than benefits [50]. For sure, discard ban should lead to more reliable data on fish stocks, supporting thus better management and improving resource-restoring efficiency. It's a lot of work on solving possible problems with discards ban implementation through adoption of technical solutions such as more selective fishing gear.

On of the main aim of the reform is to clarify roles and obligations of each actor and bring decisions closer to the fishing grounds. The idea is that EU legislators will just define the general framework with the basic principles and the overall targets, while Member States will decide regarding the actual implementing measures, and will cooperate at regional level. Since state of resources and the socioeconomic context of each EU member state vary greatly, a balanced representation of local stakeholders who knows best how to apply EU rules in their respective areas is requested. The Advisory Councils (ACs), as stakeholder-led organisations, are recognized for such task achievement. Further on, they called for a revision of long-term management plans, clearer regulations regarding landing plans and a slow introduction of the January 2015 discard ban. Advisory Councils requested for stronger EU financial assistance to allow Member States to complete collection of missing scientific data and movement beyond single stock advice. Namely, more accurate and reliable time series of landing data or official statistics are requested [11, 17, 18]. Also fundings focused on greening, innovation, coastal development, science and research are already available but they are not fully used.

Without doubt, coastal communities that depend on fishing also require specific support. The future financial instrument for fisheries will include measures beneficial to SSF and will help local economies adapt to the changes. Although SSF are in many cases actually more sustainable than large-scale fisheries, they are disadvantaged because of their typical remoteness, lack of infrastructure, and marginal political power [22, 51]. An additional opportunity is also ecolabeling [22] but small-scale fishers worldwide will never benefit from sustainable seafood initiatives because they are not well informed and organised. Such solution may lead to the elimination of harmful fisheries subsidies. Beside all aforementioned, a better framework for aquaculture is predicted by reform CFP in term to increase production and supply of seafood in the EU, reduce dependence on imported fish and rural development of coastal areas [52]. Also, new Aquaculture Advisory Council will be established to give advice on industry-related issues. The main hypotheses of reform CFP is that simplified rules and decentralised management will greatly improve fisheries in EU countries. In the line with that, results-based management, will replace top-down decisions and thus industry will be able to choose optimal ways how to best achieve the agreed outcomes. It s considered that better marketing of EU fisheries and aquaculture products will help to reduce waste and provide market feedback to producers [39].

However, one of the major objectives of the reformed CFP is to ensure that the exploitation rates of the fish populations maintain or restore marine resources at level which can produce high and long-term sustainable fishing yield, known as maximum sustainable yield (MSY) [53]. As stated in Regulation N°1380/2013 [54], all European fish stocks should be brought to a state where they can produce at maximum sustainable yield by 2020 at the latest! Namely, the positive trend observed in northern European seas needs to be replicated in the Mediterranean and Black Sea, where the assessment from the STECF showed that more than 90% of the 30 stocks analysed were being fished well above MSY levels [55]. Several measures are proposed in order to improve such situation including a reduction of fishing combined with the use of new nets, which let smaller and unwanted fish escape, catch limits for small pelagic and some demersal fish, more regionally based management plans, greater cooperation between countries and the addition of more financial resources should steer the Mediterranean and Black Sea in line with the goals of the reformed CFP [55]. In general, the state of knowledge of Mediterranean and Black Sea stocks is improving, and the recovery potential of stocks is still very high. A reconstruction of historical data in order to establish biomass baselines would help improve assessment throughout the region, as would greater access by scientists to landing and discard data. However, current assessment of western mediterranean fisheries should be taken with care due to important quantities of unreported catches in the most high-valued species and output values of the stock status indicators and the bioeconomic modelling [56] while eastern Mediterranean stocks has been inhibited due to a lack of data from some countries, leaving a number of stocks in the region unassessed [17, 57] and for that reason, several Mediterranean areas have been characterized as fisheries data poor areas [58]. The general impression is that still there is a lack of communication among the managers, fisherman and scientists, while this should be the focus of any management process [17]. Improved scientific knowledge accompanied with better flow of communication between scientists and stakeholders should significantly facilitate the desired outcomes [59].

IMPLEMENTATION OF REFORMED CFP AND ENFORCEMENT OF NEW RULES IN THE MEDITERRANEAN FISHERIES

Focusing on issues related to policy implementation and control, monitoring and enforcement as the means of ensuring compliance with regulations, are the most painful points of modern fisheries management [60]. Most EU member states manage their fisheries as a combination of biological, economic, social, and political objectives achievement [61, 62]. However, often there is uncertainty regarding definition and balance of these objectives [63] and scientific advice has often been poorly or only partially implemented [64, 65].

As a short resume, quota systems and many means of fishing effort control are generally not applied in the Mediterranean fisheries [18], it is data-poor, with short time-series and data collection has only been carried out on a continuous and standardized basis in the last decade. Most of the fisheries are managed just through a combination of effort-control and technical measures, which makes it difficult to distinguish and evaluate individual measures [59]. On the other hand, commonly utilized management provisions and their measures effects have rarely been accompanied by studies of fisheries resources [17, 66]. Although such legislative changes are relatively common, the reasons for ordaining them have remain outside the domain of scientific literature [59]. Furthermore, fishing behaviour is usually unpredictable [67] which as a consequence creates problems for fisheries management [68].

Mediterranean fishery sector, like in most EU waters, is suffering from overfishing, fleet overcapacity, heavy subsides, low economic resilience and decline in the volume and size of fish caught [69]. Although the EU Data Collection Regulation [70] was enforced in all EU Member States, data collection of many coastal fisheries resources caught by SSF are still not sufficiently systematic (high number of fishers and landing places) and thus hinder the assessment and management of SSF. All Mediterranean countries are limited in the control of fishing capacity and fishing effort and/or to the application of technical measures. Scientific advice, although it is always sought for, has rarely been really considered into the implementation of such measures. In addition, control and surveillance at sea, has not been fully efficient in enforcing the adopted measures. However, the use of fisheries catches data and catch- based methods for evaluating the state of a fishery is important for scientific and socio-economic reasons. Over 25 countries border Mediterranean Sea, belonging to three continents, all with varying levels of fisheries technologies and management regulations [17, 18, 71, 72], and exposed to environmental/climatic changes [73, 74].

As stated in the Regulation N°1380/2013 [54], the European Union (EU) should ensure that the exploitation of marine resources restores and maintains stocks above MSY by 2015 whenever possible or by 2020 at the latest [75]. A large part of the EU fisheries is managed based on single species stock assessments; however, in reality, most species are caught together within multi-species and multi-fleet fisheries [17], for which single species assessments, and obtained

MSY reference points, are often not valid [76]. As each species has specific MSYs, it is extremely difficult to regulate the fishing mortality for each of them independently [56, 76 - 78], notably when the dynamics of these species may additionally be influenced by the environmental [74] and fishing impacts [79 - 81]. Achieving the MSY goal, therefore, is not an easy task because many different contrasting socioeconomic and ecological interests need to be confronted. As consequence, that usually demands socially unacceptable management measures implying drastic reductions of fishing exploitation rates [56]. To minimize those tensions, it is essential to actively involve scientists, fisherman and policy makers in fisheries assessment and management. The objective of reformed CFP have to be achieved through a multiannual plans (MAPs) as it is already mentioned before. In the Mediterranean, the MAPs should cover multiple stocks where those are jointly exploited with definition of clear time frames and safeguard mechanisms [56, 82]. The new management framework requires co-management and collective decision-making process [37, 83]. The projections also showed that the only management scenario under which both resource sustainability and economic viability of the Mediterranean fisheries are ensured is the decrease of fleet capacity in terms of vessel numbers. In this case, however, measures to support the fisheries-dependent communities need to be implemented to prevent the collapse of local economies due to employment decrease. Scenarios assuming selectivity improvements would be also beneficial for the stocks but they showed low economic performance and their application would threaten the viability of the fleets, particularly that of the trawlers [81].

In the EU Mediterranean waters, monitoring of discards has only been regularly undertaken since 2002 within the framework of the EU data collection programmes [70]. Recently, it is estimated that discards represent 18.6% of the average annual catches [84]. Numerous reasons for discarding are reported including legal, economic, technical, biological and environmental aspects [85 - 87, 84] and are most complex in mixed-species demersal trawl fisheries [84]. Generally, the absence of discard data affects the reliability of the stock assessment process, and can result in bad scientific advice due inaccurate estimates of fishing mortality [88]. Higher market demand for smaller sizes of certain species by local consumers and tourists [89], strengthening of monitoring, control, and surveillance to enforce established minimum landing sizes [90], absence of catch quotas [91] significantly affected discarding practices. Contrary recent investigation of the effectiveness of the landings obligation in the Mediterranean Sea revealed more weaknesses and threats than strengths and opportunities [92]. Thus, in order to compile an eligible future discard management plan for the Mediterranean fisheries and apply for any exemption, a clear framework is requested.

Generally, negative perception of policy was found at the Mediterranean fishers, influencing their resilience. They feel estranged from the decision-making process and confidence in the outcomes of policy making is low. Further, they highlighted that social goals are almost completely neglected. Namely, resource and habitat protection policies (particularly those that are in competition with SSF like proclamation of new MPAs) that are introduced without consideration of socio-economic factors are ineffective. Those policies are typically associated with intense conflict and low compliance, leading to loss of confidence in future work [93]. Developing better national fisheries management strategies based on fishermen knowledge can provide invaluable practical information. Examination and determination of factors affecting fishers' satisfaction will contribute to a better understanding of their perception and will provide additional information on the effectiveness of current management systems [93].

In order to reverse overexploitation trends in the Mediterranean and the Black Seas, near future fisheries management scenarios should focus on the preservation of ecosystem health by strictly enforcing current regulations, limiting fishing effort, banning excessive subsidies, and excluding a large part of current fishing grounds from any fishing activity [94, 95]. Further on, as well in other parts, illegal, unreported and unregulated fishing (IUU) is still serious issue in the Mediterranean. So, that imposes a clear conclusion that overfishing is still the most important threat to Mediterranean underwater ecosystems, "more than pollution, invasive species, or climate change [96]." So, only restrictive management can protect the Mediterranean fishing resources, and declaring the closed season and permanent marine protected areas and zones without fishing are powerful tools for management that would allow further recovery of resources and biodiversity conservation.

CONFLICT OF INTEREST

The author (editor) declares no conflict of interest, financial or otherwise.

ACKNOWLEDGEMENTS

Special thanks to organizing committee of the 2nd International Congress of Applied Ichthyology & Aquatic Environment "HydroMedit 2016" for invitation to be thematic speaker for Marine Fisheries theme, and to present this review regarding recent findings on Mediterranean fisheries in the framework of the reformed CFP published by a number of eminent scientists.

REFERENCES

[1] Sala E. The past and present topology and structure of the Mediterranean sutidal rocky-shore food webs. Ecosystems (N Y) 2004; 7: 333-40.

[http://dx.doi.org/10.1007/s10021-003-0241-x]

[2] Coll M, Piroddi C, Steenbeek J, *et al.* The biodiversity of the Mediterranean Sea: estimates, patterns, and threats. PLoS One 2010; 5(8): e11842. [http://dx.doi.org/10.1371/journal.pone.0011842] [PMID: 20689844]

[3] Halpern BS, Walbridge S, Selkoe KA, *et al.* A global map of human impact on marine ecosystems. Science 2008; 319(5865): 948-52. [http://dx.doi.org/10.1126/science.1149345] [PMID: 18276889]

[4] Costello MJ, Coll M, Danovaro R, Halpin P, Ojaveer H, Miloslavich P. A census of marine biodiversity knowledge, resources, and future challenges. PLoS One 2010; 5(8): e12110. [http://dx.doi.org/10.1371/journal.pone.0012110] [PMID: 20689850]

[5] Mouillot D, Albouy C, Guilhaumon F, *et al.* Protected and threatened components of fish biodiversity in the Mediterranean sea. Curr Biol 2011; 21(12): 1044-50. [http://dx.doi.org/10.1016/j.cub.2011.05.005] [PMID: 21658949]

[6] Coll M, Piroddi C, Albouy C, Ben Rais Lasram F, Cheung W W L. The Mediterranean under siege: spatial overlap between marine biodiversity, cumulative threats and marine reserves. Glob Ecol Biogeogr 2012; 21: 465-81. [http://dx.doi.org/10.1111/j.1466-8238.2011.00697.x]

[7] Micheli F, Halpern BS, Walbridge S, *et al.* Cumulative human impacts on Mediterranean and Black Sea marine ecosystems: assessing current pressures and opportunities. PLoS One 2013; 8(12): e79889. [http://dx.doi.org/10.1371/journal.pone.0079889] [PMID: 24324585]

[8] Lejeusne C, Chevaldonné P, Pergent-Martini C, Boudouresque CF, Pérez T. Climate change effects on a miniature ocean: the highly diverse, highly impacted Mediterranean Sea. Trends Ecol Evol (Amst) 2010; 25(4): 250-60. [http://dx.doi.org/10.1016/j.tree.2009.10.009] [PMID: 19959253]

[9] Zenetos A, Gofas S, Morri C, *et al.* Alien species in the Mediterranean Sea by 2012. A contribution to the application of European Union's Marine Strategy Framework Directive (MSFD). Part 2. Introduction trends and pathways. Mediterr Mar Sci 2012; 13: 328-52. [http://dx.doi.org/10.12681/mms.327]

[10] Katsanevakis S, Bogucarskis K, Gatto F, Vandekerkhove J, Deriu I, Cardoso AC. Building the European Alien Species Information Network (EASIN): a novel approach for the exploration of distributed alien species data. BioInvasions Rec 2012; 1: 235-45. [http://dx.doi.org/10.3391/bir.2012.1.4.01]

[11] Farrugio H, Oliver P, Biagi F. An overview of the history, knowledge, recent and future trends in Mediterranean fisheries. Sci Mar 1993; 57: 105-19.

[12] Goñi R, Polunin NV, Planes S. The Mediterranean: marine protected areas and the recovery of a large marine ecosystem. Environ Conserv 2000; 27: 95-7. [http://dx.doi.org/10.1017/S0376892900000126]

[13] Juanes F. Mediterranean marine protected areas. Trends Ecol Evol (Amst) 2001; 16(4): 169-70. [http://dx.doi.org/10.1016/S0169-5347(01)02111-5] [PMID: 11245934]

[14] Guidetti P, Bussotti S, Pizzolante F, Ciccolella A. Assessing the potential of an artisanal fishing co-management in the marine protected area of Torre Guaceto (southern Adriatic Sea, SE Italy). Fish Res 2010; 101: 180-7. [http://dx.doi.org/10.1016/j.fishres.2009.10.006]

[15] FAO Fisheries management 2 The ecosystem approach to fisheries FAO Technical Guidelines for Responsible Fisheries 4. 2003; p. 112.

[16] FAO. The State of World Fisheries and Aquaculture: Opportunities and challenges. Rome: FAO 2014.

[17] Lleonart J, Maynou F. Fish stock assessments in the Mediterranean: State of the art. Sci Mar 2003; 67:

37-49.
[http://dx.doi.org/10.3989/scimar.2003.67s137]

[18] Papaconstantinou C, Farrugio H. Fisheries in the Mediterranean. Mediterr Mar Sci 2000; 1: 5-18. [http://dx.doi.org/10.12681/mms.2]

[19] Farrugio H. Mediterranean Fisheries Status and Management. Diplomatic Conference on Fisheries Management in the Mediterranean. Venezia (Italy). 1996.

[20] Tzanatos E, Somarakis S, Tserpes G, Koutsikopoulos C. Identifying and classifying small-scale fisheries me ′tiers in the Mediterranean: A case study in the Patraikos Gulf, Greece. Fish Res 2006; 81: 158-68.
[http://dx.doi.org/10.1016/j.fishres.2006.07.007]

[21] Tzanatos E, Castro J, Forcada A, Matić-Skoko S, Gaspar M, Koutsikopoulos CA. Métier-Sustainability-Index (MSI25) to evaluate fisheries components: assessment of cases from data-poor fisheries from southern Europe. ICES J Mar Sci 2013; 70: 78-98.
[http://dx.doi.org/10.1093/icesjms/fss161]

[22] Jacquet J, Pauly D. Funding priorities: big barriers to small-scale fisheries. Conserv Biol 2008; 22(4): 832-5.
[http://dx.doi.org/10.1111/j.1523-1739.2008.00978.x] [PMID: 18637910]

[23] Gómez S, Lloret J, Demestre M, Riera V. The decline of the artisanal fisheries in Mediterranean coastal areas: the case of cap de Creus (Cape Creus). Coast Manage 2006; 34: 217-32.
[http://dx.doi.org/10.1080/08920750500531389]

[24] Lloret J, Cowx IG, Cabral H, *et al.* Small-scale coastal fisheries in European Seas are not what they were: ecological, social and economic changes. Mar Policy 2016.
[http://dx.doi.org/10.1016/j.marpol.2016.11.007]

[25] Stagličić N, Matić-Skoko S, Pallaoro A, *et al.* Long-term trends in the structure of eastern Adriatic littoral fish assemblages: Consequences for fisheries management. Estuar Coast Shelf Sci 2011; 94: 263-71.
[http://dx.doi.org/10.1016/j.ecss.2011.07.005]

[26] Guidetti P, Fanelli G, Fraschetti S, Terlizzi A, Boero F. Coastal fish indicate human-induced changes in the Mediterranean littoral. Mar Environ Res 2002; 53(1): 77-94.
[http://dx.doi.org/10.1016/S0141-1136(01)00111-8] [PMID: 11767255]

[27] Occhipinti-Ambrogi A. Global change and marine communities: alien species and climate change. Mar Pollut Bull 2007; 55(7-9): 342-52.
[http://dx.doi.org/10.1016/j.marpolbul.2006.11.014] [PMID: 17239404]

[28] Rijnsdorp AD, Peck MA, Engelhard GH, Möllmann C, Pinnegar JK. Resolving the effect of climate change on fish populations. ICES J Mar Sci 2009; 66: 1570-83.
[http://dx.doi.org/10.1093/icesjms/fsp056]

[29] Jennings S, Kaiser MJ, Reynold JD. Marine Fisheries Ecology. Oxford, UK: Blackwell Science 2001.

[30] Pauly D, Christensen V, Dalsgaard J, Froese R, Torres F Jr. Fishing down marine food webs. Science 1998; 279(5352): 860-3.
[http://dx.doi.org/10.1126/science.279.5352.860] [PMID: 9452385]

[31] Jennings S, Greenstreet SP, Reynolds JD. Structural change in an exploited fish community: a consequence of differential fishing effects on species with contrasting life histories. J Anim Ecol 1999; 68: 617-27.
[http://dx.doi.org/10.1046/j.1365-2656.1999.00312.x]

[32] Garcia SM, Kolding J, Rice J, *et al.* Conservation. Reconsidering the consequences of selective fisheries. Science 2012; 335(6072): 1045-7.
[http://dx.doi.org/10.1126/science.1214594] [PMID: 22383833]

[33] European Environment Agency (EEA). Status of marine fish stocks. Copenhagen, Denmark 2011; p. 17.

[34] Sanchirico JN, Malvadkar U, Hastings A, Wilen JE. When are no-take zones an economically optimal fishery management strategy? Ecol Appl 2006; 16(5): 1643-59. [http://dx.doi.org/10.1890/1051-0761(2006)016[1643:WANZAE]2.0.CO;2] [PMID: 17069360]

[35] Abbott JK, Haynie AC. What are we protecting? Fisher behavior and the unintended consequences of spatial closures as a fishery management tool. Ecol Appl 2012; 22(3): 762-77. [http://dx.doi.org/10.1890/11-1319.1] [PMID: 22645809]

[36] Di Franco A, Thiriet P, Di Carlo G, *et al.* Five key attributes can increase marine protected areas performance for small-scale fisheries management. Sci Rep 2016; 6: 38135. [http://dx.doi.org/10.1038/srep38135] [PMID: 27905533]

[37] Caddy JF, Seijo JC. This is more difficult than we thought! The responsibility of scientists, managers and stakeholders to mitigate the unsustainability of marine fisheries. Philos Trans R Soc Lond B Biol Sci 2005; 360(1453): 59-75. [http://dx.doi.org/10.1098/rstb.2004.1567] [PMID: 15713588]

[38] Stefansson G, Rosenberg AA. Combining control measures for more effective management of fisheries under uncertainty: quotas, effort limitation and protected areas. Philos Trans R Soc Lond B Biol Sci 2005; 360(1453): 133-46. [http://dx.doi.org/10.1098/rstb.2004.1579] [PMID: 15713593]

[39] Policy CF. (CFP) brochure. Belgium: European Union 2014.

[40] Pullin AS, Báldi A, Can OE, *et al.* Conservation focus on Europe: major conservation policy issues that need to be informed by conservation science. Conserv Biol 2009; 23(4): 818-24. [http://dx.doi.org/10.1111/j.1523-1739.2009.01283.x] [PMID: 19627313]

[41] Jennings S, Rice J. Towards an ecosystem approach to fisheries in Europe: a perspective on existing progress and future directions. Fish Fish 2011; 12: 125-37. [http://dx.doi.org/10.1111/j.1467-2979.2011.00409.x]

[42] Gascuel D, Merino G, Doring R, *et al.* Towards the implementation of an inte- grated ecosystem fleet-based management of European fisheries. Mar Policy 2012; 36: 1022-32. [http://dx.doi.org/10.1016/j.marpol.2012.02.008]

[43] Jennings S. Indicators to support an ecosystem approach to fisheries. Fish Fish 2005; 6: 212-32. [http://dx.doi.org/10.1111/j.1467-2979.2005.00189.x]

[44] Hilborn R. Ecosystem-based fisheries management: the carrot or the stick. Mar Ecol Prog Ser 2004; 274: 275-8.

[45] van Leeuwen J, Raakjaer J, van Hoof L, van Tatenhove J, Long R, Ounanian K. Implementing the Marine Strategy Framework Directive: A policy perspective on regulatory, institutional and stakeholder impediments to effective implementation 2014. [http://dx.doi.org/10.1016/j.marpol.2014.03.004]

[46] Boyes SJ, Elliott M, Murillas-Maza A, Papadopoulou N, Uyarra MC. Is existing legislation fit-fo--purpose to achieve Good Environmental Status in European seas? Mar Pollut Bull 2016; 111(1-2): 18-32. [http://dx.doi.org/10.1016/j.marpolbul.2016.06.079] [PMID: 27377000]

[47] Damalas D. Mission impossible: discard management plans for the EU Mediterranean fisheries under the reformed Common Fisheries Policy. Fish Res 2015; 165: 96-9. [http://dx.doi.org/10.1016/j.fishres.2015.01.006]

[48] Font T, Lloret J. Improving the efficiency of MPAs as fisheries management tools and benefits from involving the small scale fisheries sector 2016.

[49] Villasante SA, Guimeráns P, Rodrigues J, Antelo Rivero Rodríguez M, Da Rocha JM, *et al.* Small-

scale fisheries and the zero discard target, Policy Departement B: Structural and Cohesion Policies. European Parliament, Fisheries Brussels 2015.

[50] Veiga P, Pita P, Rangel M, *et al.* The EU landing obligation and European small-scale fisheries: What are the odds for success? Mar Policy 2016; 64: 64-71. [http://dx.doi.org/10.1016/j.marpol.2015.11.008]

[51] Ponte S, Raakjaer J, Campling L. Swimming upstream: market access for African fish exports in the context of WTO and EU negotiations and regulation. Dev Policy Rev 2007; 25: 113-38. [http://dx.doi.org/10.1111/j.1467-7679.2007.00362.x]

[52] Nielsen M, Jensen F, Nielsen R. Increased competition for aquaculture from fisheries: Does improved fisheries management limit aquaculture growth? Fish Res 2014; 159: 25-33. [http://dx.doi.org/10.1016/j.fishres.2014.05.004]

[53] Quetglas A, Keller S, Massutí E. Can Mediterranean cephalopod stocks be managed at MSY by 2020? The Balearic Islands as a case study. Fish Manag Ecol 2015; 22: 349-58. [http://dx.doi.org/10.1111/fme.12131]

[54] Council Regulation (EC) No 1380/2013. Common Fisheries Policy EU 2013.

[55] European Commision (EC). Concerning a consultation on Fishing Opportunities for 2015 under the Common Fisheries Policy. Brussels: COM 2014.

[56] Quetglas A, Merino G, Ordines F, *et al.* Assessment and management of western Mediterranean small-scale fisheries. Ocean Coast Manage 2016; 133: 95-104. [http://dx.doi.org/10.1016/j.ocecoaman.2016.09.013]

[57] Stergiou KI, Tsikliras AC. Underrepresentation of regional ecological research output by bibliometric indices 2006.

[58] Pilling GM, Apostolaki P, Failler P, *et al.* Assessment and management of data- poor fisheries.Advances in Fisheries science: 50 years on from Beverton and Holt Blackwell Publishing. CEFAS 2008; pp. 280-305. [http://dx.doi.org/10.1002/9781444302653.ch12]

[59] Matić-Skoko S, Stagličić N, Pallaoro A, *et al.* Effectiveness of conventional management in Mediterranean type artisanal fisheries. Estuar Coast Shelf Sci 2011; 91: 314-24. [http://dx.doi.org/10.1016/j.ecss.2010.10.029]

[60] Gazelius SS, Raakjaer J, Eds. Making Fisheries Management Work: Implementation of Policies for Sustainable Fishing. Springer 2008. [http://dx.doi.org/10.1007/978-1-4020-8628-1]

[61] Mardle S, Pascoe S, Boncoeur J, Le Gallic B, Garcia-Hoyo JJ, Herrero I. Jimenez- Toribio R, Cortes C, Padilla N, Nielsen JR, Mathiesen C. Objectives of fisheries management: case studies from the UK, France, Spain and Denmark. Mar Policy 2002; 26: 415-28. [http://dx.doi.org/10.1016/S0308-597X(02)00022-2]

[62] Hilborn R. Defining success in fisheries and conflicts in objectives. Mar Policy 2007; 31: 153-8. [http://dx.doi.org/10.1016/j.marpol.2006.05.014]

[63] Dichmont CM, Pascoe S, Kompas T, Punt AE, Deng R. On implementing maximum economic yield in commercial fisheries. Proc Natl Acad Sci USA 2010; 107(1): 16-21. [http://dx.doi.org/10.1073/pnas.0912091107] [PMID: 20018676]

[64] Daw T, Gray T. Fisheries science and sustainability in international policy: a study of failure in the European Union's common fisheries policy. Mar Policy 2005; 29: 189-97. [http://dx.doi.org/10.1016/j.marpol.2004.03.003]

[65] Villasante SA, do Carme García-Negro M, González-Laxe F, Rodríguez GR. Overfishing and the Common Fisheries Policy: (un)successful results from TAC regulation? Fish Fish 2011; 12: 34-50. [http://dx.doi.org/10.1111/j.1467-2979.2010.00373.x]

[66] Battaglia P, Romeo T, Consoli PG, Scotti G, Andaloro F. Characterization of the artisanal fishery and its socio-economic aspects in the central Mediterranean Sea (Aeolian Islands, Italy). Fish Res 2010; 102: 87-97.
[http://dx.doi.org/10.1016/j.fishres.2009.10.013]

[67] Maunder MN, Punt AE. Standardizing catch and effort data: a review of recent approaches. Fish Res 2004; 70(2–3): 141-59.
[http://dx.doi.org/10.1016/j.fishres.2004.08.002]

[68] Salas S, Gaertner D. The behavioural dynamics of fishers: management implications. Fish Fish 2004; 5: 153-67.
[http://dx.doi.org/10.1111/j.1467-2979.2004.00146.x]

[69] European Commission (EC). The Common Fisheries Policy: a User's Guide. Luxembourg: Office for Official Publications of the European Communities 2009.

[70] Council Regulation (EC) No No 1543/2000. Establishing a Community Framework for the Collection and Management of Data Needed to Conduct the CFP. EU 2000.

[71] Piroddi C, Gristina M, Zylich K, *et al.* Reconstruction of Italy's marine fisheries removals and fishing capacity, 1950–2010. Fish Res 2015; 172: 137-47.
[http://dx.doi.org/10.1016/j.fishres.2015.06.028]

[72] Colloca F, Cardinale M, Giannoulaki M, *et al.* Rebuilding Mediterranean fisheries: toward a new paradigm for ecological sustainability in single species population models. Fish Fish 2011; 14: 89-109.
[http://dx.doi.org/10.1111/j.1467-2979.2011.00453.x]

[73] Golani D. Impact of Red Sea Fish Migrants through the Suez Canal on the Aquatic Environment of the Eastern Mediterranean. Bull Ser Yale School Frestry Environ Stud 1998; 103: 375-87.

[74] Pranovi F, Colla S, Valeri P, Marco Anelli M. Present and future status of artisanal fisheries in the Adriatic Sea (western Mediterranean Sea). Ocean Coast Manage 2016; 122: 49-56.
[http://dx.doi.org/10.1016/j.ocecoaman.2016.01.004]

[75] Maunder MN, Deriso RB. A stock–recruitment model for highly fecund species based on temporal and spatial extent of spawning. Fish Res 2013; 146: 96-101.
[http://dx.doi.org/10.1016/j.fishres.2013.03.021]

[76] Guillen J, Macher C, Merzéréaud M, Bertigna M, Fifas S, Guyader O. Estimating MSY and MEY in multi-species and multi-fleet fisheries, consequences and limits: an application to the Bay of Biscay mixed fishery. Mar Policy 2013; 40: 64-74.
[http://dx.doi.org/10.1016/j.marpol.2012.12.029]

[77] Rätz H-J, Charef A, Abella AJ, *et al.* A medium-term, stochastic forecast model to support sustainable, mixed fisheries management in the Mediterranean Sea. J Fish Biol 2013; 83(4): 921-38.
[http://dx.doi.org/10.1111/jfb.12236] [PMID: 24090555]

[78] Mackinson S, Deas B, Beveridge D, Casey J. Mixed-fishery or ecosystem conundrum? Multispecies considerations inform thinking on long-term management of North Sea demersal stocks. Can J Fish Aquat Sci 2009; 66(7): 1107-29.
[http://dx.doi.org/10.1139/F09-057]

[79] Mueter FJ, Megrey BA. Using multi-species surplus production models to estimate ecosystem-level maximum sustainable yields. Fish Res 2006; 81(2-3): 189-201.
[http://dx.doi.org/10.1016/j.fishres.2006.07.010]

[80] Merino G, Quetglas A, Maynou F, *et al.* Improving the performance of a Mediterranean demersal fishery towards economic objectives beyond MSY. Fish Res 2015; 161: 131-44.
[http://dx.doi.org/10.1016/j.fishres.2014.06.010]

[81] Tserpes G, Nikolioudakis N, Maravelias C, Carvalho N, Merino G. Tsikliras AC. Viability and Management Targets of Mediterranean Demersal Fisheries: The Case of the Aegean Sea. PLoS ONE.

2016; 11: p. e0168694.

[82] Simmonds EJ, Döring R, Daniel P, Angot V. The role of fisheries data in the development evaluation and impact assessment in support of European fisheries plans. ICES J Mar Sci 2011; 68: 1689-98. [http://dx.doi.org/10.1093/icesjms/fsr067]

[83] Linke S, Dreyer M, Sellke P. The Regional Advisory Councils: what is their potential to incorporate stakeholder knowledge into fisheries governance? Ambio 2011; 40(2): 133-43. [http://dx.doi.org/10.1007/s13280-010-0125-1] [PMID: 21446392]

[84] Tsagarakis K, Mytilineou C, Haralabous J, Lorance P, Politou C-Y, Dokos J. Mesoscale spatio-temporal dynamics of demersal assemblages of the Eastern Ionian Sea in relationship with natural and fisheries factors. Aquat Living Resour 2013; 26(4): 381-97. [http://dx.doi.org/10.1051/alr/2013067]

[85] Alverson DL, Hughes SE. Bycatch: from emotion to effective natural resource management. Rev Fish Biol Fish 1996; 6: 443-62. [http://dx.doi.org/10.1007/BF00164325]

[86] Stratoudakis Y, Fryer RJ, Cook RM. Discarding practices for commercial gadoids in the North Sea. Can J Fish Aquat Sci 1998; 55: 1632-44. [http://dx.doi.org/10.1139/f98-029]

[87] Rochet M-J, Trenkel VM. Factors for the variability of discards: assumptions and field evidence. Can J Fish Aquat Sci 2005; 62: 224-35. [http://dx.doi.org/10.1139/f04-185]

[88] Diamond B, Beukers-Stewart BD. Fisheries Discards in the North Sea: Waste of Resources or a Necessary Evil? Rev Fish Sci 2011; 19: 231-45. [http://dx.doi.org/10.1080/10641262.2011.585432]

[89] Sacchi J. The use of trawling nets in the Mediterranean 2008.

[90] Borges L. The evolution of a discard policy in Europe. Fish Fish 2015; 16(3): 534-40. [http://dx.doi.org/10.1111/faf.12062]

[91] Bellido JM, García-Rodriguez M, García-Jiménez T, González-Aguilar M, Carbonell-Quetglas A. Could the obligation to land undersized individuals increase the black market for juveniles: evidence from the Mediterranean? Fish Fish 2017; 18: 185-94. [http://dx.doi.org/10.1111/faf.12166]

[92] García-Rivera S, Lizaso JL, Millán JM. A quantitative and qualitative assessment of the discard ban in European Mediterranean waters. Mar Policy 2015; 53: 149-58. [http://dx.doi.org/10.1016/j.marpol.2014.12.003]

[93] Marsall NA. Can policy perception influence social resilience to policy change? Fish Res 2007; 86: 216-27. [http://dx.doi.org/10.1016/j.fishres.2007.06.008]

[94] Pauly D, Christensen V, Guénette S, *et al.* Towards sustainability in world fisheries. Nature 2002; 418(6898): 689-95. [http://dx.doi.org/10.1038/nature01017] [PMID: 12167876]

[95] Tsikliras AC, Dinouli A, Tsalkou E. Exploitation trends of the Mediterranean and Black Sea fisheries. Acta Adriat 2013; 54: 273-82.

[96] Guidetti P, Baiata P, Ballesteros E, *et al.* Large-scale assessment of Mediterranean marine protected areas effects on fish assemblages. PLoS One 2014; 9(4): e91841. [http://dx.doi.org/10.1371/journal.pone.0091841] [PMID: 24740479]

CHAPTER 5

Antibiotic Resistance in Fish

Dragana B. Ljubojević*

Scientific Veterinary Institute "Novi Sad", Novi Sad, Republic of Serbia

Abstract: Foodborne transmission of antibiotic resistant strains of bacteria from contaminated food has been recognized as an important hazard for human health in the past few decades and pathogenic strains of aquatic bacteria have long been considered as serious zoonotic hazards. Relatively low and competitive price of fish meat, the lack of religious and cultural barriers and the nutritional quality are the main reasons for the fact that fish meat is very attractive for consumers worldwide, so the measures to preserve the safety of fish meat are very important issue. The widespread consumption of different antimicrobial drugs in aquaculture could lead to the evolution and transmission of resistance determinates from fish to humans *via* the food chain. The link between the utilization of such drugs and the existence of antibioric resistance in infective either in commensal microorganism which can be found in fish and in aquatic environment and also residues of those drugs in fish due to improper utilization of veterinary drugs is discussed in this manuscript. The need for prudent use of antibiotics, particularly those which are used in human medicine is highlighted. The aim is also to point toward the present difficulty in the aquaculture management and to show the possible ecological and economical impacts. Monitoring and recording of occurrence should be the main operations in the upcoming years which should moot questions of how to make better and put into actions the established directives.

Keywords: Antibiotic resistance, Aquaculture, Bacterial pathogens, Commensal bacteria, Safety management systems, Zoonotic hazards.

INTRODUCTION

Fish meat has a high nutritive value which is of a great importance for human nutrition [1 - 4]. Relatively low and competitive price of fish meat, the lack of religious and cultural barriers and the nutritional quality are the main reasons for the fact that fish meat is very attractive for consumers worldwide [5 - 8], so the measures to preserve the safety of fish meat are very important issues.

The apparent emergence of resistance to antibiotics in microorganisms from diff-

* **Corresponding author Dragana B. Ljubojević:** Scientific Veterinary Institute "Novi Sad", Novi Sad, Republic of Serbia; Tel: +381648185486; E-mail: dragana@niv.ns.ac.rs

Panagiotis Berillis (Ed)

erent kinds of livestock farming as well as from the environment throughout the past decades and its potential involvements in problems related to human health have resulted in comprehensive examination of resistance in microorganisms [9, 10]. Dissemination of pathogens *via* the food chain has been recognized as a risk and antibiotic resistant strains of pathogens represent an important threat to public health [9 - 11]. The dissemination might happen throughout the whole process of production including stages before and after harvest, and also by eating of animal products carrying antibiotic resistant pathogens [12, 13]. The outbreaks of foodborne diseases are still very common despite the fact that plenty of food safety management measures have been put into practice in order to control the pathogens [14, 15]. Widespread and unlimited utilization of prophylactic drugs in aquaculture farming, in particular in developing countries in order to prevent microbial infection as result of poor hygienic conditions during fish farming lead to the emergence of antibiotic resistance in bacteria isolated from fish as well as from surrounding environment [16]. The great proportion of antibiotics are utilized in aquaculture for preventive and therapeutic measures due to bad sanitary and stressful conditions on fish farms, so the hazard of infections in cultured fish is notable. Appearance of resistant microorganisms makes more height worry about safety of fish and fishery products. The outbreaks of foodborne diseases are likely to happen, particularly in the developing counties due to conditions in all stages of the food chain, from fish farms, transport, fish processing facilities, distribution and cultural habits during preparation of fish at home [17]. The presence of antibiotic resistant microorganisms on fish rearing facilities in any part of the world has an international importance due to risk of their dissemination *via* farm workers, animals, insects, food products or water [18]. The fish originated from fish ponds are eaten by humans and can present a medium for spreading of resistance due to the fact that determinants of resistance are carried on transferable plasmids [19]. Since multiple antibiotic resistant of pathogenic and commensal bacteria represent a public health concerns, the occurrence of resistant bacteria on fish farms will be presented in this paper. The aims of the present work are the analysis and comparison of the results from available literature regarding the prevalence of antibiotic resistance in different bacteria collected from fish in different countries. Moreover, the most important international policies on use of antibiotics in fish have been reviewed. Also, some recommendations for preventive measures and monitoring of the use of antibiotics are proposed.

THE UTILIZATION OF VETERINARY DRUGS IN AQUACULTURE

Different types of veterinary drugs have been widely utilized in aquaculture for decades [20]. Notwithstanding the fact that they are used therapeutically in order to improve health and welfare of fish, they often were given for prophylactic purposes and as growth promoters in order to improve growth rate and feed conversion efficiency. Potential outcomes of the utilization of antibiotic in fish feeds are the evolution of antibiotic resistant microorganisms, the occurrence of multi-resistant strains, transmission of resistance to pathogens and shortcomings of antibiotics therapy in humans or in animals.

Fish are commonly treated as groups and antibiotics are commonly administered through the feed. In some cases, antibiotics can be used in fish in fish hatcheries in bath treatments added to the water. Brood stock fish can be treated with intraperitoneal infectious of antibiotics. The fact that the process of producing the medicated feed is commonly performed by the same feed equipment that also make the regular feed for fish representing a huge problem [21]. Another problem is the fact that usually larger amount of antibiotics have to be administered in seawater to secure a sufficient level of active antibiotic substances because the high level of salt may reduce the effect of the antibiotics [22]. Furthermore, some shortcomings in fish rearing such as overpopulation, chemical treatments and artificial diets may affect antibiotic resistance. The antibiotics used in fish production are often the same, or belong to the same class, as those used in human medicine [23]. The main consequences of the use of antibiotics, particularly the same ones or those from the same class in both humans and fish can cause propagation of antibiotic resistance due to a continuous positive selection for drug resistant clones in pathogens and commensal microorganisms [24]. Antibiotic resistance frequently occurs on fish farms and surrounding environment and can be spread to humans *via* food or water chain as well as by routes such as environmental contamination and direct contact with animals [9, 10]. Kumar and coworkers [25] and Sayadi and coworkers [26] noted that up to 90% of the antibiotic administered orally may pass through unaltered in fish excrement, so regular use of antibiotics in fish feed increases antibiotic resistant bacteria including pathogens being excreted by fish.

The utilization of veterinary drugs in fish production may lead to the evolution of resistance and many investigations have shown that antibiotic resistance in microorganisms is consequences of their utilization in fish production [27 - 31]. Moreover, there is strong evidence from around the world which indicate that fish meat may represent a vehicle of resistant bacteria [32].

During the last two decades, the use of antibiotic in aquaculture has changed due to the fact that their use for growth promotion was severely restricted or banned in some countries. Local, national and international management campaigns and policies have been developed to encourage prudent use of and limit unnecessary exposure to antibiotics, with the final goal of preserving their effectiveness for serious diseases [33]. This has led to divergence among different parts of the world regarding to incidence of antibiotic resistance [29, 30, 34 - 36].

The determinants of resistance in microorganisms is particularly frequently encountered following the introduction of some antibiotics in clinical practice and treatment of human or animal pathogens with antibiotics has a direct effect on the occurrence of resistance in microorganisms [37]. On the other hand, Rolain and coworkers [38] suggested that the appearance and rice of antibiotic resistance cannot be explained just by increasing use of antibiotics and that a complex interaction in an ecosystem is involved.

Tetracyclines represent a family of broad-spectrum antibiotics and tetracycline, chlortetracycline, and oxytetracycline are used in the treatment of bacterial, chlamydial, rickettsial and protozoal infections [39]. They are among the most popular and the most frequently used antibiotics in aquaculture but this group of antibiotics is also still used in human medicine due to its effectiveness, relatively reasonable price and relatively the lack of adverse effects. Tetracycline, chlortetracycline and oxytetracycline belong to the first generation, while minocycline and doxycycline belong to second generation of tetracyclines. Tetracyclines from first generation were usually used for growth promotion in animals, while tetracyclines from second generation were more used for therapeutic purposes in humans and pets [40]. Tetracyclines are frequently given to fish as first line drugs and they are commonly given before determination of resistant profile in bacteria. Thus, such extensive utilization of tetracycline has led to evolution of resistance in bacteria and could result in horizontal transfer of resistance determinants from fish to humans as well as to the environment. Indubitably, the improper utilization of tetracyclines has led to increasing of prevalence of resistance to tetracyclines among bacteria. Worrying health risk represents a fact that tetracycline resistant isolates of bacteria from fish have a high chance to become resistant to another antibiotics [29 - 31]. Moreover, oxytetracycline has a bed absorption rate in the fish intestinal tract [41], so it is usually given at a high dosage rate causing a high elimination rate of oxytetracycline which consequently increased the selective pressure and consequently can lead to selection of oxytetracycline- resistant microorganisms in fecal as well as in gastrointestinal microbiota of fish following treatment with oxytetracycline. Furthermore, tetracycline resistance can be protect and preserved in populations of microorganism over time, despite of selection pressure, so that can led to a general enhancement in antibiotic resistance over the period of time [42]. Tetracycline resistance is mediated by plasmid which carry lot of different genetic determinants which increased the possibility of bacteria to receive those factors of resistance by conjugation [40, 43] or by transformation [44]. The rapid dissemination of genetic determinants among population of microorganisms is a consequence of the fact that the genes are placed on mobile elements [45].

Besides tetracyclines, sulfonamides have been predominantly given for treatment of diseases caused by bacteria and protozoa in fish production [46] which results in the existence of residues of the above mentioned drugs in aquaculture and consequently to the prevalent resistance in bacteria to both tetracycline and sulfonamide at aquaculture facilities [29 - 31, 47 - 49]. Sulfonamides are very important due to their considerable excretion rate and solubility as well as their persistence in the surroundings [50]. Enne and coworkers [51] noted that sulfonamide-resistant bacteria can remain stable in the aquatic environment for 5 or ten years even without selective pressure and the bacteria resistant to sulfonamide can be more persistent than the sulfonamide. Luo and coworkers [52] found the high prevalence of sulfonamide resistance genes (sul 1 and sul 2) in the Haihe River in China which most likely resulted from its widespread use in fish farming.

THE ANTIBIOTIC RESISTANCE IN AQUATIC ENVIRONMENT

The spread of antibiotic resistance represents a great public health concern worldwide. Pathogenic and commensal microorganisms that are resistant to antibiotics are very important from the viewpoint of human health and the environmental reservoirs for resistance genetic determinants are still very badly known. Because of that, the origin of antibiotic resistance in nature is very important. In aquaculture production, antibiotics are released in the water environment after the therapy of infections in fish caused by bacteria or after the prophylactic use and that represent a serious public health concern since it have an impact on future therapeutic measures and also on environment. The utilization of antibiotic in prophylactic purposes in fish farming is certainly one of factors which leads to the emergence and dissemination of antibiotic resistant bacteria in the water [16]. The situation is different in comparison with the situation with terrestrial animals because the environment surrounding the threated fish is directly exposed to the antibiotic which is dissolved in the water. Moreover, it is very difficult to assess the effect of antibiotics on the resident microbiota having in mind the complexity of the aquatic environment. Antibiotic resistant microorganism come in the water from both human and animal sources and might transfer the genes of resistance to the microorganism which live in water and have their own genes of resistance. Antibiotics originated from industry can also be present in water and can affect ecosystems of existing microorganism. Akinbowale and coworkers [35] suggested that effluent from fish production contributes to the resistance to antibiotics and metals in surroundings. Also, antibiotic contamination in environment may have a key role in the transmission of resistance in aquaculture. Miranda and Zemelman [53] found the high occurrence of antibiotic resistant bacteria in wild fish so they supposed that there is probability that antibiotic resistant fecal microorganisms of home sewage released into the marine waters can transfer their resistant genes to indigenous microorganisms of fish, rising their spread in the surroundings. The antibiotic resistant bacteria can come to seawater *via* domestic sewage [54]. Moreover, the occurrence of antibiotic resistant microorganisms in coastal waters worldwide can be result of selective pressure generated by high discarding of pharmaceutical wastes into water [53]. Schmidt and coworkers [29] highlighted the fact that aquaculture facilities mainly discharge their wastewaters without any treatment and as a result antibiotic residues persist in that waters and around facilities after the utilization of antibiotic. Furthermore, results obtained by DePaola and coworkers [55], McPhearson and coworkers [56], Spanggaard and coworkers [57] and Guardabassi and coworkers [58] suggested a link between increasing resistance levels in microorganisms on and toward aquaculture facilities and the utilization of veterinary drugs during fish farming. Chelossi and coworkers [59] examined the effects of fish rearing on possibility of their connection with resistance to antibiotics in costal sediments of the Ligurian Sea. Their results demonstrated a high prevalence of antibiotic resistance among microorganisms originated from control and affected sediment and suggested high levels of resistance to antibiotic around aquaculture facilities. The study conducted by Chelossi and coworkers [59] also reported a similarity in antibiotic resistance patterns between the benthic community of microorganisms and microorganisms from fish farms. The common introduction of antibiotic supplemented feed and the absence of treatment of wastewater from fish farms enable to antibiotics to come into environment and make a persistent selective pressure on pathogenic and environmental bacterial communities. Antibiotic residues can affect the bacterial population of sediment and promote proliferation of resistant strains of bacterial pathogens. Also, the occurrence of enteric microorganisms in waters can contaminate fish, so they represent a medium for dissemination of pathogens to people. Antibiotic resistant enteropathogenic bacteria have been isolated from different aquatic environments [60 - 65]. The utilization of veterinary drugs in fish farming, in particular no biodegradable rises selective pressure of antibiotic in aquatic environment, makes easier the transfer of determinants of resistance among microorganisms encompass pathogenic strains and enables the occurrence of drug residues in cultured aquaculture products [16]. Samuelsen and coworkers [66] reported that resistant microorganisms persisted in sediment from fish farms certainly eighteen months following therapy. Miranda and Zemelman [53] also observed that bacteria isolated from demersal fish (fish which live near the bottom) showed higher levels of resistance to more than two antibiotics than those from pelagic habitats and that that difference may be attributed to the high level of resistant microorganisms originated from sediment in contaminated water.

Chelossi and coworkers [59] examined bacterial strains for resistance to antibiotics and found that 96% of examined bacteria were resistant to at least on antibiotic and that 100% of Gram negative bacteria were not susceptible to ampicillin and sulphamethoxazole and 61.5% were resistant to chloramphenicol and 76.9% were resistant to streptomycin, while Gram-positive bacteria were resistant to ampicillin (84.6%), nalidixic acid (80%) and sulphamethoxazole (96.1%), while they reported low level of resistance to tetracycline (23.1%). They further reported quinolone resistance in 70% of examined bacteria. The resistance to nalidixic acid was higher around facilities for rearing fish and that approximately 90% of bacteria were Gram positive bacteria that can be resistant to nalidixic acid by nature but, they also found high level resistance to nalidixic acid (quinolone) in Gram negative bacteria in sediment originated from farm indicating a presumable utilization of quinolone in aquaculture practices. According to results obtained by Chelossi *et al.* [59], Gram negative bacteria were highly resistant to ampicillin and streptomycin in affected sediments and also the changes in bacterial population is evidently associated with resistance in Gram negative bacteria from sediments. Ampicillin was routinely used on examined farms so a high level of Gram negative bacteria in sediment under the cages may be result of adaptation of microbial population. Chelossi and coworkers [59] noted that the greater occurrence of multi-resistant strains in fish-farm sediment in their research represents a significant concern of veterinary drugs use. They suggested that the high occurrence of antibiotic resistance in examined microorganisms may be connected to the fact that some bacteria from the genus *Bacillus* produced antibiotics.

The results obtained by Economou and coworkers [67] highlighted the complexity of the interactions between pathogens and commensal microorganisms. They collected water samples from chosen areas in Acheron and Kalamas Rivers and coast of the Ionian Sea in north western Greece in order to examine the presence and incidence of resistance in *Salmonella* strains and their possible interactions with pathogens and commensal microorganism from examined areas. They detected twenty eight serovars of *Salmonella* spp., including *Salmonella Enteritidis*, *Salmonella Thompson* and *Salmonella Virchow*. They found that more than 30% of *Salmonella Enteritidis* and all isolates of *Salmonella Virchow* were resistant to more than three antibiotics. Wastewaters from livestock farms contribute to the contamination of the rivers with antibiotic resistant microorganisms, both directly *via* release of such waters without treatment or *via* the manuring the field or fishponds. The theory is supported by the isolation of serovars of *Salmonella,* resistant to more than two antibiotics by Economou and coworkers [67] which are zoonotic serotypes.

Gao and coworkers [31] examined the occurrence of resistance to sulfonamides and tetracyclines in the water and sediment samples collected from fish farms in China. They identified the incidence of antibiotic resistance in microorganisms and genes of resistance at six fish farms and found the highest prevalence of genes related to resistance to sulfonamides which are the most widely used on examined fish farms. They reported that genes of resistance to sulfamethoxazole and tetracycline including tetM, tetO, tetT, tetW, sul1 and sul 2 were isolated from *Bacillus cereus* which suggested the occurrence of multi-resistance in the examined area.

Matyar and coworkers [22] reported the link between resistance to antibiotic and heavy metals in Gram negative microorganisms isolated from shrimps, sediment and seawater. They performed their study in contaminated Iskenderun Bay, on the south coast of Turkey. They isolated thirty one bacteria species and found in 11.4% of samples *Escherichia coli*, in 9.7% *Aeromonas hydrophila* and in 9.3% *Stenotrophomonas maltophilia.* They found that resistance to ampicillin, streptomycin and cefazolin was 93.2, 90.2 and 81.3%, respectively. Further, resistance to imipenem, meropenem and cefepime was 16.5, 13.9 and 8.0%. Furthermore, they reported that more than 50% of examined microorganisms were multi-resistant and showed resistance to more than seven antibiotics. They observed that the resistance patterns were different in metal resistant bacteria from seawater and sediment compared to those from shrimp and showed resistance to different groups of antibiotics. They showed high levels of resistance to antibiotics and heavy metals in examined microorganism from Iskenderun Bay.

CO-SELECTION FOR THE RESISTANCE TO THE ANTIBIOTICS

The phenomenon of co-selection contributes to the massive persistence of multi-resistant bacteria. Nielsen and coworkers [68] suggested that co-selection remarkably contributed to the occurrence of bacterial resistance to more than one antibiotic. Schmidt and coworkers [29, 30] reported that tetracyclines provoke resistance to other groups of antimicrobial drugs and vice versa. The high occurrence of sulphonamide resistance was recorded by many authors [31, 46] and that might be due to slow degradation rate of sulphonamides [69]. Diarrassouba and coworkers [70] hypothesized the maintance of genes related to resistance to sulphonamides as result of their interactions with other genes. That is confirmed by results obtained by Chelossi and coworkers [59]. They noted resistance to amoxicillin and tetracycline in sulphonamide resistant bacteria. That implicated that resistance to antibiotics which are not routinely used in clinical medicine could be potential threat to humans due to co-selection for the resistance to the used antibiotics. Adams and coworkers [71] and Rhodes and coworkers [19] suggested that this is probably due to the fact that resistance determinants are often found on mobile genetic elements. Also, mobile elements may carry both tetracycline and other antibiotic related resistance genes [72]. Schmidt and coworkers [30] further observed that isolates with chloramphenicol acetyltransferase gene (catl) have been found to be resistant to ampicillin and tetracycline which could be explain by the fact that chloramphenicol acetyltransferase gene (catl) is plasmid mediated and the tetracycline resistant gene is also plasmid mediated then the two genes can be mediated by one plasmid. Goñi-Urriza and coworkers [73] and Hatha and coworkers [74] confirmed that and showed that the isolates which were resistant to chloramphenicol showed resistance to ampicillin and tetracycline which may implicated that continuous use of tetracycline in fish could resulting in the persistence of chloramphenicol resistance in bacteria without its use. Broughton and Walker [36] and Ryu *et al.* [32], further observed that the greater number of the ampicillin resistant microorganisms also showed resistance to tetracycline and Adelaide and coworkers [75] reported that ampicillin and its analogs are rarely used in in Kenya so there was no specific selection pressure resulting directly from ampicillin while the animals were continuously exposed to tetracycline. Adelaide and coworkers [75] noted that selective pressure for tetracycline was present in all the farms examined and that the selective pressure played a key role in the selection of the other antibiotics and it could be that the resistance factors for the antibiotics are harbored on the same mobile element. Scaletsky and coworkers [76] described a relationship between class 1 integron and resistance to ampicillin, sulphamethoxazole, tetracycline, streptomycin, kanamycin and chloramphenicol. Koo and Woo [40] further reported that the co-selection of resistance to tetracycline and other antibiotics is influenced by the quantity of used veterinary drugs.

MECHANISM OF ANTIBIOTIC RESISTANCE IN AQUACULTURE

Antibiotic resistance is result of plasmid-mediated horizontal gene transfer or chromosomal mutation. It can be a result of exposition of water polluted with effluent from fecal wastes of livestock and humans and can be related with resistance to metals. Jones and coworkers [77] claimed that the presence of antibiotic and metal resistance in microbial community is environment phenomena on natural selection for survival. The existence of transferable R-plasmids is also associated with antibiotic resistance. R-plasmids encode resistance to different antibiotics and they were detected in *Aeromonas salmonicida* [78]. Further, resistance which is medicated by plasmids was detected in a majority of fish pathogens such as *Aeromonas salmonicida*, *Aeromonas hydrophila*, *Edwardsiella tarda, Pasteurella piscicida*, *Pseudomonas fluorescens*, *Yersinia ruckeri* and *Vibrio anguillarum*. Aoki and Takahashi [79], DePaola and coworkers [55], DePaola and coworkers [80], Andersen and Sandaa [81], Adams and coworkers [71] and Rhodes and coworkers [19] described different tetracycline related resistance genes on R-plasmids in the genus *Aeromonas*. It is well known that tetracycline resistance determinants can be carried by transposons or other mobile genetic elements, which can enhance the ability to move [82]. The occurrence of R-plasmids in various bacteria from fish farms and sediment was also reported by Chelossi and coworkers [59]. Acquired resistance to beta-lactams and tetracyclines are often plasmid-coded or transposon-coded [83]. R plasmids were detected in *Yersinia ruckeri* [84, 85] while Schmidt and coworkers [29] did not found resistance to antibiotic in isolates of *Yersinia ruckeri* and stated that it is not clear why antibiotic resistance is not often in *Yersinia ruckeri* but it is highly present in the other microorganisms under the same conditions.

It was thought that the transfer of R-plasmids has a key role in spreading of resistance to oxytetracycline resistance around fish farms [19, 79, 86] but likely not the sole mechanism of transfer of genes of resistance. According to results obtained by Schmidt and coworkers [30] a relationship existed between oxytetracycline resistant *Aeromonas* spp. carrying plasmids and occurrence of tetA gene. On the other hand De Paola and coworkers [55] and DePaola and coworkers [80] found that the majority of donors in successful mattings had unknown tetracycline resistance genes. Goni-Urriza and coworkers [73] reported probable chromosomally mediated resistance and noted increased levels of resistance to tetracycline in *Aeromonas* spp. isolated from water before and after point of wastewater release. Schmidt and coworkers [29] did not find R plasmids or similar structures in *Flavobacterium psychrophilum*, so they supposed that chromosomally determined mechanisms of resistance could be predominant. The occurrence of a truncated or completed transposon Tn1721 was documented on R-plasmids and indicated one more mechanism of tetracycline resistance [19, 30, 82].

THE LINK BETWEEN THE UTILIZATION OF VETERINARY DRUGS ON THE FISH FARMS AND THE LEVEL OF RESISTANCE IN THE MICROORGANISMS AND THE POSSIBLE EFFECTS ON HUMAN HEALTH

The bacterial population on fish ponds is affected by production practices, especially by nutrition and utilization of veterinary drugs on the fish ponds. The antibiotic of choice for medical treatment of *E. coli* infections is trimethoprim-sulphamethoxazole [88], so the increased level of resistance to this antibiotic in cultured fish obtained by Sarter and coworkers [87] represent a serious health problem. Also, cephalosporins are used for the treatment of infection caused by *Salmonella* spp. in children [47] as well as for the treatment of animals, so the reported resistance to broad-spectrum cephalosporins detected in the bacteria from fish and sediment in Australia by Akinbowale *et al.* [35], represent a public health concern.

The high resistance to ampicillin, oxytetracycline, trimethoprim-sulphamethoxazole, nalidixic acid, chloramphenicol and nitrofurantoin found by Sarter and coworkers [87] in isolates from commercial fish farms indicate that the utilization of medicated feeds with more than one veterinary drug might be a reason for spreading resistance. The increased levels of resistance to the mentioned as well as other antibiotics in bacterial isolates from aquaculture were also observed by Miranda and Zemelman [89], Chelossi and coworkers [59] and Furushita and coworkers [90]. Overall, the results suggest that one of the possibilities to reduce the spread of antibiotic resistance could be by decreasing the number of veterinary drugs in the feed. Besides, the bacteria can acquired mechanisms of resistance before drug utilization, however drug application can lead to shift in genetic structure of microorganisms toward their adaptation to recently circumstances [91]. Continual utilization of veterinary drugs hence contributes to preserving resistance in microorganisms in surroundings. Further, they can continue to live in the drugs presence in different ecosystems and to spread mobile determinants of resistance to other microorganisms. Many fish farms are in vicinity of livestock farms and antibiotic resistance could be results of run-off from areas populated with animals treated with antibiotics. Also, there is possibility that the resistance has emerged as a consequence of improper utilization of veterinary drugs [92]. The use of large amounts of a different types of veterinary drugs, especially no biodegradable provides that they can exist in water, producing selective pressure [16]. Further, that leads to the occurrence of resistant microorganisms in environment, the changes in microbial community in water and in sediment, the emergence of resistance in pathogenic bacteria of fish and the spread of resistant determinants to human and animal microorganisms. The utilization of veterinary drugs in fish feed also lead to the possibility for the occurrence of drug residues in fish flesh. Veterinars or fish farmers sometimes use antibiotics to control bacterial diseases in the aquaculture production. In many countries, there is either no, or no productive, control on the utilization of veterinary drugs in cultured fish species [93]. Besides that, Alderman and Hastings [94] noted that the use of antibiotics in ornamental fish, especially in some exporting countries is significant and there is evidence that multi-resistant bacteria may be frequent in such fish which can be a health concern because ornamental fish are in close contact in humans although they are not eaten. Antibiotic resistant microorganisms have been described both in cultured fish [32, 56, 57, 87] and in wild fish [53, 95, 96]. Miranda and Zemelman [53] examined resistant microorganisms from fish caught in the Concepcion Bay in Chile and observed that the level of resistance of microorganisms obtained from gills and contents of intestine was significant to ampicillin, nitrofurantoin, streptomycin, tetracycline and very low to amikacin, cotrimoxazol and gentamycin. They noted that the most resistant microorganisms were resistant to three or four groups of veterinary drugs and were members of families *Vibrionaceae* or *Enterobacteriaceae*. Their results suggested that fish living in areas near the disposal of domestic sewage may also be a reservoir of resistant microorganisms and consequently represent a serious health concern. Pathak and Gopal [95] determined resistance for antibiotics and metals in arborescent organ, gill, kidney, liver and spleen of *Clarias batrachus*. They observed high prevalence of antibiotic and metal resistance which can be a public health threat and found that resistance to colistin, ampicillin, gentamycin, carbenicilin, tetracycline, streptomycin and ciprofloxacin was 98, 82, 34, 28, 20, 12 and 2%, respectively. Sarter and coworkers [87] examined the resistance of fish bacteria to antibiotics from 3 different catfish farms of *Pangasius* sp. in Viet Nam and reported high prevalence of resistance to antibiotic in fish indigenous microorganisms. Also they evaluated the selective pressure of antibiotics exerted in water environment. They determined that 49.1, 35.2 and 15.7% of the examined microorganisms belonged to families *Enterobacteriaceae*, *Pseudomonads* and *Vibrionaceae,* respectively. They determined the high levels of resistance to ampicillin, oxytetracycline, trimethoprim-sulphamethoxazole, nalidixic acid, chloramphenicol and nitrofurantoin.

According to results obtained by Bruun and coworkers [97] and Schmidt and coworkers [29, 30] the patterns of resistance of fish pathogenic microorganisms can be a result of wide utilization of antibiotics. DePaola [27] and DePaola and coworkers [28] suggested existence of a link between administration of oxytetracycline in fish feed and the incidence of resistance to oxytetracycline resistance in catfish intestinal microbiota and they supposed that the fish digestive tract can be a reservoir for selection of resistant bacteria. High levels of oxytetracycline resistant *Aeromonas* spp. from farmed catfish and surroundings as a consequence of regularly utilization in fish feed was observed by DePaola and coworkers [55]. They found tetA and tetE genes in more than 90% of tetracycline resistant *Aeromonas hydrophila*. The application of oxytetracycline for preventing and controlling diseases in salmon rearing is common in Chile, so Miranda and Zemelman [89] investigated the level of resistance to oxytetracycline in microorganisms isolated from fish fry, feed and water collected on fish farms without data of recently use of oxytetracycline. They found the highest proportion of resistant bacteria in the effluent samples and the lowest proportion in influent samples. They identified *Pseudomonas fluorescens, Aeromonas hydrophila, Stenotrophomonas maltophilia, Sphingomonas paucimobilis*, *Acinetobacter lwoffii*, *Pseudomonas putida*, *Burkholderia cepacia*, *Brevundimonas vesicularis*, *Acinetobacter johnsonii*, *Pantoea* sp. and *Moraxella* sp. in 28.2, 9.7, 5.8, 5.8, 4.8, 4.8, 3.9, 3.9, 2.9, 2.9 and 2.9% of samples, respectively. They observed that *Acinetobacter lwoffii* and *Stenotrophomonas maltophilia* were the most prevalent in fish feed, while *Pseudomonas fluorescens* and *Aeromonas hydrophila* were the most prevalent in fish fry. They reported the high occurrence of resistance to oxytetracycline. Their results suggested the areas around salmon farms can be a notable source of microorganisms which harboured mobile genetic elements for resistance to oxytetracycline and so represent a serious threat for people employed on farms. McPhearson and coworkers [56] compared the prevalence of antibiotic resistance in Gram negative bacteria originated from the sediment, water and digestive tracts of catfish from rivers and fish farms in the USA. They determined resistance to ampicillin, chloramphenicol, kanamycin, nitrofurantoin, tetracycline and oxytetracycline. They observed that *Plesiomonas shigelloides* and *Aeromonas hydrophila* were the most prevalent bacteria and that various *Enterobacteriaceae* spp. and pseudomonads were also present but in lower numbers. Resistance to one or more antibiotics were linked to the utilization of antibiotics and it was more prevalent on fish farms with a recent use of veterinary drugs than in those without their recent use. Furthermore, in microorganisms originated from rivers the occurrence of resistance was the lowest. They observed that the incidence of microorganisms resistant to one or more than one group of antibiotics was significantly different among the examined samples originated from farms with or without history of antibiotic utilization and samples from river. They found that the proportion of resistance to antibiotic in general was higher in bacteria from farms which used antibiotic than from farms without their use while the lowest proportion was in samples from rivers. Resistance to tetracycline was more than 50% in microorganisms from farms which used antibiotics, while it was only 1% in the microorganisms from rivers. They noticed increasing of resistance levels in microorganisms from rivers, to those from farms without antibiotic use and to treated farms for each antibiotic. Alderman and Hastings [94] reported that the resistance to the 4quinolones, such as oxolinic acid and flumequine, authorized for aquaculture use in fish farming arises as a consequence of the selection of resistant mutants and that this selection can be demonstrated to occur both *in vitro* and *in vivo* in the farm environment. On the other hand, Kerry and coworkers [98] did not provide argument for a selection of resistant bacteria in digestive tract of smolts of Atlantic salmon after treatment with oxytetracycline. Also, in research conducted by Schmidt and coworkers [29] was not shown the existence of correlation between utilization of veterinary drugs and occurrence of resistance. One of the reasons could be the fact that the time of collecting samples was not always in correlation with the use of antibiotics or diseases outbreaks. Results obtained by Schmidt and coworkers [29] showed the high influence of trout rearing on microorganisms in environment and pathogenic bacteria of fish whereas incidence of resistance was higher in samples originated from ponds or outlet that those from inlets. They also noted an unexpected occurrence of resistance to oxytetracicline which was 69% in *Aeromonas* spp. and 72% in *Flavobacterium* spp. Namely, oxytetracycline was not be commonly used antibiotic on examined farms. Schmidt and coworkers [30] also pointed out that oxytetracycline was rarely used for diseases treatment on fish farms in Denmark. On the contrary, sulfadiazine-trimethoprim and oxolinic acid have been widely utilized of fish farms in Denmark. They further compared the *Flavobacterium psychrophilum* sampled during their study to a samples of Danish clinical isolates from 1994 and noted that the prevalence of resistance increased from 50 to 100% for oxolinic acid and from 0 to 36% for amoxicillin [97] and that such prompt emergence of antibiotic resistant microorganisms on trout farms represent a serious public health concern. Results obtained by Schmidt *et al.* [23], also highlighted an important influence of fish farming on the *Aeromonas* spp. in the examined farms whereas high occurrence such as 48% of multi-resistance suggested that the dissemination occurred by the horizontal gene transfer. Also, findings of class 1 integrons in 45% of examined bacteria contributed to this theory. Furthermore, the clonal spread of resistance in *Aeromonas* spp. was not supported by the determined genes and the high diversity of bacteria. Schmidt *et al.* [30], demonstrated by experiment of conjugation that class 1 integrons were cotransferred to *E. coli* recipients on oxytetracycline resistance plasmids and that there was possible existence of other mechanisms of transfer resistance. They also suggested that extensive antibiotic resistance within motile *Aeromonas* species could be a reservoir of resistance determinants able to spread to other microorganisms. Furushita and coworkers [90] isolated tetracycline-resistant bacteria from fish obtained from three farms in Japan and found the same origins of tet genes from examined fish and those from clinical isolates. Their result indicated that tetracycline resistant microorganisms isolated from fish may transfer their tetracycline resistance determinants to pathogenic bacteria even to human pathogens.

THE RESISTANCE TO ANTIBIOTICS IN FISH PATHOGENIC AND COMMENSAL MICROORGANISMS

The *Aeromonas* spp. are ubiquitous bacteria in water environment and some species are primary pathogens to fish [86]. In general, they are not pathogenic for humans, but some of them were found in them [99]. Investigations conducted by Schmidt and coworkers [29] indicated that trout rearing seems to have effect on *Aeromonas* spp in environment. Schmidt *et al.* [29], reported that the predominant multiresistance phenotype was oxytetracycline – sulfadiazine - trimethoprim resistance which was observed in 89 isolates (28%), followed by oxytetracycline - oxolinic acid resistance which was found in 31 isolates (10%) and oxytetracycline - oxolinic acid – sulfadiazine - trimethoprim which was detected in 28 isolates (9%). The motile *Aeromonas* spp. were often selected as bacterial indicators due to their ubiquitous presence in the water, sediment and fish. Besides, some species of the *Aeromonas* spp. develop resistance to one or more antibiotics [89]. Also, R- plasmids were commonly found [19, 55, 71, 79 - 81]. Schmidt and coworkers

[30] isolated motile *Aeromonas* spp. from fish farms in order to determine resistant determinants associated with high occurrence of antibiotic resistance. They found *Aeromonas hydrophila*, *Aeromonas bestiarum*, *Aeromonas veronii* biovar *sobria* and unknown bacteria in 35.3, 19, 15.3 and 15% of samples, respectively. They reported that antibiotic resistance patterns did not vary significantly. They found that the most abundant resistance phenotype was a combined oxytetracycline and sulphadiazine - trimethoprim resistance in 37% of isolates and it was associated with the determination of class 1 integron. They noted occurrence of four various gene cassettes in bacteria with integrons. They reported that there were no gene cassettes in twenty three bacteria containing integrons. They found a relationship between the presence of tetA and R-plasmids among bacteria resistant to oxytetracycline while tetE and unknown oxytetracycline genes and class 1 integrons were not related with the presence of plasmids. Results obtained by Schmidt *et al.* [30], indicated the involvement of other gene transfer mechanisms in the spreading of resistance to antibiotics among *Aeromonas* spp. in environment. They reported that patterns of resistance were not different among *Aeromonas* species. Schmidt and coworkers [29] noted that 69% of *Aeromonas* isolates showed resistance to oxytetracycline, 43% to sulfadiazine-trimethoprim and 20% to oxolinic acid. Among 69% of the oxytetracycline resistant bacteria the tetracycline resistance genes were found only in 30% of them. Schmidt and coworkers [30] found out that sulfadiazine - trimethoprim resistance was highly associated with the existence of class 1 integrons. They noted that a majority of integron-positive bacteria were without inserted gene cassettes. They found a relationship between the tetA and plasmids in oxytetracycline resistant *Aeromonas*. The dominant gene in transconjugants was tetA. Contrary, the presence of class 1 integrons was not correlated with the presence of large plasmids. *Aeromonas* from eel farms showed high resistance to antibiotics [49]. They are also commonly resistant to heavy metals [100]. Hatha and coworkers [74] isolated *Aeromonas* spp. from the intestines of cultured freshwater fish (*Catla catla, Labeo rohita* and *Ctenopharyngodon idella*) and identified *Aeromonas hydrophila* (61%), *Aeromonas caviae* (30%) and *Aeromonas sobris* (7%). Further, they found high resistance to ampicillin (100%), novobiocin (92.6-100%), oxytetracycline (about 40%), amoxycillin (33.3-52.7%) and polymixin-B (0-41.8%), implicating possible utilization of mentioned drugs on fish farms in India. Chloramphenicol resistance was less than 20% (5.5-18.5%). They further noted that the presence of residues of chloramphenicol in fish produced in China and India lead to a ban of import of fish from some producers from those countries by European Union. *Aeromonas salmonicida* can cause various diseases in different fish species. *Aeromonas salmonicida* subspecies *salmonicida* causes furunkulosis in salmonid fish. Sulfonamides were primarily used for the therapy of furunculosis in fish hatcheries in the United

States in 1950s [101]. Snieszko and coworkers [102] isolated *Aeromonas salmonicida* from trout in the United States in 1959 which was resistant to sulfathiazole and tetracycline and that was the first report on acquired antibiotic resistance in fish pathogen. Snieszko isolated a strain from diseased brook trout in 1959 which was designated the type strain of the species (ATCC 14174) which was later reported as a strain which contain transferable resistance to both tetracycline and sulfonamides. Goñi-Urriza and co-workers [73] reported R-plasmids were the responsible genetic factors for the transfer of resistance in *Aeromonas salmonicida* and 104 isolates of *Aeromonas salmonicida* in France was 100% resistant to sulfonamides, 11.5% to tetracycline and/or chloramphenicol. Furthermore, the resistance to tetracyclines and /or chloramphenicol was transferred to *Escherichia coli* K-12, *A. salmonicida* and *Aeromonas hydrophila* together with resistance to sulfonamides [103]. It is interesting that the resistance to sulfonamides was not transferred as a single characteristic. Beta-lactamases occur widely among *Aeromonas* spp. and their occurrence has been reported in *Aeromonas salmonicida* ssp. *achromogenes* [104]. Gram-negative bacterium *Aeromonas hydrophila* is often present in water, including fish farms. These bacteria are cause of the septicaemia, which occurs in many species of fresh water fish [105]. Infection caused by *A. hydrophila* (motile aeromonad septicemia) is one of the biggest problems in carp aquaculture. Currently, the most widely used method to control the motile aeromonad septicaemia is use of antibiotics. In the absence of appropriate vaccines to control this economically significant infection, the use of the proper antibiotics is of great importance [105]. Despite the fact that there are several alternatives to the treatment with antibiotics, including vaccination, immunostimulators or probiotics [106], treatment with antibiotic is usually used for therapy of diseases caused by bacteria. Intensive aquaculture has resulted in a growing problem of bacteria diseases, leading to the widespread use of antibiotics for their treatment, and increased antimicrobial resistance in aquatic bacteria [89]. Motile aeromonad septicaemia can still be controlled by the proper application of antibiotics such as flumequine, oxytetracycline or trimethoprim-sulfonamide. The emergence of resistance in *A. hidrophila* can be consequence of improper and subtherapeutic utilization of antibiotics. During the treatment of viral diseases, the antibiotics are used in order to solve difficulties caused by bacteria. Resistance to more than two antibiotics in *Aeromonas* spp. was described worldwide [107]. Sulfonamides were first used in fish feed and after that the utilization of oxytetracline and 4-quinolones was widely described [105]. DePaola and coworkers [55] and DePaola and Roberts [108] noted wide prevalence of tetracycline resistant *A. hydrophila* in catfish. Vivekanandhan and coworkers [92] isolated *Aeromonas hydrophila* from fish and prawns. They found high resistance to methicillin and rifampicin of 100 and 99% respectively. The resistance to chloramphenicol was detected in 3% of examined bacteria.

Li and coworkers [34] examined the resistance to antibiotics and the plasmid occurrence in *Vibrio* spp. isolated from unhealthy silver sea bream (*Sparus sarba*) in Hong Kong. They isolated fifty one *Vibrio* spp. from ill fish and identified 7 species including 24 strains of *Vibrio alginolyticus*, 12 of *Vibrio vulnificus*, 7 of *Vibrio parahaemolyticus*, 4 of *Vibrio logei*, and 2 of *Vibrio pelagius II*, one of *Vibrio fluvialis* and one strain of *Vibrio medittterranei*. They reported that all examined bacteria showed no resistance to ceftriaxone, streptomycin, nalidixic acid and rifampicin, while most of them showed no resistance to ceftazidime, netilimicin, chloramphenicol and sulfamethoxazole. On the other hand, they observed that most isolates were resistant to trimethoprim, cefuroxime, ampicillin, kanamycin and amikacin, 76.5, 66.7, 60.8, 58.8 and 55% respectively. High resistance can be a consequence of uncontrolled use of the mentioned antibiotics in fish. They noted that one to four plasmids were carried by fifteen isolates. Further they proved by conjugation that plasmids with resistance to related antibiotic can be transferred. Their results indicated chromosomal mechanism of resistance to these related antibiotics (ampicillin, cefuroxime and trimethoprim).

E. coli are refer as emerging pathogen just as the majority of pathogens involved in diseases caused by food. They can be involved in many diseases of humans including gastroenteritis, diseases of urinary tract, meningitis in babies, haemolytic uremic syndrome and rarely septicaemia [109]. *E. coli* represent ubiquitous microorganisms in fish intestines, but sometimes cause systemic infection. All in all, *E. coli* have a great ability to adapt and live in many different conditions [91]. Ryu *et al.* [32], characterized the resistance genes in *E. coli* from fish and seafood in Seoul, Korea. They observed high levels of resistance to the tetracycline, streptomycin, cephalothin, ampicillin and ticarcillin, which were 30.7, 12.8, 11.7, 6.7and 6.1%, respectively. All bacteria were susceptible to amikacin, amoxicillin-clavulanic acid and cefoxitin. Further they found the tetB and tetD in 41.4% of isolates and 20% of isolates, respectively. The blaTEM gene was detected in 21.4% bacteria and the gene for resistance to aminoglycoside, aadA in 25.7%. They found class 1 and class 2 integrons in 41.4 and 2.9% of the bacteria, respectively. Further, they found two different gene cassettes in isolates containing class 1 integron. Such results indicate that fish and seafood represent a significant reservoir for resistant microorganisms.

It is also important fact that the majority of the veterinary drugs which are utilized in fish are not completely absorbed in the fish intestinal tract and about ninety percent of the given amount may be excreted in aquatic environment [25].

The antibiotic resistance in *E. coli* represents severe clinical threats to humans. It is well documented that *E. coli* as commensal microorganisms might be reservoirs of resistance genes for potentially pathogenic microorganisms [110]. It can be transferred between humans and animals [111] and also can be spread through water, food or soil. The antibiotic resistance in isolates of *E. coli* from faecal samples of animals may be anthropogenic in origin. Moreover, *E. coli* is important faecal indicator bacteria and it is commonly used in order to determine contaminated waters. Antibiotic resistant *E. coli* have been described in livestock [112, 113] and in humans [114] and it can spread to the environment, infecting fish which becoming new reservoir of resistant *E. coli* in the environment. Recently, the emergence of *E. coli* strains resistant to broad spectrum cephalosporin's as well as *E. coli* strains resistant to fluoroquinolones became great public health concerns and represent a serious problem for the therapy of human infections [115, 116]. Skurnik and coworkers [117] suggested that the level of antibiotic resistance in faecal microorganisms from animals can be in correlation with the level of contact of these animals with humans. The existence of direct exchange of resistant determinants between animals and human pathogens is not completely clarified as well as the extent to which the use of antibiotics in veterinary practice participate in selective pressure which could be the important contributor in spreading of antibiotic resistance in human isolates [118]. The resistance to tetracycline, streptomycin and ampicillin is common in farm and wild animals which live in human vicinity [119]. Importantly, antibiotics are used for growth promotions on fish farms worldwide, and that contributed to the dissemination of resistance to antibiotics. The utilization of veterinary drugs during fish rearing can lead to appearance of resistant microorganisms in fish, their excrement and hence in the surroundings [120]. Fish and their surroundings are potent reservoirs of both resistant microorganisms and genes of resistance that could be transferred to people by direct contact or indirectly as a result of the fertilisation or by food chain [121].

E. coli could be a source of contamination of fish meat during evisceration after slaughter and processing, during contact with contaminated water or during food handling [122]. Thus, antibiotic resistant faecal *E. coli* from fish can affect people either directly or *via* food. Besides that, *E. coli* are commensal bacteria *E. coli* are also very important pathogen and are connected to extra-intestinal pathogenic *E. coli* strains in humans [109], such as urinary tract infections [123]. The majority of infections caused by *E. coli* in humans are not harmful, but some infections may be lethal. It is well established that the level of antibiotic resistance in commensal *E. coli* could be used as a great indicator of the selection pressures due to the utilization of veterinary drugs and moreover the patterns of resistance that could be expected in pathogens. Monitoring of antibiotic resistance among commensal *E. coli* is of great importance not only in humans but also in fish and meat thereof in order to detect the possible route of transfer of resistant microorganisms and resistant determinants from fish to humans.

Akinbowale *et al.* [35], examined isolates of *Pseudomonas* spp. and *Aeromonas* spp. from nine trout (*Oncorhynchus mykiss*) farms in Australia and found high levels of resistance in samples of sediment and fish. They isolated *Pseudomonas* spp. from fish and sediment including *Pseudomonas aeruginosa*, *Pseudomonas stutzeri*, *Pseudomonas fluorescens* and *Burkholderia pseudomallei* in 45, 29, 20 and 6%, respectively. Besides that, they isolated 90 *Aeromonas* spp. including *Aeromonas hydrophila*, *Aeromonas veronii* biovar *sobria*, *Aeromonas caviae* and *Aeromonas veronii* biovar *veronii*, in 77, 11, 10 and 2%, respectively. They reported that no resistance to all studied antibiotic was observed just in one *Aeromonas* and absence of resistance to cefotaxime, oxolinic acid, ciprofloxacin and gentamycin was observed in all bacteria from fish and sediment. They found that the incidence of resistance in sediment and fish bacteria were 85.7% and 97.2% to amoxicillin, 83.3% and 85.4% to cefalothin, 31% and 39.6% to ticarcillin, 30.9% and 10.4% to oxytetracycline, 26.2% and 33.3% to streptomycin, 14.3% and 18.8% to sulfamethoxazole, 4.8% and 8.3% to ceftiofur, 4.8% and 2.1% to florfenicol, respectively. Further, they found resistance to chloramphenicol, nitrofurantoin and trimethoprim in 2.1% of the fish isolates. They reported no resistance to ciprofloxacin and gentamicin in all *Pseudomonas* isolates both from sediments and fish as well as to oxytetracycline in all isolates from fish. Also, two isolates of *Pseudomonas* showed no resistance to all studied antibiotics. They further found that one isolate from sediment was resistant to tetracycline and sulfamethoxazole. According to their report incidence of antibiotic resistance in bacteria from sediment and fishes was 96.5% and 97.7% to cefalothin; 95.3% and 90.9% to amoxicillin; 92.9% and 90.9% to florfenicol; 88.2% and 86.6% to ticarcillin; 87.1% and 95.5% to nitrofurantoin and trimethoprim; 87.1% and 90.9% to ceftiofur, 78.8% and 65.9% to chloramphenicol; 50.6% and 43.2% to streptomycin; 10.6% and 38.6% to cefotaxime and 9.4% and 4.6% to oxolinic acid, respectively.

Broughton and Walker [36] examined fish from fish facilities in Guangdong Province, China, in order to examine the incidence of resistance to antibiotics in *Salmonella* spp. They sampled cultured freshwater fish from stands at two wholesale and four retail markets in 2008. They collected 35 catfish, 19 tilapia, 20 perch and 18 silver carp and found the prevalence of 5% for *Salmonella*, which were isolated from one perch, two catfish and two silver carp. They reported that all *Salmonella* showed no resistance to neomycin, cefotaxime, and cefepime, while were resistant to erythromycin and penicillin. *Salmonella* from perch was resistant to erythromycin, nitrofurantoin, penicillin, compound sulfanomides and tetracycline and intermediate to nalidixic acid. One isolate from silver carp was resistant to erythromycin and penicillin and intermediate to nitrofurantoin and streptomycin while the other was resistant to erythromycin and penicillin and intermediate to nalidixic acid. One isolate from catfish was resistant to erythromycin, nalidixic acid and penicillin while the other was resistant to ampicillin, cotrimoxazole, erythromycin; gentamycin, nalidixic acid, penicillin, streptomycin, compound sulfanomides, tetracycline and trimethoprim and intermediate to chloramphenicol, ciprofloxacin and nitrofurantoin. Broughton and Walker [36] have not found resistance to cephalosporines of third and fourth generation and on the other hand, Chao *et al.* [124], detected 11.6% of *Salmonella* in food from markets in China were resistant to cephalosporines of first and second generation. The interesting fact is finding of intermediate susceptibility to nitrofuran, because the use of nitrofuran in food producing livestock was prohibited in China and some other countries in 1995 because it have mutagenic and carcinogenic properties [125]. Moreover, the results that three *Salmonella* were resistant to nitrofurantoin indicate that nitrofurans are still used. Undoubtedly, worldwide efforts in order to encourage prudent utilization of veterinary drugs in fish farming are necessary.

THE IMPLEMENTATION OF GOOD PRACTICES AND MONITORING PROGRAMS ON THE FISH FARMS

Today, fish farms are commonly in vicinity of human settlements. That can be a serious concern for public health. In spite of the fact that the farming practices in aquaculture has been changing in developing parts of the world, there is not enough awareness regarding the utilization of veterinary drugs in aquaculture. It should have in mind that food safety starts on the farm or for fish on fish ponds. The progressive improvement in the known methods of production which are implemented in order to control pathogens, particularly antibiotic-resistant strains, include the development of new antibiotics, vaccines, as well as the new methods for microorganisms detection in livestock and surroundings. Biosecurity measures for aquaculture facilities improved methods for the detection of infection or contamination in the animals and their environment. Different biosecurity measures that put stop to transfer of pathogenic bacteria among animals and surroundings can be implemented. It should have in mind that absence of clinical symptoms does not mean absence of microorganisms and such carriers help in amplifications of pathogens. Mainly, manufacturers are not even aware that pathogens are in fish and their environment. Manufacturers should be encouraged to introduce preventive measures and risk management systems with the aim to prevent the occurrence of pathogens in aquaculture. Good hygienic practices, good farming practices, management of animal wastes in order to prevent transfer of pathogenic microorganisms, management of vectors, management of feed and water *via* different treatments, utilization of probiotics and vaccine are some of existed strategies. The Directive on zoonosis [126] set up the key principle for monitoring programs.

THE PREVENTIVE MEASURES

Antibiotic resistance undoubtedly represents one of the most serious public health threat. Using all the presented results together, it could be concluded that fish as well as aquatic environment are very important as potential source of antibiotic resistance. The transfer of antibiotics utilized in fish could be the cause for high levels of antibiotic resistance. Further, overpopulation in aquaculture as well as poor sanitation may contribute to the dissemination of antibiotic resistance genes among fish. Also, infections of urinary tract can be connected with retail meat consumption and the acquisition of antimicrobial resistant *E. coli* [127] which endorses the necessity for further investigation.

The monitoring of antibiotic resistance in both pathogenic and commensal bacterial isolated from fish and elected as prototype of overall community gives important data on the resistance occurring in the community as well as data for investigations the link with the selective pressure as a consequence of utilization of veterinary drugs on the bacterial community in fish and their environment.

Because fish meat can be an important reservoir of antibiotic-resistant bacteria, the necessity for education of people regarding good hygienic practices should be highlighted. Preventive measures at home include carefully cooking of fish, prevention of cross-contamination from fish by regular hands washing and cleaning surfaces in kitchen throughout cooking, hand washing after contact with animals, their feces or environments. Education of food handlers is also significant in order to decrease the potential risk regarding resistance to antibiotics. Moreover, the education of fish farmers, veterinarians, public, and health practitioners is very important in order to ensure the suitable utilization of veterinary drugs and to ensure that they will not be used without need, such as treatment of infections caused by viruses, prophylactic use and the use as growth promoter. Undoubtedly, the utilization of veterinary drugs in fish feed must be firmly regulated. WHO suggested investigation, monitoring, suitable utilization of medicaments, trainings and application of antibiotic alternatives as measures which should be used in order to decrease resistance levels in pathogenic microorganisms [128]. A number of studies have highlighted the use of various probiotic, phytogenic additives, herbs, species and their essential oils into animal feed as alternatives to antibiotics [106, 129 - 131].

CONCLUSION

Undoubtedly, antibiotic use in aquaculture production cannot be stopped. They represent a valuable tool for reducing fish diseases as well as suffering. The use of antibiotics concerning fish nutrition and as growth promoters is undoubtedly beneficial for the improvement production performance and prevention of diseases. However, because of threats for human and animals health which comc from emerging resistance of pathogens and also the accumulation of residues of antibiotic in fish meat and products as well as in the environment there is a global need to remove antibiotics as growth promoters from fish feed. Having that in mind, it should be very careful when some veterinary drugs should be used. The intensive aquaculture production industry is challenged to optimize production performance and minimize economic loses as a result of antibiotic growth promoters removal, as well as ensuring the safety of fish meat *via* the control and elimination of food borne hazards. Phytogenic additives, herbs, species and their essential oils as a feed supplements present promising alternatives to antibiotic growth promoters in aquaculture industry. Conserving the antibiotic effectiveness is crucial to protecting human health as well as animal health. The need for prudent use of antibiotics, particularly because of the fact that some of them are still used in human medicine should be highlighted. Continuous monitoring and reporting on incidences in the future should improve the current regulations.

CONFLICT OF INTEREST

The author (editor) declares no conflict of interest, financial or otherwise.

ACKNOWLEDGEMENT

This work was supported by the Ministry of Education, Science and Technological Development of the Republic of Serbia (TR31011).

REFFERENCES

[1] Ćirković M, Ljubojević D, Đorđević V, Novakov N, Petronijević R, Matekalo-Sverak V, *et al.* The breed effect on productivity and meat nutrient compsition of fish. Kafkas Univ Vet Fak Derg 2012; 18(5): 775-80.

[2] Ljubojević D, Ćirković M, Novakov N, Jovanović R, Janković S, Đorđević V, *et al.* Productivity and meat nutrient in fish: the diet effect. Kafkas Univ Vet Fak Derg 2013; 19(1): 43-9.

[3] Ljubojević D, Ćirković M, Đorđević V, Puvača N, Trbović D, Vukadinov J, *et al.* Fat Quality of Marketable Fresh Water Fish Species in the Republic of Serbia. Czech J Food Sci 2013; 31(5)

[4] Ljubojevic D, Trbovic D, Lujic J, Bjelic-Cabrilo O, Kostic D, Novakov N, *et al.* Fatty acid composition of fishes from inland waters. Bulg J Agric Sci 2013; 19(1): 62-71.

[5] Ljubojević D, Ćirković M, Novakov N, Puvača N, Aleksić N, Lujić J, *et al.* Comparison of meat quality of tench, Tinca tinca, reared in extensive and semi□intensive culture systems. J Appl Ichthyology 2014; 30(s1): 50-7. [http://dx.doi.org/10.1111/jai.12425]

[6] Ljubojević D, Radosavljević V, Puvača N, Baloš MŽ, Đorđević V, Jovanović R, *et al.* Interactive effects of dietary protein level and oil source on proximate composition and fatty acid composition in common carp (Cyprinus carpio L.). J Food Compos Anal 2015; 37: 44-50. [http://dx.doi.org/10.1016/j.jfca.2014.09.005]

[7] Ljubojević D, Radosavljević V, Pelić M, Đorđević V, Živkov-Baloš M, Ćirković M. Fatty acid composition, chemical composition and processing yield of traditional hot smoked common carp (Cyprinus carpio, L). Iran J Fish Sci 2016; 15(4): 1293-306.

[8] Župan B, Ljubojević D, Pelić M, Ćirković M, Đorđević V, Bogut I. Epidemiological significance of poultry litter for spreading the antibiotic-resistant strains of *Escherichia coli*. World's Poultry Science Journal 2016; 72(03): 485-94.

[9] Ljubojević D, Puvača N, Pelić M, Todorović D, Pajić M, Milanov D, *et al.* Epidemiological significance of poultry litter for spreading the antibiotic-resistant strains of Escherichia coli. Worlds Poult Sci J 2016; 72(03): 485-94. [http://dx.doi.org/10.1017/S004393391600043X]

[10] Ljubojević D, Radosavljević V, Milanov D. The role of gulls (Laridae) in the emergence and spreading of antibiotic resistance in the environment. Worlds Poult Sci J 2016; 72(4): 853-64. [http://dx.doi.org/10.1017/S0043933916000659]

[11] Nyachuba DG. Foodborne illness: is it on the rise? Nutr Rev 2010; 68(5): 257-69. [http://dx.doi.org/10.1111/j.1753-4887.2010.00286.x] [PMID: 20500787]

[12] Aarestrup FM, Wegener HC, Collignon P. Resistance in bacteria of the food chain: epidemiology and control strategies. Expert Rev Anti Infect Ther 2008; 6(5): 733-50. [http://dx.doi.org/10.1586/14787210.6.5.733] [PMID: 18847409]

[13] Sofos JN. Challenges to meat safety in the 21st century. Meat Sci 2008; 78(1-2): 3-13. [http://dx.doi.org/10.1016/j.meatsci.2007.07.027] [PMID: 22062090]

[14] Feldhusen F. The role of seafood in bacterial foodborne diseases. Microbes Infect 2000; 2(13): 1651-60. [http://dx.doi.org/10.1016/S1286-4579(00)01321-6] [PMID: 11113384]

[15] Ljubojevic D, Pelic M, Djordjevic V, Milojevic L, Cirkovic M. Bacterial hazards in fish meat: The aetiologic agents of foodborne diseases. Meat Technology. Tehnol Mesa 2016; 57(1)

[16] Cabello FC. Heavy use of prophylactic antibiotics in aquaculture: a growing problem for human and animal health and for the environment. Environ Microbiol 2006; 8(7): 1137-44. [http://dx.doi.org/10.1111/j.1462-2920.2006.01054.x] [PMID: 16817922]

[17] Skovgaard N. New trends in emerging pathogens. Int J Food Microbiol 2007; 120(3): 217-24. [http://dx.doi.org/10.1016/j.ijfoodmicro.2007.07.046] [PMID: 17976849]

[18] Davies J, Davies D. Origins and evolution of antibiotic resistance. Microbiol Mol Biol Rev 2010; 74(3): 417-33. [http://dx.doi.org/10.1128/MMBR.00016-10] [PMID: 20805405]

[19] Rhodes G, Huys G, Swings J, *et al.* Distribution of oxytetracycline resistance plasmids between aeromonads in hospital and aquaculture environments: implication of Tn1721 in dissemination of the tetracycline resistance determinant tet A. Appl Environ Microbiol 2000; 66(9): 3883-90. [http://dx.doi.org/10.1128/AEM.66.9.3883-3890.2000] [PMID: 10966404]

[20] Wegener HC. Antibiotics in animal feed and their role in resistance development. Curr Opin Microbiol 2003; 6(5): 439-45. [http://dx.doi.org/10.1016/j.mib.2003.09.009] [PMID: 14572534]

[21] Kennedy DG, Cannavan A, McCracken RJ. Regulatory problems caused by contamination, a frequently overlooked cause of veterinary drug residues. J Chromatogr A 2000; 882(1-2): 37-52. [http://dx.doi.org/10.1016/S0021-9673(00)00320-4] [PMID: 10895931]

[22] Matyar F, Kaya A, Dinçer S. Antibacterial agents and heavy metal resistance in Gram-negative bacteria isolated from seawater, shrimp and sediment in Iskenderun Bay, Turkey. Sci Total Environ 2008; 407(1): 279-85. [http://dx.doi.org/10.1016/j.scitotenv.2008.08.014] [PMID: 18804847]

[23] Schwarz S, Chaslus-Dancla E. Use of antimicrobials in veterinary medicine and mechanisms of resistance. Vet Res 2001; 32(3-4): 201-25. [http://dx.doi.org/10.1051/vetres:2001120] [PMID: 11432414]

[24] Wassenaar TM. Use of antimicrobial agents in veterinary medicine and implications for human health. Crit Rev Microbiol 2005; 31(3): 155-69. [http://dx.doi.org/10.1080/10408410591005110] [PMID: 16170906]

[25] Kumar K, Gupta SC, Chander Y, Singh AK. Antibiotic use in agriculture and its impact on the terrestrial environment. Adv Agron 2005; 87: 1-54. [http://dx.doi.org/10.1016/S0065-2113(05)87001-4]

[26] Sayadi M, Trivedy R, Pathak R. Pollution of pharmaceuticals in environment. I Control Pollution 2015; 2010.

[27] Depaola A. Tetracycline resistance by bacteria in response to oxytetracycline-contaminated catfish feed. J Aquat Anim Health 1995; 7(2): 155-60. [http://dx.doi.org/10.1577/1548-8667(1995)007<0155:TRBBIR>2.3.CO;2]

[28] DePaola A, Peeler JT, Rodrick GE. Effect of oxytetracycline-medicated feed on antibiotic resistance of gram-negative bacteria in catfish ponds. Appl Environ Microbiol 1995; 61(6): 2335-40. [PMID: 7793953]

[29] Schmidt AS, Bruun MS, Dalsgaard I, Pedersen K, Larsen JL. Occurrence of antimicrobial resistance in fish-pathogenic and environmental bacteria associated with four danish rainbow trout farms. Appl Environ Microbiol 2000; 66(11): 4908-15. [http://dx.doi.org/10.1128/AEM.66.11.4908-4915.2000] [PMID: 11055942]

[30] Schmidt AS, Bruun MS, Dalsgaard I, Larsen JL. Incidence, distribution, and spread of tetracycline resistance determinants and integron-associated antibiotic resistance genes among motile aeromonads from a fish farming environment. Appl Environ Microbiol 2001; 67(12): 5675-82. [http://dx.doi.org/10.1128/AEM.67.12.5675-5682.2001] [PMID: 11722922]

[31] Gao P, Mao D, Luo Y, Wang L, Xu B, Xu L. Occurrence of sulfonamide and tetracycline-resistant bacteria and resistance genes in aquaculture environment. Water Res 2012; 46(7): 2355-64. [http://dx.doi.org/10.1016/j.watres.2012.02.004] [PMID: 22377146]

[32] Ryu SH, Park SG, Choi SM, *et al.* Antimicrobial resistance and resistance genes in Escherichia coli strains isolated from commercial fish and seafood. Int J Food Microbiol 2012; 152(1-2): 14-8. [http://dx.doi.org/10.1016/j.ijfoodmicro.2011.10.003] [PMID: 22071288]

[33] Landers TF, Cohen B, Wittum TE, Larson EL. A review of antibiotic use in food animals: perspective, policy, and potential. Public Health Rep 2012; 127(1): 4-22. [http://dx.doi.org/10.1177/003335491212700103] [PMID: 22298919]

[34] Li J, Yie J, Foo RW, Ling JM, Xu H, Woo NY. Antibiotic resistance and plasmid profiles of Vibrio isolates from cultured silver sea bream, Sparus sarba. Mar Pollut Bull 1999; 39(1): 245-9. [http://dx.doi.org/10.1016/S0025-326X(99)00062-4]

[35] Akinbowale OL, Peng H, Barton MD. Diversity of tetracycline resistance genes in bacteria from aquaculture sources in Australia. J Appl Microbiol 2007; 103(5): 2016-25. [http://dx.doi.org/10.1111/j.1365-2672.2007.03445.x] [PMID: 17953612]

[36] Broughton EI, Walker DG. Prevalence of antibiotic-resistant Salmonella in fish in Guangdong, China. Foodborne Pathog Dis 2009; 6(4): 519-21. [http://dx.doi.org/10.1089/fpd.2008.0196] [PMID: 19292686]

[37] Allen HK, Donato J, Wang HH, Cloud-Hansen KA, Davies J, Handelsman J. Call of the wild: antibiotic resistance genes in natural environments. Nat Rev Microbiol 2010; 8(4): 251-9. [http://dx.doi.org/10.1038/nrmicro2312] [PMID: 20190823]

[38] Rolain JM, Canton R, Cornaglia G. Emergence of antibiotic resistance: need for a new paradigm. Clin

Microbiol Infect 2012; 18(7): 615-6.
[http://dx.doi.org/10.1111/j.1469-0691.2012.03902.x] [PMID: 22703444]

[39] Chopra I, Roberts M. Tetracycline antibiotics: mode of action, applications, molecular biology, and epidemiology of bacterial resistance. Microbiol Mol Biol Rev 2001; 65(2): 232-60.
[http://dx.doi.org/10.1128/MMBR.65.2.232-260.2001] [PMID: 11381101]

[40] Koo HJ, Woo GJ. Distribution and transferability of tetracycline resistance determinants in Escherichia coli isolated from meat and meat products. Int J Food Microbiol 2011; 145(2-3): 407-13.
[http://dx.doi.org/10.1016/j.ijfoodmicro.2011.01.003] [PMID: 21324543]

[41] Lunestad BT, Goksøyr J. Reduction in the antibacterial effect of oxytetracycline in sea water by complex formation with magnesium and calcium. Dis Aquat Organ 1990; 9(1): 67-72.
[http://dx.doi.org/10.3354/dao009067]

[42] Sayah RS, Kaneene JB, Johnson Y, Miller R. Patterns of antimicrobial resistance observed in Escherichia coli isolates obtained from domestic- and wild-animal fecal samples, human septage, and surface water. Appl Environ Microbiol 2005; 71(3): 1394-404.
[http://dx.doi.org/10.1128/AEM.71.3.1394-1404.2005] [PMID: 15746342]

[43] Sunde M, Norström M. The prevalence of, associations between and conjugal transfer of antibiotic resistance genes in *Escherichia coli* isolated from Norwegian meat and meat products. J Antimicrob Chemother 2006; 58(4): 741-7.
[http://dx.doi.org/10.1093/jac/dkl294] [PMID: 16931539]

[44] Miles TD, McLaughlin W, Brown PD. Antimicrobial resistance of *Escherichia coli* isolates from broiler chickens and humans. BMC Vet Res 2006; 2(1): 7.
[http://dx.doi.org/10.1186/1746-6148-2-7] [PMID: 16460561]

[45] Roberts MC, Roberts MC. Tetracycline therapy: update. Clin Infect Dis 2003; 36(4): 462-7.
[http://dx.doi.org/10.1086/367622] [PMID: 12567304]

[46] Kim SR, Nonaka L, Suzuki S. Occurrence of tetracycline resistance genes tet(M) and tet(S) in bacteria from marine aquaculture sites. FEMS Microbiol Lett 2004; 237(1): 147-56.
[http://dx.doi.org/10.1111/j.1574-6968.2004.tb09690.x] [PMID: 15268950]

[47] Alcaide E, Blasco M-D, Esteve C. Occurrence of drug-resistant bacteria in two European eel farms. Appl Environ Microbiol 2005; 71(6): 3348-50.
[http://dx.doi.org/10.1128/AEM.71.6.3348-3350.2005] [PMID: 15933039]

[48] Dang H, Zhang X, Song L, Chang Y, Yang G. Molecular determination of oxytetracycline-resistant bacteria and their resistance genes from mariculture environments of China. J Appl Microbiol 2007; 103(6): 2580-92.
[http://dx.doi.org/10.1111/j.1365-2672.2007.03494.x] [PMID: 18045442]

[49] Penders J, Stobberingh EE. Antibiotic resistance of motile aeromonads in indoor catfish and eel farms in the southern part of The Netherlands. Int J Antimicrob Agents 2008; 31(3): 261-5.
[http://dx.doi.org/10.1016/j.ijantimicag.2007.10.002] [PMID: 18160266]

[50] Lamshöft M, Sukul P, Zühlke S, Spiteller M. Metabolism of 14C-labelled and non-labelled sulfadiazine after administration to pigs. Anal Bioanal Chem 2007; 388(8): 1733-45.
[http://dx.doi.org/10.1007/s00216-007-1368-y] [PMID: 17619182]

[51] Enne VI, Livermore DM, Stephens P, Hall LM. Persistence of sulphonamide resistance in *Escherichia coli* in the UK despite national prescribing restriction. Lancet 2001; 357(9265): 1325-8.
[http://dx.doi.org/10.1016/S0140-6736(00)04519-0] [PMID: 11343738]

[52] Luo Y, Mao D, Rysz M, *et al.* Trends in antibiotic resistance genes occurrence in the Haihe River, China. Environ Sci Technol 2010; 44(19): 7220-5.
[http://dx.doi.org/10.1021/es100233w] [PMID: 20509603]

[53] Miranda CD, Zemelman R. Antibiotic resistant bacteria in fish from the Concepción Bay, Chile. Mar Pollut Bull 2001; 42(11): 1096-102.

[http://dx.doi.org/10.1016/S0025-326X(01)00093-5] [PMID: 11763221]

[54] Reinthaler FF, Posch J, Feierl G, *et al.* Antibiotic resistance of *E. coli* in sewage and sludge. Water Res 2003; 37(8): 1685-90.
[http://dx.doi.org/10.1016/S0043-1354(02)00569-9] [PMID: 12697213]

[55] DePaola A, Flynn PA, McPhearson RM, Levy SB. Phenotypic and genotypic characterization of tetracycline- and oxytetracycline-resistant Aeromonas hydrophila from cultured channel catfish (Ictalurus punctatus) and their environments. Appl Environ Microbiol 1988; 54(7): 1861-3. [PMID: 3046493]

[56] McPhearson RM, DePaola A, Zywno SR, Motes ML, Guarino AM. Antibiotic resistance in Gram-negative bacteria from cultured catfish and aquaculture ponds. Aquaculture 1991; 99(3-4): 203-11.
[http://dx.doi.org/10.1016/0044-8486(91)90241-X]

[57] Spanggaard B, Jørgensen F, Gram L, Huss HH. Antibiotic resistance in bacteria isolated from three freshwater fish farms and an unpolluted stream in Denmark. Aquaculture 1993; 115(3): 195-207.
[http://dx.doi.org/10.1016/0044-8486(93)90136-M]

[58] Guardabassi L, Dijkshoorn L, Collard J-M, Olsen JE, Dalsgaard A. Distribution and *in-vitro* transfer of tetracycline resistance determinants in clinical and aquatic Acinetobacter strains. J Med Microbiol 2000; 49(10): 929-36.
[http://dx.doi.org/10.1099/0022-1317-49-10-929] [PMID: 11023190]

[59] Chelossi E, Vezzulli L, Milano A, Branzoni M, Fabiano M, Riccardi G, *et al.* Antibiotic resistance of benthic bacteria in fish-farm and control sediments of the Western Mediterranean. Aquaculture 2003; 219(1): 83-97.
[http://dx.doi.org/10.1016/S0044-8486(03)00016-4]

[60] Sapkota AR, Curriero FC, Gibson KE, Schwab KJ. Antibiotic-resistant enterococci and fecal indicators in surface water and groundwater impacted by a concentrated Swine feeding operation. Environ Health Perspect 2007; 115(7): 1040-5.
[http://dx.doi.org/10.1289/ehp.9770] [PMID: 17637920]

[61] Papadopoulou C, Economou V, Sakkas H, *et al.* Microbiological quality of indoor and outdoor swimming pools in Greece: investigation of the antibiotic resistance of the bacterial isolates. Int J Hyg Environ Health 2008; 211(3-4): 385-97.
[http://dx.doi.org/10.1016/j.ijheh.2007.06.007] [PMID: 17728184]

[62] Lugo-Melchor Y, Quiñones B, Amézquita-López BA, León-Félix J, García-Estrada R, Chaidez C. Characterization of tetracycline resistance in Salmonella enterica strains recovered from irrigation water in the Culiacan Valley, Mexico. Microb Drug Resist 2010; 16(3): 185-90.
[http://dx.doi.org/10.1089/mdr.2010.0022] [PMID: 20704512]

[63] Abhirosh C, Sherin V, Thomas AP, Hatha AA, Mazumder A. Potential public health significance of faecal contamination and multidrug-resistant *Escherichia coli* and Salmonella serotypes in a lake in India. Public Health 2011; 125(6): 377-9.
[http://dx.doi.org/10.1016/j.puhe.2011.03.015] [PMID: 21620427]

[64] Akinyemi KO, Iwalokun BA, Foli F, Oshodi K, Coker AO. Prevalence of multiple drug resistance and screening of enterotoxin (stn) gene in Salmonella enterica serovars from water sources in Lagos, Nigeria. Public Health 2011; 125(2): 65-71.
[http://dx.doi.org/10.1016/j.puhe.2010.11.010] [PMID: 21277605]

[65] Ham YS, Kobori H, Kang JH, Matsuzaki T, Iino M, Nomura H. Distribution of antibiotic resistance in urban watershed in Japan. Environ Pollut 2012; 162: 98-103.
[http://dx.doi.org/10.1016/j.envpol.2011.11.002] [PMID: 22243853]

[66] Samuelsen OB, Torsvik V, Ervik A. Long-range changes in oxytetracycline concentration and bacterial resistance toward oxytetracycline in a fish farm sediment after medication. Sci Total Environ 1992; 114: 25-36.
[http://dx.doi.org/10.1016/0048-9697(92)90411-K] [PMID: 1594922]

[67] Economou V, Gousia P, Kansouzidou A, Sakkas H, Karanis P, Papadopoulou C. Prevalence, antimicrobial resistance and relation to indicator and pathogenic microorganisms of Salmonella enterica isolated from surface waters within an agricultural landscape. Int J Hyg Environ Health 2013; 216(4): 435-44.
[http://dx.doi.org/10.1016/j.ijheh.2012.07.004] [PMID: 22901425]

[68] Nielsen HU, Hammerum AM, Ekelund K, Bang D, Pallesen LV, Frimodt-Møller N. Tetracycline and macrolide co-resistance in Streptococcus pyogenes: co-selection as a reason for increase in macrolide-resistant S. pyogenes? Microb Drug Resist 2004; 10(3): 231-8.
[http://dx.doi.org/10.1089/mdr.2004.10.231] [PMID: 15383167]

[69] García-Galán MJ, Díaz-Cruz MS, Barceló D. Identification and determination of metabolites and degradation products of sulfonamide antibiotics. TrAC Trends in Analytical Chemistry 2008; 27(11): 1008-22.
[http://dx.doi.org/10.1016/j.trac.2008.10.001]

[70] Diarrassouba F, Diarra MS, Bach S, *et al.* Antibiotic resistance and virulence genes in commensal *Escherichia coli* and Salmonella isolates from commercial broiler chicken farms. J Food Prot 2007; 70(6): 1316-27.
[http://dx.doi.org/10.4315/0362-028X-70.6.1316] [PMID: 17612058]

[71] Adams CA, Austin B, Meaden PG, McIntosh D. Molecular characterization of plasmid-mediated oxytetracycline resistance in Aeromonas salmonicida. Appl Environ Microbiol 1998; 64(11): 4194-201.
[PMID: 9797265]

[72] Olasz F, Fekete PZ, Blum-Oehler G, Boldogkői Z, Nagy B. Characterization of an F18+ enterotoxigenic *Escherichia coli* strain from post weaning diarrhoea of swine, and of its conjugative virulence plasmid pTC. FEMS Microbiol Lett 2005; 244(2): 281-9.
[http://dx.doi.org/10.1016/j.femsle.2005.01.057] [PMID: 15766780]

[73] Goñi-Urriza M, Capdepuy M, Arpin C, Raymond N, Caumette P, Quentin C. Impact of an urban effluent on antibiotic resistance of riverine Enterobacteriaceae and Aeromonas spp. Appl Environ Microbiol 2000; 66(1): 125-32.
[http://dx.doi.org/10.1128/AEM.66.1.125-132.2000] [PMID: 10618213]

[74] Hatha M, Vivekanandhan AA, Joice GJ, Christol . Antibiotic resistance pattern of motile aeromonads from farm raised fresh water fish. Int J Food Microbiol 2005; 98(2): 131-4.
[http://dx.doi.org/10.1016/j.ijfoodmicro.2004.05.017] [PMID: 15747443]

[75] Adelaide OA, Bii C, Okemo P. Antibiotic resistance and virulence factors in *Escherichia coli* from broiler chicken slaughtered at Tigoni processing plant in Limuru, Kenya. East Afr Med J 2008; 85(12): 597-606.
[PMID: 19413216]

[76] Scaletsky IC, Souza TB, Aranda KR, Okeke IN. Genetic elements associated with antimicrobial resistance in enteropathogenic *Escherichia coli* (EPEC) from Brazil. BMC Microbiol 2010; 10(1): 25.
[http://dx.doi.org/10.1186/1471-2180-10-25] [PMID: 20105329]

[77] Jones JG, Gardener S, Simon BM, Pickup RW. Antibiotic resistant bacteria in Windermere and two remote upland tarns in the English Lake District. J Appl Bacteriol 1986; 60(5): 443-53.
[http://dx.doi.org/10.1111/j.1365-2672.1986.tb05090.x] [PMID: 3722030]

[78] Sørum H, L'Abée-Lund TM, Solberg A, Wold A. Integron-containing IncU R plasmids pRAS1 and pAr-32 from the fish pathogen Aeromonas salmonicida. Antimicrob Agents Chemother 2003; 47(4): 1285-90.
[http://dx.doi.org/10.1128/AAC.47.4.1285-1290.2003] [PMID: 12654659]

[79] Aoki T, Takahashi A. Class D tetracycline resistance determinants of R plasmids from the fish pathogens Aeromonas hydrophila, Edwardsiella tarda, and Pasteurella piscicida. Antimicrob Agents Chemother 1987; 31(8): 1278-80.

[http://dx.doi.org/10.1128/AAC.31.8.1278] [PMID: 3631946]

[80] DePaola A, Hill WE, Harrell FM. Oligonucleotide probe determination of tetracycline-resistant bacteria isolated from catfish ponds. Mol Cell Probes 1993; 7(5): 345-8. [http://dx.doi.org/10.1006/mcpr.1993.1051] [PMID: 8264667]

[81] Andersen SR, Sandaa RA. Distribution of tetracycline resistance determinants among gram-negative bacteria isolated from polluted and unpolluted marine sediments. Appl Environ Microbiol 1994; 60(3): 908-12. [PMID: 8161183]

[82] Sørum H. Antibiotic resistance in aquaculture. Acta Vet Scand Suppl 1999; 92: 29-36. [PMID: 10783715]

[83] Sobecky PA, Mincer TJ, Chang MC, Helinski DR. Plasmids isolated from marine sediment microbial communities contain replication and incompatibility regions unrelated to those of known plasmid groups. Appl Environ Microbiol 1997; 63(3): 888-95. [PMID: 9055407]

[84] De Grandis SA, Stevenson RM. Antimicrobial susceptibility patterns and R plasmid-mediated resistance of the fish pathogen Yersinia ruckeri. Antimicrob Agents Chemother 1985; 27(6): 938-42. [http://dx.doi.org/10.1128/AAC.27.6.938] [PMID: 4026266]

[85] Klein BU. Investigations on transferable antibiotic resistance through R-plasmids between obligate and facultative fish pathogenic bacteria. Bull Eur Assoc Fish Pathol 1996; 16: 138-42.

[86] Son R, Rusul G, Sahilah AM, Zainuri A, Raha AR, Salmah I. Antibiotic resistance and plasmid profile of Aeromonas hydrophila isolates from cultured fish, Telapia (Telapia mossambica). Lett Appl Microbiol 1997; 24(6): 479-82. [http://dx.doi.org/10.1046/j.1472-765X.1997.00156.x] [PMID: 9203404]

[87] Sarter S, Nguyen HN, Hung LT, Lazard J, Montet D. Antibiotic resistance in Gram-negative bacteria isolated from farmed catfish. Food Control 2007; 18(11): 1391-6. [http://dx.doi.org/10.1016/j.foodcont.2006.10.003]

[88] Karlowsky JA, Kelly LJ, Thornsberry C, Jones ME, Sahm DF. Trends in antimicrobial resistance among urinary tract infection isolates of *Escherichia coli* from female outpatients in the United States. Antimicrob Agents Chemother 2002; 46(8): 2540-5. [http://dx.doi.org/10.1128/AAC.46.8.2540-2545.2002] [PMID: 12121930]

[89] Miranda CD, Zemelman R. Bacterial resistance to oxytetracycline in Chilean salmon farming. Aquaculture 2002; 212(1): 31-47. [http://dx.doi.org/10.1016/S0044-8486(02)00124-2]

[90] Furushita M, Shiba T, Maeda T, *et al.* Similarity of tetracycline resistance genes isolated from fish farm bacteria to those from clinical isolates. Appl Environ Microbiol 2003; 69(9): 5336-42. [http://dx.doi.org/10.1128/AEM.69.9.5336-5342.2003] [PMID: 12957921]

[91] Ljubojević D, Velhner M, Todorović D, Pajić M, Milanov D. Tetracycline resistance in *Escherichia coli* isolates from poultry. Arh Vet Med 2016; 9(1): 21.

[92] Vivekanandhan G, Savithamani K, Hatha AA, Lakshmanaperumalsamy P. Antibiotic resistance of Aeromonas hydrophila isolated from marketed fish and prawn of South India. Int J Food Microbiol 2002; 76(1-2): 165-8. [http://dx.doi.org/10.1016/S0168-1605(02)00009-0] [PMID: 12038573]

[93] Kümmerer K. Antibiotics in the aquatic environment--a review--part I. Chemosphere 2009; 75(4): 417-34. [http://dx.doi.org/10.1016/j.chemosphere.2008.11.086] [PMID: 19185900]

[94] Alderman D, Hastings T. Antibiotic use in aquaculture: development of antibiotic resistance–potential for consumer health risks. Int J Food Sci Technol 1998; 33(2): 139-55. [http://dx.doi.org/10.1046/j.1365-2621.1998.3320139.x]

[95] Pathak SP, Gopal K. Occurrence of antibiotic and metal resistance in bacteria from organs of river fish. Environ Res 2005; 98(1): 100-3.
[http://dx.doi.org/10.1016/j.envres.2004.05.012] [PMID: 15721889]

[96] Ozaktas T, Taskin B, Gozen AG. High level multiple antibiotic resistance among fish surface associated bacterial populations in non-aquaculture freshwater environment. Water Res 2012; 46(19): 6382-90.
[http://dx.doi.org/10.1016/j.watres.2012.09.010] [PMID: 23039919]

[97] Bruun MS, Schmidt AS, Madsen L, Dalsgaard I. Antimicrobial resistance patterns in Danish isolates of Flavobacterium psychrophilum. Aquaculture 2000; 187(3): 201-12.
[http://dx.doi.org/10.1016/S0044-8486(00)00310-0]

[98] Kerry J. NicGabhainn S, Smith P. Changes in oxytetracycline resistance of intestinal microflora following oral administration of this agent to Atlantic salmon (Salmo salar L.) smolts in a marine environment. Aquaculture 1997; 157(3): 187-95.
[http://dx.doi.org/10.1016/S0044-8486(97)00159-2]

[99] Castro-Escarpulli G, Figueras MJ, Aguilera-Arreola G, *et al.* Characterisation of Aeromonas spp. isolated from frozen fish intended for human consumption in Mexico. Int J Food Microbiol 2003; 84(1): 41-9.
[http://dx.doi.org/10.1016/S0168-1605(02)00393-8] [PMID: 12781953]

[100] Baker-Austin C, Wright MS, Stepanauskas R, McArthur JV. Co-selection of antibiotic and metal resistance. Trends Microbiol 2006; 14(4): 176-82.
[http://dx.doi.org/10.1016/j.tim.2006.02.006] [PMID: 16537105]

[101] Starliper CE, Cooper RK. Biochemical and conjugation studies of Romet-resistant strains of Aeromonas salmonicida from salmonid rearing facilities in the Eastern United States. J Aquat Anim Health 1998; 10(3): 221-9.
[http://dx.doi.org/10.1577/1548-8667(1998)010<0221:BACSOR>2.0.CO;2]

[102] Snieszko S, Dunbar C, Bullock G. Resistance to ulcer disease and furunculosis in eastern brook trout, Salvelinus fontinalis. Prog Fish-Cult 1959; 21(3): 111-6.
[http://dx.doi.org/10.1577/1548-8659(1959)21[111:RTUDAF]2.0.CO;2]

[103] Aoki T, Kitao T, Arai T. R plasmids in fish pathogens Plasmid: medical and theoretical aspects Avicenum Czechoslovak Medical Press. Prague: Czechoslovakia 1977; pp. 39-45.

[104] Rossolini GM, Walsh T, Amicosante G. The Aeromonas metallo-β-lactamases: genetics, enzymology, and contribution to drug resistance. Microbial drug resistance 1996; 2(2): 245-52.

[105] Janda JM, Abbott SL. The genus Aeromonas: taxonomy, pathogenicity, and infection. Clin Microbiol Rev 2010; 23(1): 35-73.
[http://dx.doi.org/10.1128/CMR.00039-09] [PMID: 20065325]

[106] Ljubojević D, Ćirković M, Miščević M. Probiotics in carp fish nutrition. Vet Glas 2013; 67(5-6): 429-39.
[http://dx.doi.org/10.2298/VETGL1306429L]

[107] Holmström K, Gräslund S, Wahlström A, Poungshompoo S, Bengtsson BE, Kautsky N. Antibiotic use in shrimp farming and implications for environmental impacts and human health. Int J Food Sci Technol 2003; 38(3): 255-66.
[http://dx.doi.org/10.1046/j.1365-2621.2003.00671.x]

[108] DePaola A, Roberts MC. Class D and E tetracycline resistance determinants in gram-negative bacteria from catfish ponds. Mol Cell Probes 1995; 9(5): 311-3.
[http://dx.doi.org/10.1016/S0890-8508(95)91572-9] [PMID: 8569770]

[109] Ewers C, Antão E-M, Diehl I, Philipp H-C, Wieler LH. Intestine and environment of the chicken as reservoirs for extraintestinal pathogenic *Escherichia coli* strains with zoonotic potential. Appl Environ Microbiol 2009; 75(1): 184-92.

[http://dx.doi.org/10.1128/AEM.01324-08] [PMID: 18997030]

[110] Wray C, Gnanou J-C. Antibiotic resistance monitoring in bacteria of animal origin: analysis of national monitoring programmes. Int J Antimicrob Agents 2000; 14(4): 291-4. [http://dx.doi.org/10.1016/S0924-8579(00)00139-4] [PMID: 10794949]

[111] Velhner M, Milanov D. Resistance to tetracycline in *Escherichia coli* and Staphylococcus aureus: brief overview on mechanisms of resistance and epidemiology. Arh Vet Med 2015; 8: 27-36.

[112] Guerra B, Junker E, Schroeter A, Malorny B, Lehmann S, Helmuth R. Phenotypic and genotypic characterization of antimicrobial resistance in German *Escherichia coli* isolates from cattle, swine and poultry. J Antimicrob Chemother 2003; 52(3): 489-92. [http://dx.doi.org/10.1093/jac/dkg362] [PMID: 12888584]

[113] Lanz R, Kuhnert P, Boerlin P. Antimicrobial resistance and resistance gene determinants in clinical *Escherichia coli* from different animal species in Switzerland. Vet Microbiol 2003; 91(1): 73-84. [http://dx.doi.org/10.1016/S0378-1135(02)00263-8] [PMID: 12441233]

[114] Kang HY, Jeong YS, Oh JY, *et al.* Characterization of antimicrobial resistance and class 1 integrons found in *Escherichia coli* isolates from humans and animals in Korea. J Antimicrob Chemother 2005; 55(5): 639-44. [http://dx.doi.org/10.1093/jac/dki076] [PMID: 15761064]

[115] Paterson DL, Bonomo RA. Extended-spectrum β-lactamases: a clinical update. Clin Microbiol Rev 2005; 18(4): 657-86. [http://dx.doi.org/10.1128/CMR.18.4.657-686.2005] [PMID: 16223952]

[116] Strahilevitz J, Jacoby GA, Hooper DC, Robicsek A. Plasmid-mediated quinolone resistance: a multifaceted threat. Clin Microbiol Rev 2009; 22(4): 664-89. [http://dx.doi.org/10.1128/CMR.00016-09] [PMID: 19822894]

[117] Skurnik D, Ruimy R, Andremont A, *et al.* Effect of human vicinity on antimicrobial resistance and integrons in animal faecal *Escherichia coli*. J Antimicrob Chemother 2006; 57(6): 1215-9. [http://dx.doi.org/10.1093/jac/dkl122] [PMID: 16581916]

[118] Carattoli A. Animal reservoirs for extended spectrum β-lactamase producers. Clin Microbiol Infect 2008; 14(s1) (Suppl. 1): 117-23. [http://dx.doi.org/10.1111/j.1469-0691.2007.01851.x] [PMID: 18154535]

[119] Costa D, Poeta P, Sáenz Y, *et al.* Detection of *Escherichia coli* harbouring extended-spectrum β-lactamases of the CTX-M, TEM and SHV classes in faecal samples of wild animals in Portugal. J Antimicrob Chemother 2006; 58(6): 1311-2. [http://dx.doi.org/10.1093/jac/dkl415] [PMID: 17023496]

[120] Heuer H, Smalla K. Manure and sulfadiazine synergistically increased bacterial antibiotic resistance in soil over at least two months. Environ Microbiol 2007; 9(3): 657-66. [http://dx.doi.org/10.1111/j.1462-2920.2006.01185.x] [PMID: 17298366]

[121] Marshall BM, Levy SB. Food animals and antimicrobials: impacts on human health. Clin Microbiol Rev 2011; 24(4): 718-33. [http://dx.doi.org/10.1128/CMR.00002-11] [PMID: 21976606]

[122] Schroeder CM, White DG, Meng J. Retail meat and poultry as a reservoir of antimicrobial-resistant *Escherichia coli*. Food Microbiol 2004; 21(3): 249-55. [http://dx.doi.org/10.1016/S0740-0020(03)00074-1]

[123] Johnson JR, Kuskowski MA, Smith K, O'Bryan TT, Tatini S. Antimicrobial-resistant and extraintestinal pathogenic *Escherichia coli* in retail foods. J Infect Dis 2005; 191(7): 1040-9. [http://dx.doi.org/10.1086/428451] [PMID: 15747237]

[124] Chao G, Zhou X, Jiao X, Qian X, Xu L. Prevalence and antimicrobial resistance of foodborne pathogens isolated from food products in China. Foodborne Pathog Dis 2007; 4(3): 277-84. [http://dx.doi.org/10.1089/fpd.2007.0088] [PMID: 17883311]

[125] Chang C, Peng DP, Wu JE, Wang YL, Yuan ZH. Development of an indirect competitive ELISA for the detection of furazolidone marker residue in animal edible tissues. J Agric Food Chem 2008; 56(5): 1525-31.
[http://dx.doi.org/10.1021/jf0726684] [PMID: 18260630]

[126] Directive 2003/99/EC of The European Parliament and of the Council of 17 November 2003 on the monitoring of zoonoses and zoonotic agents, amending Council Decision 90/424/EEC and repealing Council Directive 92/117/EEC. Official Journal of the European Union L 2003; 325: 31-40.

[127] Manges AR, Smith SP, Lau BJ, *et al.* Retail meat consumption and the acquisition of antimicrobial resistant *Escherichia coli* causing urinary tract infections: a case-control study. Foodborne Pathog Dis 2007; 4(4): 419-31.
[http://dx.doi.org/10.1089/fpd.2007.0026] [PMID: 18041952]

[128] Mayrhofer S, Paulsen P, Smulders FJ, Hilbert F. Antimicrobial resistance profile of five major food-borne pathogens isolated from beef, pork and poultry. Int J Food Microbiol 2004; 97(1): 23-9.
[http://dx.doi.org/10.1016/j.ijfoodmicro.2004.04.006] [PMID: 15527915]

[129] Miščević M, Ćirković M, Jovanović R, Ljubojević D, Novakov N, Mašić Z, *et al.* Effect of probiotics to the production of one-year old tench and common carp. Arch Zootech 2012; 15(4): 69-76.

[130] Puvača N, Ljubojević D, Kostadinović L, Lukač D, Lević J, Popović S, *et al.* Spices and herbs in broilers nutrition: Effects of garlic (Allium sativum L.) on broiler chicken production. Worlds Poult Sci J 2015; 71(03): 533-8.
[http://dx.doi.org/10.1017/S0043933915002214]

[131] Puvača N, Ljubojević D, Kostadinović L, Lević J, Nikolova N, Miščević B, *et al.* Spices and herbs in broilers nutrition: hot red pepper (Capsicum annuum L.) and its mode of action. Worlds Poult Sci J 2015; 71(04): 683-8.
[http://dx.doi.org/10.1017/S004393391500241X]

CHAPTER 6

Stress and Fish Health: Towards an Understanding of Allostatic Load

Hijran Yavuzcan Yıldız* and **F. Sertel Seçer**

Ankara University, Faculty of Agriculture, Department of Fisheries and Aquaculture, 06110, Ankara, Turkey

Abstract: In aquaculture, stress related disease problems which are leading to serious limitations in sustainability of the sector are of major importance however, stressors in aquaculture are unavoidable. In terms of characteristics of interaction of stress and health in aquaculture, stress physiology has been the center of attention. In fish, the physiological stress response to stressors causes primary response involving neuro-hormonal stimulation, resulting in an increase in corticosteroid and catecholamine secretions. In turn, these primary effects cause a number of physiological changes known as a secondary response. Tertiary stress response involves the effects on immunocompetence and potential susceptibility to diseases. The effects of stress on the fish physiology have been extensively studied for many years. It is known that acute stress can have beneficial effects in short-term, however, chronic stress hinders the immune system from functioning in teleost fish. Thus, the recent approach to explain the negative effects of stress, allostasis theory and allostatic load is presumed, concerning with how stability of the body's is achieved through change. The correlation between the stressors and disease was reported for various fish species cultured and the diseases. To characterize the stress response, the immunocompetence state may provide better approach to understand the disease emergence. Stress response measurements are mainly based on the levels of cortisol nevertheless, it might be better to characterize the stress response with an index of immunological parameters and/or allostatic load.

Keywords: Allostatic load, Aquaculture, Fish, Fish health, Physiological stress response.

INTRODUCTION

Stress can be defined as internal or external challenges threatening homeostasis which is the maintenance of equilibrium within an internal environment in response to external changes. Fish as all living organisms maintain homeostasis, which is constantly challenged by internal or external adverse effects, called stres-

* **Corresponding author Hijran Yavuzcan Yıldız:** Ankara University, Faculty of Agriculture, Department of Fisheries and Aquaculture, Ankara, Turkey; Tel: +903125961638; E-mail: yavuzcan@ankara.edu.tr

Panagiotis Berillis (Ed)

sors. In intensive fish culture, a variety of physiological changes in fish's organism in response to stressors may occur at three different levels: i) the primary stress response involves neuroendocrine response such as catecholamine's release and activation of corticotrophin *via* interrenal axis; ii) secondary stress response involves haematological, metabolic, blood enzymatic and osmoregulatory changes under the effect of primary stage; iii) tertiary stress reponse involves increase in susceptibility to pathogens, disruption in growth and reproduction, and death. It is presumed that the stress-triggered neuroendocrine hormones lead to immune dysregulation, which ultimately results in disease emergence by altering normal immune response. This chapter covers the terms of "allostasis and allostatic load", which are relatively new concepts proposed to explain physiological responses to stress. Allostasis is a process by which an organism changes its internal milieu to meet environmental challenges. The term "allostatic load" can explain the stress results in fish organism regarding fish health thus, higher allostatic load is associated with worse health outcomes.

Concept of Stress

Stress can generally be described as disruption in the state of optimum balance of organ function in fish. Any factor which has negative effects on the normal pyhsiological functions of organs in fish, is a stressor and the fish's response to this stressor is the stress response [1]. Health expresses the ability of an animal to fulfil normal physiological functions and to maintain homeostatis, hence resistance ability of fish against infectious and non-infectious stressors is a part of health [2]. The existence of close and mutual links between stress and health is obvious with the definitions of stress and health.

The study by Selye [3] on the facts of stress and General Adaptation Syndrome (G.A.S.) in 1950 elucidated the stress response in mammalians. In Selye's G.A.S. three stages were identified: i) Alarm Reaction, ii) the Stage of Resistance, and iii) the Stage of Exhaustion (Fig. **1**).

In the first phase-the stage of alarm, body sources mobilize under the effect of releasing stress hormones; catecholamines and corticosteroids to confront the threat. In the second phase-the stage of resistance, physiological forces are activated to resist the damage by noxious stimulus. Thus, blood glucose levels remain high, cortisol and adrenalin continue to circulate at elevated levels, but outward appearance of organism seems normal. This stage, often results in adaptation or the disappearance of symptoms and does not progress to the third stage. The last phase-the stage of exhaustion occurs if stressor continues beyond body's capacity, organism exhausts resources and becomes susceptible to disease. It is important to note that adaptation energy is limited by individual's genetic and the magnitude of adaptive energy alters by genetic factors.

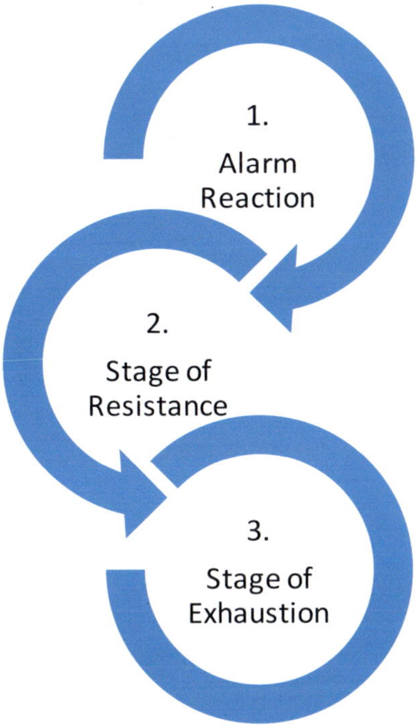

Fig. (1). Stages of general adaptation syndrome.

Although Selye's study was on the mammalians, he emphasized that "all living organisms can respond to stress as such, and that in this respect the basic reaction pattern is always the same, irrespective of the agent used to produce stress. This response is called the general adaptation syndrome, and it derails the diseases of adaptation [3]." This idea was adopted for fish and the first researches on stress in fish began with Snieszko and Wedemeyer in 1960's. However, due to high diversity in fish species and being cold-blooded animal, the stress studies in fish differed from mammalian studies. Thus, it is understood that an analogy between the mammalian pattern and fish response against stress exists however, not complete. Studies showed that pyhsiological stress response may vary with fish species and stressor type. Currently, producing physiological data on stress response is still of major concern to understand the stress in fish.

Stress Response in Fish

Stress response is an important component of fish's defence against stressors as in other vertebrates. Fish is able to develop 'coping' strategies to challenge the stress through the neuroendocrine, physiological and behavioural responses. The role of stress response is to enhance the individual's possibility of survival when confronted with an instant threat, existing or perceived [4]. A stress response can be observed at different levels [5] and classified as primary, including neural and neuroendocrine response; secondary, such as physiological consequence of such primary response; and tertiary, such as changes in behavior, growth rate, increased susceptibility to diseases and change in population (Fig. **2**).

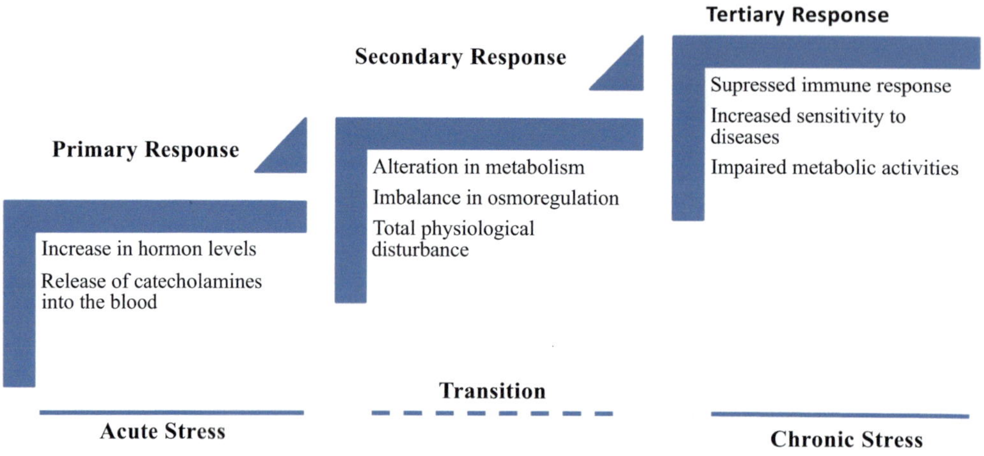

Fig. (2). Physiological alteration in fish organism by the stress levels (adapted from Nardocci *et al.* [6],).

1. The primary neuroendocrine response – The primary neuroendocrine response is characterised by the activation and the secretion of the hormones, corticosteroids (cortisol) and catecholamines into the blood. The peak of neuroendocrine response leads to the 2° response and behavioural differences occur [5]. The response is initiated within the limbic system of the brain. At the primary level, the stress response has two major neuroendocrine components. The first is a rapid activation of the sympathetic nervous system whose endpoint is the release of catecholamines into the blood from the chromaffin tissue of the head kidney [4]. Second one is the activation of the hypothalamus, the pituitary and the interrenal (HPI) axis, which ends in releasing of the stress hormone cortisol into the blood. Cortisol is produced in fish by interenal bodies in the head kidney under the effect of adrenocorticotrophic hormone (ACTH). Cortisol is the steroid hormone and lipid soluble.

2. The secondary stress response – the biochemical and physiological changes resulted from the effects of the factors released during the primary response (*e.g.* changes in blood circulation, the oxygen carrying capacity of blood, increase in plasma glucose). Production of heat shock proteins (Hsp) is included in secondary level as an intracellular stress response. In case of prolonged stress response 3° response becomes.

3. Tertiary stress response – Chronic exposure to stressors provokes tertiary stress responses that lead to a number of pathological changes and reduction in reproductive success, depression of growth and decrease in disease resistance. The tertiary stress response represents whole animal and population level changes associated with stress [7].

Acute and Chronic Stress

Acute stress responses are originated from short-term and sudden discomfort such as sharp changes in dissolved oxygen or chemical exposure (Fig. **3**). Acute response can be lethal in case that the stress exceeds the tolerance of organisms.

Chronic stress is more ubiquitous and considered as sublethal stress. Chronic stress is generally caused by continuous or periodic exposure to stressors for the long periods. Chronic stress has an impact on whole physiology of the fish with reducing growth, increasing sensitivity to the diseases, impairing reproduction and reduce the capability of fish to overcome additional stress. It is accepted that chronic stress depresses the immune response and lowers the disease resistance.

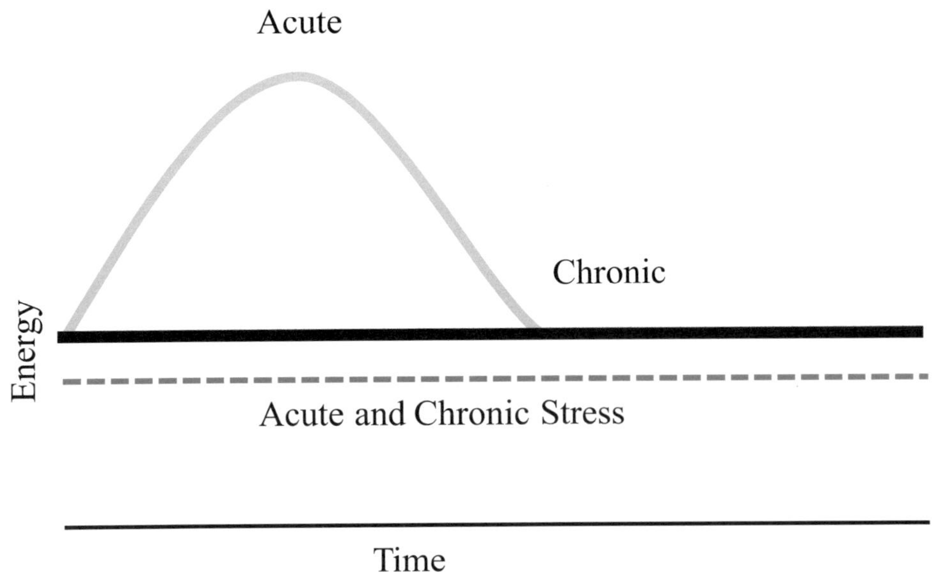

Fig. (3). Energy use in acute and chronic stress.

Stressors

The factor that leads to stress called "stressors". Fish in aquaculture expose to various stressors [8]. Wedemeyer [9] classified the stressors in fish culture as chemical, physical, biological and procedural stressor (Fig. **4**).

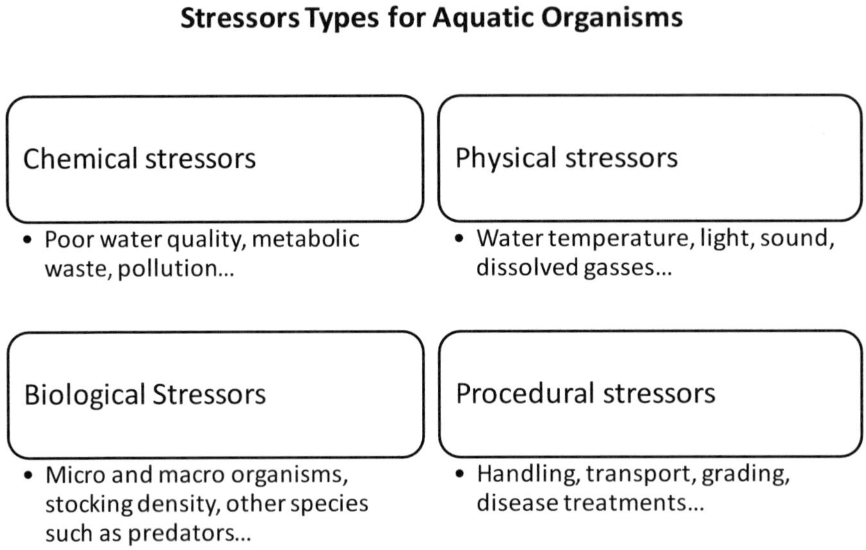

Fig. (4). Type of stressors in aquaculture (adapted from Barton [10]).

Stress Hormones

The activation of neuro-endocrine system is the primary response to stress, leading to the elevation of of stress hormones; corticosteroids and catecholamines (Fig. **5**).

Catecholamines

The catecholamine hormones; adrenaline and noradrenaline are released into the circulation from chromaffin cells. The secretion of catecholamines into the blood and total effects on physiological systems is termed the acute humoral adrenergic stress response [11]. Catecholamines are mobilized into the circulation of fish during the stressful situations. This is generally related with the modulation of cardio-respiratory function or mobilization of energy reserves [12]. Hyperglycemia after stress due to glycogenolysis stimulation from catecholamines to meet increased energy demands for a flight or fight response [13]. Regarding catecholamines effects on the cardiovascular and respiratory functions, increased blood flow, gill permeability, and lamellar recruitment occur [14]. Increases in gill permeability to water and some ions result in disruption in the balance between water and ion influx both in freshwater fishes and marine fishes [14 - 16]. The magnitude of the elevation in plasma catecholamine levels can vary depending upon the type of stress, the severity of the stress and the species of fish [12].

Cortisol

Cortisol is one of the major corticosteroids produced by the interrenal tissue of teleost fish with glucocorticoid and mineralocorticoid functions. This hormone has physiological effects on osmoregulation, intermediary metabolism, growth, stress, and immune function [17]. Releasing of cortisol to the blood is initiated by the hypothalamo-pituitary-interrenal (HPI) axis. HPI axis is activated by higher brain centres. Stress stimulates the expression and synthesis of corticotropin releasing hormone (CRH) which moves from the hypothalamus to the pituiatry *via* direct neural contact in fish. Cortisol is produced from cholesterol when the interrenal cells in head kidney are stimulated by HPI. The secretion of cortisol is slower than catecholamines, but its effects are more prolonged [18, 19].

In unstressed conditons, basal cortisol values in teleosts are estimated <10 ng mL^{-1}, however, apart from stress, corticosteroid levels change by various factors, sex and maturity, time of day, feeding, and season [20]. The peak values are reported to be in the range of 30-300 ng mL^{-1} depending on the duration or severity of the stressor [10]. The decline of cortisol levels to normal values following to stress requires some hours thus, recovery from stress takes, at least several hours rather than minutes considering the return of cortisol levels. Nonetheless, measurement of cortisol or any other stress indicator usually involves subsequent maintenance and manipulation of fish under conditions that may themselves be perceived as stressful. Hence, the stress indicators should be evaluated by the fact that the sampling for assessing the stress effects, itself may affect the values.

Cortisol plays an important role in osmoregulation by the maintenance of salt and water balance. It has long been known that cortisol acts of seawater acclimation in teleost fishes. Cortisol is also involved in ion uptake and can interact positively with prolactin, indicating that cortisol has a dual osmoregulatory function in teleosts [17]. High levels of plasma cortisol in fish transferred from fresh water to sea water help sea water acclimation with stimulating the proliferation and differentiation of ion-transporting chloride cells within the gill epithelium, and with stimulating Na+- K+-ATPase activity within osmoregulatory organs including the gill, intestine and kidney [21]. Cortisol is also involved in ionic/osmotic regulation of freshwater fish. Cortisol treatment triggers chloride cell proliferation and increases branchial ion (Na^{+}, Ca^{2+}, Cl^{-}) influx in rainbow trout [21 - 23].

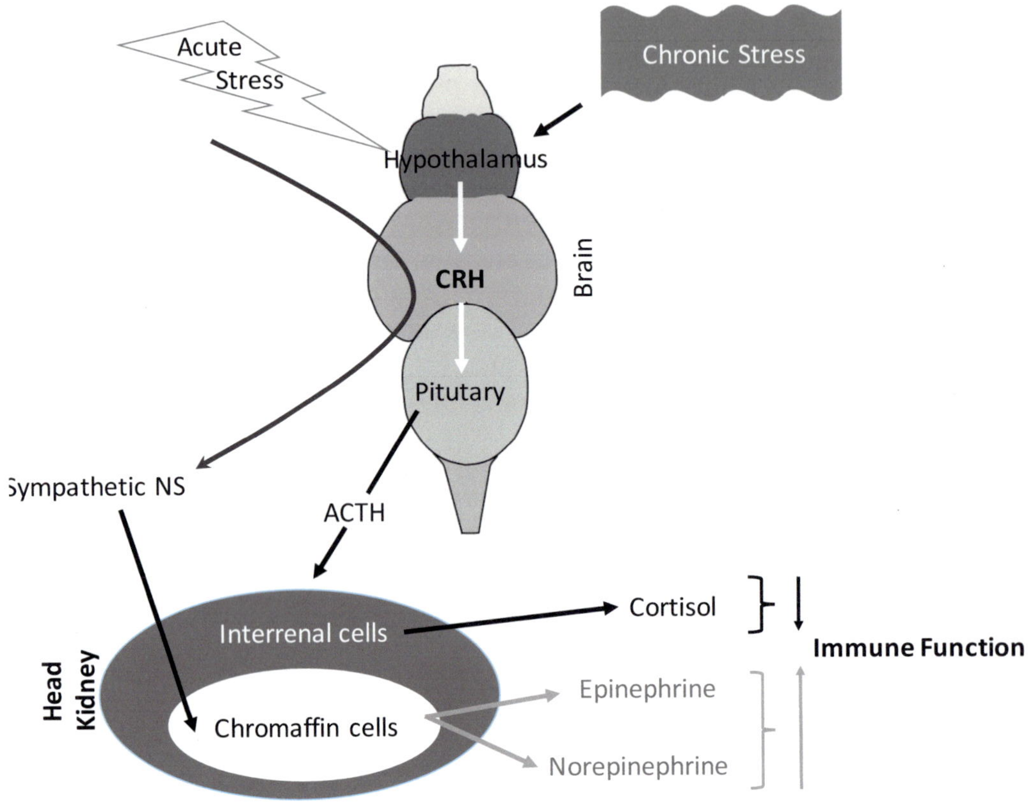

Fig. (5). Schematic representation of stress effects on HPI axis in fish (adapted from Nardocci *et al.* [6],).

Cortisol affects the protein metabolism: increased protein synthesis *via* enhanced activity of messenger RNA; the proteins synthesized seem to be mostly proteolitic. Inhibition of growth and mobilization of protein from thymus, spleen and liver are under the effects of cortisol.

Cortisol activates glycogenolysis and gluconeogenesis processes in fish; gluconeogenesis means the production of carbohydrate from non-carbohydrate sources. Thus, increased glucose production from tissue protein occurs. Under high levels of cortisol in fish, carbohydrate use is reduced and glycogen is deposited in the liver [24]. Cortisol has also an impact on the circulatory system. White blood cells are mobilized and inflammatory response is reduced due to high level of cortisol [1].

Under normal conditions, the role of cortisol as a regulatory factor for the physiological function and rapid physiological readjustments in stress response is of significance to maintain homeostatis. Thus, the key point is that the stress response is fundamentally a regulatory response to return the animal to 'homeostatic norms' [25]. However, due to the inhibitory effects of cortisol on growth, reproduction and disease resistance metabolic effects of cortisol seem to be complicated, in general [26]. Nevertheless, inconsistent results exist in terms of the immunosuppressive effects of cortisol in fish. For example, Narnaware and Baker [27] reported that a lack of depression of macrophage phagocytic activity *in vitro* following to 200 nM cortisol injected trout. By contrast, corticosteroid treatment resulted in a reduction of circulating B-cells while an increased number of neutrophils [28]. Hence, to ascertain whether cortisol is immunosuppressive, one should consider that cortisol actions are not always inhibitory but depend on the stressor so that some non-specific responses are enhanced while specific responses are inhibited [29].

Immune Response of Fish and Stress

A well functioning immune system is a sensitive indicator of a healthy fish and can be used to understand the status of fish health in aquaculture. Immune response in fish can be examined in two categories: innate and adaptive immunity by cellular and humoral factors (Table **1**).

Table 1. Humoral and cellular elements of immune system in fish. (adapted from Toledo-Ibarra GA [30].

	Innate	Adaptative
Humoral	Antimicrobial peptides Lectins Lysozyme Complement	Antibodies: IgM IgD IgT/IgZ
Cellular	Macrophages Neutrophils Eosinophils	B-lymphocytes T-lymphocytes

Innate Immunity

Innate immunity takes place in early defence against pathogen infection regarding its mechanisms. Innate immunity has a basilar role in initiating the adaptive response. The innate immune response of teleosts is highly developed and the components such as lysozyme, C3, 2 macroglobulin and C-reactive protein may enhance immune recognition. Teleosts, as lower poikilothermic vertebrates, the non-specific type may be the centre of immune response [31].

The innate humoral components that are mainly found in fishes are antibacterial peptides, lysozymes, lectins, acute-phase proteins, and molecules of the complement system. The cells of the innate immune system are activated by pathogen recognition receptors (PRR) [30].

In fishes, the cells of the innate immune response are the macrophages, neutrophils, and eosinophils. Macrophages are strong phagocytic cells in fishes that are of significance in early antimicrobial responses. Neutrophils have also phagocytic properties and neutrophils are the first cells to migrate to the site of infection; their bactericidal capacity is high relating with the release of proteolytic enzymes, antimicrobial peptides, and reactive oxygen species (ROS). In fishes, neutrophils have the capacity of releasing neutrophil extracellular traps (NET), which are complex structures consisting of DNA, histones, and proteins from granules. These structures are extremely important for the immunity of fish as functioning for trapping and extracellular killing of bacteria, fungi, parasites, and also for inactivating viruses. Fish eosinophils are responsible for the parasites by releasing cytoplasmic granules against extracellular parasites [30].

Phagocytic response is one of the significant components of innate immunity in fish and can be detected using several indicators: the respiratory burst reaction, as measured by the levels of oxygen radicals produced; phagocytic indicators, such as the reduction of nitroblue tetrazolium (NBT) caused by the oxygen radicals; cell adherence; and the ability of macrophages to engulf particles. Another indicator of cell response are the white blood cells: the total and differential white cells numbers as well as their function and distribution. The interaction of stress and phagocytes, mainly phagocytic capacity of macrophages has been studied for various fish species. However, few studies provide a complete picture of the status of white cells in fish during stress, *i.e.* determination of the number and type of cells (leukocytes or lymphocytes) both in blood and in immune related organs (spleen, head kidney and thymus) [30].

Adaptive Immunity

Adaptive immunity in fishes are mediated by T- and B-lymphocytes and antibodies. Adaptive immunity plays a critical role in protection against recurrent infections through the cooperation of cellular and humoral mechanisms, which generate immunological memory. In fish, the memory response is infirm and slower, with taking some time to reach a maximum [31]. The hemapoietic tissue of head kidney in fish is analog to the mammalian bone marrow due to the production of B- and T-lymphocytes, and phagocytic cells in the head kidney [6]. In teleost fish, activated B-cells, plasma blasts, and plasma cells exist, which are responsible for the production of antibodies. Soluble antibodies that have been identified in fishes are primarily (IgM), which are tetrameric and present in high concentrations in plasma and IgD, as in mammals, are expressed on the surface of B-lymphocytes. These antibodies are present in the first line defence barrier; mucosa of intestine, skin, and gills. Innate and adaptive mechanisms in fishes, similar to those occurring in mammals, are regulated and inter-connected by cytokines. Interleukin-1β, mainly produced by macrophages, has been characterised in bony fish and cartilaginous fish [31]. Other cytokines reported in fish include IL-2, IL-4, IL-8,IL-10, IL-12, and tumor growth factor beta (TGF-*β*) [30]. In addition to cytokines, the immune system of teleost fishes is regulated by neuroendocrine interactions, primarily *via* the hypothalamic pituitary-interrenal (HPI) axis, thus, the head kidney with being lymphoid tissue is also an important endocrine organ, functional homologous to mammalian adrenal glands [6]. Moreover, the head kidney with its hematopoietic and endocrine tissue makes the direct paracrine interaction easier.

Under normal conditions, specific and nonspecific humoral and cellular components are in harmony with each other. Nevertheless, the immune system can be disturbed by internal or external factors [32]. It is known that immune function is modulated by stress. Stressors are known to affect immune system functions.

The immunological potency of the immune cells may enhance or decrease in case of any stressors existence. The distribution of the immune cells in the body is also affected by hormones that are released during stress (Fig. **6**).

The immune system and response of fish can be influenced by various external factors like temperature, light, water quality, salinity and different stress factors [34]. Decreases in temperature, related with the poikilothermic nature of fish affect the rate of their physiological functions [35]. Decline in the number of circulating leukocytes in the rainbow trout exposed to photoperiod has been indicated. An increase in the photoperiod leads to increases in the activity of lysozyme and the circulating levels of IgM. The oxygen concentrations in the water may modulate the immune response; hypoxia suppresses the respiratory burst activity of macrophages and lowers the levels of circulating antibodies, which in turn, are elevated by hyperoxia [32, 36]. High levels of suspended solids in the fish environment induce to increase hematocrit levels to compensate for the diminished ability of the gills to take oxygen. Excess amount of suspended solids in the water also increases the activity of lysozyme and the levels of circulating IgM as a response to elevated levels of pathogens that are expected in an environment rich in particles [37]. The short-term holding effects on fish are associated with the supressed immune systems [16]. Increased salinity also affects immune parameters by increasing the activity of lytic enzymes, the respiratory burst of macrophages and the circulating levels of IgM. Changes in environmental pH levels show opposing results for immune system parameters, such as levels of lysozyme and IgM in the circulation. Stress in fish as a result of high stocking density by management practices can increase circulating cortisol levels and weaken the specific and nonspecific immunity, eventually, making the fish open to opportunistic pathogens [16, 37].

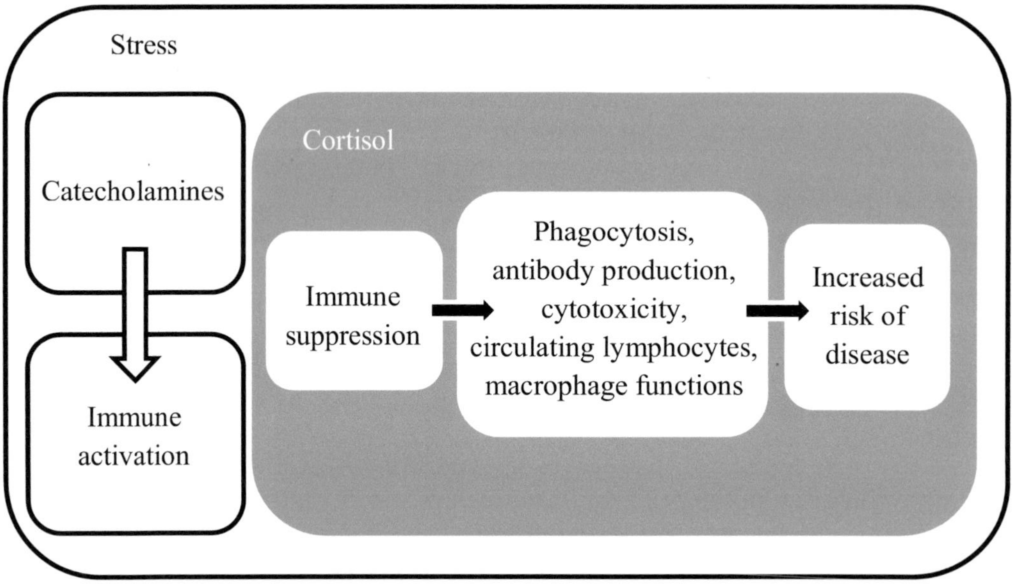

Fig. (6). Stress and immun response in fish (adapted from Tort [33]).

In evaluating the stress effects on immunological components, the duration of stress should be taken into consideration. Stress in trout caused decrease in the number of circulating lymphocytes/trombocytes, increase in the number of circulating granulocytes and depression in the phagocytic capacity of macrophages in the spleen and pronephros 3 h following to stressor exposure [27]. The adrenergic agents may be responsible for the reduced phagocytic activity immediately afterwards stress as the cortisol activation becomes later. In the longer term, cortisol has harmful effects on the immune system. In the short term, stress has depressing impact on the immune system that is not resulted from cortisol. Tort [33] explains that the early responses activated by cortisol may help to maintain key defense mechanisms, at least during the short term, whereas other actions of cortisol over longer periods of time result in suppression.

The relation between a stressor and disease outbreak can be indicated by immune indicators. Hence, identifying the appropriate indicator to monitor and understand the stress effects is of importance. Immunosupression can be analyzed by nonspecific immunological indicator assays (Table **2**) or specific immune response assays in fish.

Table 2. Nonspecific or specific assays used with *in vivo* or *in vitro* antigenic stimulation (adapted from Anderson [38]).

Assay	Sensitivity	Specific immune impairment	Statistical variation
Phagocytic index	Good	Phagocytosis	Wide
Pinocytosis	Good	Phagocytosis	Wide
Chemiluminescence (oxidative burst)	Excellent	Phagocytosis	Medium
Blastogenesis (T and B cell mitogenesis)	Fair	Lymphocytes	Wide
Blastogenesis (radioimmunoassay)	Good	Lymphocytes	Medium

Nonspecific immune assays cover hematocrit and leucocrit which are easy and quick to measure. Hematocrit values indicate hemapoietic activation of fish. Leucocrit gives an indication of the immunopoiteic system and resistance to diseases. Although hematocrit and leucocrit are not sensitive indicators due to wide variability with environmental factors, sex, feeding, season *etc.* high values in hematocrit, hemoglobin, and red blood cells have been also reported and may represent a coping strategy for increasing the blood's oxygen carrying ability during stressful periods [39].

Table 3. Specific immune response assays (adapted from Anderson [38]).

Assay	Sensitivity	Specific immune impairment	Statistical variation
Melanomacrophage center antigen accumulation	Good	Hemopoesis	Medium
Passive hemolytic plaque assay	High	Antibody production	Narrow
Agglutination (bacterial particle)	Good	Antibody production	Medium

(Table 3) contd.....

Assay	Sensitivity	Specific immune impairment	Statistical variation
Passive hemaglutination	Excellent	Antibody production	Narrow
Precipitin (gel)	Fair	Antibody production	Narrow
Counterimmunoelectrophoresis	Good	Antibody production	Narrow
ELISA	Excellent	Antibody production	Narrow
Neutralization	Good	Antibody production	Narrow
Fluorescent antibody	Excellent	Antibody production	Narrow
Challenge with pathogen	Excellent	Protection against disease	Narrow

Specific immune assays may give more sensitive indication of immune response to stress (Table **3**). The indicators of protective mechanisms in fish such as phagocytic index, macrophage migration and antibody production tests could be used to demonstrate immonusuppression. Disease challenge tests are one of the immune assays to evaluate the resistance of fish to specific disease thus, it is known that there is a link among disease occurence, immune suppression and environmental stressors.

Allostatis and Diseases

Allostasis has been described "stability through change" by Sterling and Eyer [40]. The stress response can also be defined as allostasis, which explains the struggle to maintain homeostasis through alterations in physiological systems. Allostasis enables fish to configure the physiological status to resist against stress [2]. Allostasis also clarifies an inherent ambiguity in the term "homeostasis" and distinguishes between the systems that are essential for life ("homeostasis") and those that maintain these systems in balance ("allostasis") as environment and life history stage change [41]. Stress response and resisting to a stressor is an energetically costly process, including costs associated with allostasis, attempting to reset homeostatic norms [42]. The main target in allostasis is to maintain homeostasis thus, this requires energy, imposing energetic load which refers the allostatic load [25, 42]. Allostatic load also expresses the cumulative result of allostatic states. This can be regarded as the cost an animal has to pay to realise the homeostasis *via* allostasis [41]. Allostatic load may be a part of the adaptive response against the threat and may refer the fish's competent however, when the allostatic load turns into allostatic overload, the stress pass to the phase of detrimental effects [2, 43]. Although being an adaptive response at first, prolonged and/or exposure to cumulative stress leads to a situation named allostatic overload, in which pathologies and diseases occur [41, 44]. Ultimately, as allostatic load increases the risk of diseases increases (Fig. **7**).

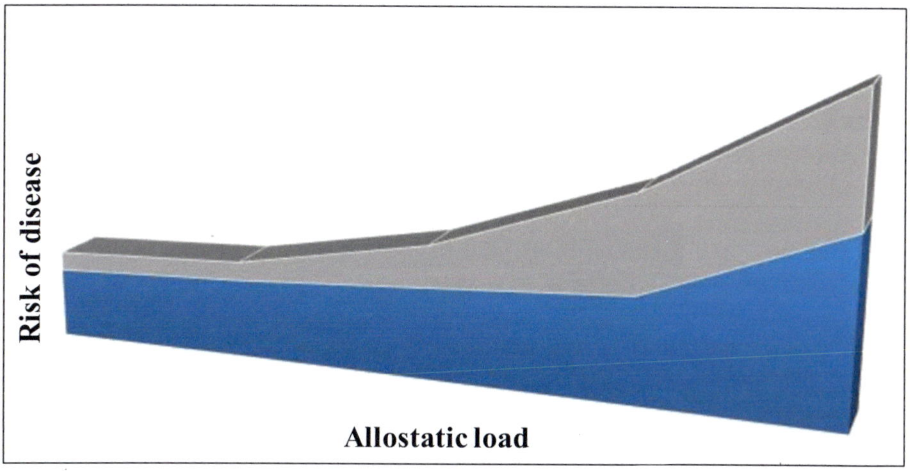

Fig. (7). Intreaction of allostatic load and disease risk (adapted from Segner *et al.* [2]).

Fish pathogens exist in aquaculture systems however, fish usually are able to resist them unless allostatic load gets to the level of overload. The relationship between stress and disease is discussed in the publication including [45 - 48]. It is known that stress makes fish more susceptible to infectious diseases, by elevated secretion of corticosteriods, which suppress the number and function of circulating lymphocytes [49].

Disease is a set of interactions among the host (fish), the pathogen, and the environment (water). The factors such as the species of fish, size or age, strain or stock, immune status, and general physiological condition of fish affect the host robustness to pathogen and resistance to diseases. For the pathogen, factors include the concentration of infectious particles present in water, their physiological properties and the virulence of the strain. The sensitive balance between the pathogen and the host is linked to disease and health, and that balance is mainly influenced by environmental factors, particularly by water quality. Fish in intensive culture are frequently exposed to adverse environmental conditions (*e.g.* poor water quality) and to management practices (*e.g.* crowding or handling) that can impose stressors and affect the immune defence reactions of the fish. The potential of opportunistic bacterial pathogens existing in the aquatic environment to disease occurence is depending on environmental quality or the defence reactions of the fish. The immune response of cold-blooded animals depends strictly on environmental temperature. Lowering the water temperature below the optimal values usually reduces or delays the onset of the immune response, particularly the humoral immune response. As an example of the low temperature effects, Tort *et al.* [50] reported the research on gilt head sea bream (*Sparus aurata*) indicates that severe immunosuppression in winter months occurs. Physiology of the fish is a crucial factor in immune response of the fish Critical physiological states of fish (*e.g.* active reproduction period) can reduce the magnitude or efficiency of the immune response. The age of the fish is also an important factor. Hence, the fish specific immune system performs its functions after several weeks of age following to hatching [51].

The agents of primary infections of fish are parasitic, fungal, bacterial and viral. Diseases of fish are mostly species and system specific.

Parasites are common in fish and identified as the significant health problem in aquaculture. It is considered that secondary bacterial infections due to presence of ectoparasites affects aquaculture production. Parasites have an interaction with the other stressors in various ways. Parasites themselves are one of the stressors and in case of overlapping the parasitism and stressors the negative impact on fish health magnifies as stressors ameliorate the effects of parasitism. Hence, stressors suppress the immune system of the fish, it is normally expected that parasites abundance and their pathogenity increase. Poor water quality, particularly decrease the immunological capabilities of fish, rendering them more susceptible to some parasites. Gill ciliates of fish are a good example of this effect [52]. Regarding immune response, ciliates create breaks in the first line of immune protection system by impairing mucus production [53]. Stress response of *Salmo salar* to *Caligus rogercresseyi* copepodids indicated that heavy infectations of *C. rogercresseyi* result in an acute stress response, metabolic alteration and excessive mucus production [54].

Svobodova *et al.* [55] reported the interaction of stress and pathogen in the aquatic environment by different disease agents. The presence or introduction of specific viral agents into ponds, reservoirs and streams is related with the outbreaks of infectious pancreatic necrosis in salmonids, viral haemorrhagic septicaemia in rainbow trout, spring viraemia of carp, infectious swim-bladder inflammation, ulcerative dermal necrosis in salmonids, pox and other viral diseases. Outbreaks of diseases are frequently in conjunction with the environmental problem, thus in diagnosing and controlling a disease any suboptimal environmental conditions should be taken into account. Where water quality is reduced, a more complicated course of the disease together with a higher mortality of fish are usually found. In particular, a reduced water quality is an important stress factor in viral haemorrhagic septicaemia. A low dissolved oxygen content and extreme changes in the pH of the water, together with an increased accumulation of metabolites, are other important factors associated with outbreaks of infectious haemopoetic necrosis in salmonids [55].

The presence of specific bacterial agents results in an infection in aquaculture. Much more of these agents can survive naturally in the environment (*e.g. Aeromonas punctata*, *Aeromonas salmonicida*) and/or in the digestive tract of clinically healthy fish; with an increase in their virulence and/or weakening of the host organism (*e.g.* due to a stressors such as polluted aquatic environment) these agents can act as causative factors in the outbreak of a bacterial disease. Many bacteria act as opportunist or facultative pathogens. These are found free-living in nature, sometimes on the skin and gill surface, and only affect fish under special circumstances. Non-infectious health problems may act synergistically with the infectious agents, resulting in high losses exceptionally in aquaculture production. In some cases, the main reason of the disease is masked by the secondary infections, the pathogenic bacteria can cross the epithelial barrier of the skin, the gills or the digestive tract and lead to septicaemia *i.e.* septicaemia of *Aeromonas spp*. and *Pseudomonas spp*. although the reason for the disease is the suboptimal water quality [55].

A direct relationship between branchiomycosis and organic pollution of water is well known in freshwater aquaculture. The disease is endemic in ponds and reservoirs; cyprinid fish species, whitefish, pike, but also wells and rainbow trout can be affected. The outbreak and duration of the disease depend on ambient environmental factors. Abnormal water temperature act as stressor thus, the disease occurs most frequently when the water temperature is above 20°C and provoked by organic pollution and unappropriate concentrations of the dissolved oxygen. Physical and/or chemical damage of the protective mucus layer of the skin, fins and gills are suitable entry for the disease outbreak. Such damage is also a precondition for the secondary development of saprolegnia, particularly in unhealthy fish. Under aquaculture conditions with high stocking density, oomycetes cause serious problems. Of the fungal species, *Saprolegnia parasitica* is typically a secondary invader of tissues damaged by trauma, infectious disease, or poor environmental conditions [55].

Wedemeyer [9, 45] reported that stress effects of aquacultural procedures which are one of the main components of intensive rearing systems, leading to compromise the immune system and reduce the resistance to the infectious diseases, emphasizing the economic dimension of loss due to diseases. Some examples of stress induced diseases in aquaculture are presented in Table **4**.

Table 4. Stress induced dieases (adapted from Wedemeyer [9, 56], Khan and Thulin [57]).

Fish Diseases Problems	Stressor
Infectious pancreatic necrosis (IPN)	Water hardness

(Table 4) contd.....

Fish Diseases Problems	**Stressor**
Bacterial gill disease *(Flavobacterium sp.)*	Crowding; chronic low oxygen (4 mg/l for salmonids); elevated ammonia (more than 0.02 mg/l for salmonids); suspended particulate matter
Blue sac, hydrocele	Temperature; ammonia; crowding
Columnaris *(Flexibacter columnaris)*	Crowding or handling during warm-water periods if carrier fish are present
Environmental gill disease	Excess organic load in the water
Epithelial tumors, ulceration	Chronic, sublethal contaminant exposure
Fin erosion	Crowding; low level of dissolved oxygen; nutritional imbalances; chronic exposure to trace contaminants; high total suspended solids; secondary bacterial invasion
Furunculosis *(Aeromonas salmonicida)*	Low oxygen (<5 mg/l for salmonids); crowding; temperature; handling when pathogen carriers are present
Hemorrhagic septicemias, red-sore disease *(Aeromonas, Pseudomonas)*	External parasite infestations; ponds not cleaned; crowding; elevated ammonia; low oxygen; stress due to elevated water temperatures; handling after overwintering at low temperatures
Kidney disease *(Renibacterium salmoninarum)*	Water hardness less than about 100 mg/l (as $CaCO_3$); diet composition; crowding; temperature
Nephrolithiasis	Water high in phosphates and carbon dioxide
Parasite infestations	High stocking density; low oxygen; excessive size variation among fish in ponds, cages and canals
Spring viremia of carp	Handling after overwintering at low temperatures
Vibriosis *(Vibrio anguillarum)*	Handling; oxygen <6 mg/l, especially at water temperatures of 10-15°C (50-59°F); salinity 10-15%
White-spot, coagulated-yolk disease	Environmental stress: air supersaturation >102-103%, temperature, metabolic wastes, chronic trace contaminant exposure
Swim bladder stress syndrome	Oil films; hypoxia; salinity; other water quality factors
Edwardsiella tarda infection	High level of ammonia
Saprolegnia infection	Confinement, crowding
Enteric septicemia, Enteric red-mouth diseases	Unsuitable temperatures

(Table 4) contd.....

Fish Diseases Problems	Stressor
Ichtyobodosis	Low level of pH
Increase in gill parasites prevalance *i.e.* Gyrodactylus sp.	Low level of oxygen

CONCLUDING REMARKS

Stress is the activation of responsive physiological systems, having both the protective and damaging effects against the stressors. However, each time the stress response is activated, the accumulated 'wear and tear effect' takes place in fish body. In the new approach to stress and fish health, the conceptions of allostasis and allostatic load are proven. Generally, the concept of allostasis refers homeostatic flexibility and allostatic physiological mechanisms are related with maintaining the fish health. Allostatic load is resulted from repeated or prolonged stress. Alongside, the transition from allostatic load to allostatic overload may be critical as the "allostatic overload" may cause to the disruption of fish health. Key issues for the future to operationalise the definition of allostatic load/overload are to measure allostatic load by various biomarkers representing neuro-endocrine and immune systems of fish and ultimate modelling of allostatic load.

CONFLICT OF INTEREST

The author (editor) declares no conflict of interest, financial or otherwise.

ACKNOWLEDGEMENTS

Declared none.

REFERENCES

[1] Smith LS. Introduction to Fish Physiology Argent Lab. USA: Publishing 1991.

[2] Segner H, Sundh H, Buchmann K, *et al.* Health of farmed fish: its relation to fish welfare and its utility as welfare indicator. Fish Physiol Biochem 2012; 38(1): 85-105. [http://dx.doi.org/10.1007/s10695-011-9517-9] [PMID: 21681416]

[3] Selye H. Stress and the general adaptation syndrome. BMJ 1950; 1(4667): 1383-92. [http://dx.doi.org/10.1136/bmj.1.4667.1383] [PMID: 15426759]

[4] Pottinger TG. The Stress Response in Fish-Mechanisms, Effects and Measurement.Fish Welfare. Oxford, UK: Blackwell Publishing Ltd 2008. [http://dx.doi.org/10.1002/9780470697610.ch3]

[5] Ellis T, Yildiz HY, López-Olmeda J, *et al.* Cortisol and finfish welfare. Fish Physiol Biochem 2012; 38(1): 163-88. [http://dx.doi.org/10.1007/s10695-011-9568-y] [PMID: 22113503]

[6] Nardocci G, Navarro C, Cortés PP, *et al.* Neuroendocrine mechanisms for immune system regulation during stress in fish. Fish Shellfish Immunol 2014; 40(2): 531-8. [http://dx.doi.org/10.1016/j.fsi.2014.08.001] [PMID: 25123831]

[7] Barman D, Kumar V, Mandal SC, Roy S, Kumar K. Stress and stressors: The role of heat shock proteins. World Aquaculture 2012; 43(2): 62-4.

[8] Kumar P, Thirunavukkarasu AR, Subburaj R, Thiagarajan G. Concept of Stress and Its Mitigation in Aquaculture. 2015. [http://dx.doi.org/10.1007/978-81-322-2271-2_10]

[9] Wedemeyer GA, McLeay OJ. Methods for determining the tolerance of fishes to enviromental stressors. 1981.

[10] Barton BA. Stress in fishes: a diversity of responses with particular reference to changes in circulating corticosteroids. Integr Comp Biol 2002; 42(3): 517-25. [http://dx.doi.org/10.1093/icb/42.3.517] [PMID: 21708747]

[11] Perry SF, Bernier NJ. The acute humoral adrenergic stress response in fish: facts and fiction. Aquaculture 1999; 177: 285-95. [http://dx.doi.org/10.1016/S0044-8486(99)00092-7]

[12] Reid SG, Bernier NJ, Perry SF. The adrenergic stress response in fish: control of catecholamine storage and release. Comp Biochem Physiol C Pharmacol Toxicol Endocrinol 1998; 120(1): 1-27. [http://dx.doi.org/10.1016/S0742-8413(98)00037-1] [PMID: 9827012]

[13] Mazeaud M, Mazeaud F. Adrenergic responses to stress in fish.Stress and Fish. New York: Academic Press 1981; pp. 49-75.

[14] Wendelaar Bonga SE. The stress response in fish. Physiol Rev 1997; 77(3): 591-625. [PMID: 9234959]

[15] Mazeaud MM, Mazeaud F, Donaldson FM. Primary and secondary effects of stress in fish: some new data with a general review. Trans Am Fish Soc 1977; 106: 201-12. [http://dx.doi.org/10.1577/1548-8659(1977)106<201:PASEOS>2.0.CO;2]

[16] Portz DE, Woodley CM. Cech Jr.JJ. Stress-associated impacts of short-term holding on fishes. Rev Fish Biol Fish 2006; 16: 125-70. [http://dx.doi.org/10.1007/s11160-006-9012-z]

[17] McCormick SD. Endocrine control of osmoregulation in teleost fish. Am Zool 2001; 41: 781-94.

[18] Gamperl AK, Vijayan MM, Boutilier RG. Experimental control of stress hormone levels in fishes: Techniques and applications. Rev Fish Biol Fish 1994; 4: 215-55. [http://dx.doi.org/10.1007/BF00044129]

[19] Waring CP, Stagg RM, Poxton MG. Physiological responses to handling in the turbot. J Fish Biol 1996; 48: 161-73. [http://dx.doi.org/10.1111/j.1095-8649.1996.tb01110.x]

[20] Pankhurst NW. The endocrinology of stress in fish: an environmental perspective. Gen Comp Endocrinol 2011; 170(2): 265-75. [http://dx.doi.org/10.1016/j.ygcen.2010.07.017] [PMID: 20688064]

[21] Sloman KA, Desforges PR, Gilmour KM. Evidence for a mineralocorticoid-like receptor linked to branchial chloride cell proliferation in freshwater rainbow trout. J Exp Biol 2001; 204(Pt 22): 3953-61. [PMID: 11807113]

[22] Perry SF, Laurent P. Adaptational responses of rainbow trout to lowered external NaCl concentration: contribution of the branchial chloride cell. J Exp Biol 1989; 147: 147-68.

[23] Laurent P, Perry SF. Effects of cortisol on gill chloride cell morphology and ionic uptake in the freshwater trout, Salmo gairdneri. Cell Tissue Res 1990; 259: 429-42. [http://dx.doi.org/10.1007/BF01740769]

[24] Martínez-Porchas M, Martínez-Córdova LR, Ramos-Enriquez R. Cortisol and Glucose: Reliable indicators of fish stress? Pan-Am J Aquat Sci 2009; 4(2): 158-78.

[25] Schreck CB. Stress and fish reproduction: the roles of allostasis and hormesis. Gen Comp Endocrinol 2010; 165(3): 549-56. [http://dx.doi.org/10.1016/j.ygcen.2009.07.004] [PMID: 19596332]

[26] Pickering AD. Factors affecting the susceptibility of Salmonid fish to disease 1989.

[27] Narnaware YK, Baker BI. Evidence that cortisol may protect against the immediate effects of stress on circulating leukocytes in the trout. Gen Comp Endocrinol 1996; 103(3): 359-66. [http://dx.doi.org/10.1006/gcen.1996.0131] [PMID: 8812406]

[28] Weyts FA, Verburg-van Kemenade BM, Flik G, Lambert JG, Wendelaar Bonga SE. Conservation of apoptosis as an immune regulatory mechanism: effects of cortisol and cortisone on carp lymphocytes. Brain Behav Immun 1997; 11(2): 95-105. [http://dx.doi.org/10.1006/brbi.1997.0484] [PMID: 9299059]

[29] Weyts FA, Cohen N, Flik G, Verburgvan KB. Interactions between the immune system and the hypothalamo pituitary interrenal axis in fish. Fish Shellfish Immunol 1999; 9: 1-20. [http://dx.doi.org/10.1006/fsim.1998.0170]

[30] Toledo-Ibarra GA, Rojas-Mayorquín AE. Influence of the Cholinergic System on the Immune Response of Teleost Fishes: Potential Model in Biomedical Research Clinical and Developmental Immunology Volume 2013; Article ID 536534, 9 pages.

[31] Tort L, Balasch JC, MacKenzie S. Fish immune system. A crossroads between innate and adaptive responses. Inmunologia 2003; 22(3): 277-86.

[32] Watts M, Munday BL, Burke CM. Immune responses of teleost fish. Aust Vet J 2001; 79(8): 570-4. [http://dx.doi.org/10.1111/j.1751-0813.2001.tb10753.x] [PMID: 11599820]

[33] Tort L. Stress and immune modulation in fish. Dev Comp Immunol 2011; 35(12): 1366-75. [http://dx.doi.org/10.1016/j.dci.2011.07.002] [PMID: 21782845]

[34] Magnadottir B. Immunological control of fish diseases. Mar Biotechnol (NY) 2010; 12(4): 361-79. [http://dx.doi.org/10.1007/s10126-010-9279-x] [PMID: 20352271]

[35] Hayward CJ, Bott NJ, Nowak BF. Seasonal epizootics of sea lice, Caligus spp., on southern bluefin tuna, Thunnus maccoyii (Castelnau), in a long-term farming trial. J Fish Dis 2009; 32(1): 101-6. [http://dx.doi.org/10.1111/j.1365-2761.2008.01010.x] [PMID: 19245634]

[36] Bowden TJ. Modulation of the immune system of fish by their environment. Fish Shellfish Immunol 2008; 25(4): 373-83. [http://dx.doi.org/10.1016/j.fsi.2008.03.017] [PMID: 18562213]

[37] Uribe C, Folch H, Enriquez R, Moran G. Innate and adaptive immunity in teleost fish: a review. Vet Med (Praha) 2011; 56(10): 486-503.

[38] Anderson DP. Immunological Indicators: Effects of Environmental Stress on Immune Protection and Disease Outbreaks (in Biological Indicators of Stress in Fish, Ed by M. Adams). American Fisheries Symposium 8. Bethesda, Maryland.

[39] Montero D, Izquierdo MS, Tort L, Roaina L, *et al.* High stocking density produces crowding stress altering some physiological and biochemical parameters in gilthead seabream, Sparus aurata, juveniles. Fish Physiol Biochem 1999; 20: 53-60. [http://dx.doi.org/10.1023/A:1007719928905]

[40] Sterling P, Eyer J. Allostasis: a new paradigm to explain arousal pathology.Handbook of Life Stress, Cognition and Health. New York: Wiley 1988; pp. 629-49.

[41] McEwen BS, Wingfield JC. The concept of allostasis in biology and biomedicine. Horm Behav 2003; 43(1): 2-15. [http://dx.doi.org/10.1016/S0018-506X(02)00024-7] [PMID: 12614627]

[42] Schreck CB, Contreras-Sanchez W, Fitzpatrick MS. Effects of stress on fish reproduction, gamete

quality, and progeny. Aquaculture 2001; 197: 3-24.
[http://dx.doi.org/10.1016/S0044-8486(01)00580-4]

[43] Moberg GP. Biological responses to stress: implications for animal welfare.The biology of animal stress: basic principles and implications for biology of animal stress:basicprinciples andimplications for animal welfare. Wallingford: CAB International 2000; pp. 1-21.
[http://dx.doi.org/10.1079/9780851993591.0001]

[44] McEwen BS, Wingfield JC. What is in a name? Integrating homeostasis, allostasis and stress. Horm Behav 2010; 57(2): 105-11.
[http://dx.doi.org/10.1016/j.yhbeh.2009.09.011] [PMID: 19786032]

[45] Wedemeyer G. Effects of rearing conditions on the health and physiological quality of fish in intensive culture.Fish Stress and Health in Aquaculture. Cambridge: Cambridge University Press 1997; pp. 35-72.

[46] Post G. Text Book of Fish Diseases. Neptune, New Jersey: TFH Publications 1987.

[47] Toranzo AE, Magarinos B, Romalde JL. A review of the main bacterial fish diseases in mariculture systems. Aquaculture 2005; 246: 37-61.
[http://dx.doi.org/10.1016/j.aquaculture.2005.01.002]

[48] Nougayrede P. Major infectious and parasitic diseases in intensive farmed trout. Rev Med Vet (Toulouse) 1995; 146: 245-50.

[49] Tucker Jr. JW Marine Fish Culture. Springer 1998.

[50] Tort L, Padros F, Rotllant J, Crespo S. Winter syndrome in the gilthead sea bream Sparus aurata. Immunological and histopathological features. Fish Shellfish Immunol 1998; 8: 37-47.
[http://dx.doi.org/10.1006/fsim.1997.0120]

[51] Anonymouse, Welfare and health in sustainable aquaculture. 2008; 66p. http://wealth.imr.no/__d ata/page/6068 /Wealth_final_report_final_17042008.pdf

[52] Lafferty KD, Kuris AM. How environmental stress affects the impacts of parasites. Limnol Oceanogr 1999; 44(3, part 2): 925-31.
[http://dx.doi.org/10.4319/lo.1999.44.3_part_2.0925]

[53] Khan RA. Crude oil and parasites of fish. Parasitol Today (Regul Ed) 1987; 3(4): 99-100.
[http://dx.doi.org/10.1016/0169-4758(87)90043-3] [PMID: 15462922]

[54] González MP, Vargas-Chacoff L, Marín SL. Stress response of Salmo salar (Linnaeus 1758) when heavily infested by Caligus rogercresseyi (Boxshall & Bravo 2000) copepodids. Fish Physiol Biochem 2016; 42(1): 263-74.
[http://dx.doi.org/10.1007/s10695-015-0134-x] [PMID: 26394864]

[55] Svobodova Z, Lloyd R, Machova J, Vykusova B. Water quality and fish health FAO Eifac Technical Paper 54. 1993; 59.

[56] Wedemeyer GA. Physiology of fish in intensive culture systems. New York: Chapman & Hall, ITP 1996.

[57] Khan RA, Thulin J. Influence of pollution on parasites of aquatic animals. Adv Parasitol 1991; 30: 201-38.
[http://dx.doi.org/10.1016/S0065-308X(08)60309-7] [PMID: 2069073]

CHAPTER 7

Fish Histopathology as Biomarker in Ecotoxicology

Božidar Rašković* and **Vesna Poleksić**

University of Belgrade - Faculty of Agriculture, Institute of Animal Science, Belgrade - Zemun, Serbia

Abstract: Fish are regarded as valuable organisms in monitoring aquatic ecosystems. They are established as indicator organisms and widely used in both laboratory studies and ecotoxicology. A range of biomarkers are developed for monitoring effects of xenobiotics on fish. Histopathology is classified as an important biomarker in terms of toxicological significance and highly relevant when chronic or sublethal pollutants are present in the water. In this chapter, a short review of histopathological methods is given, with basic guidelines for conducting field studies. The main focus is on different methods for quantifying alterations, namely: semi-quantitative scoring system, histomorphometry and stereology. Advantages and shortcomings of each method are briefly discussed.

Keywords: Ecotoxicology, Fish, Histomorphometry, Quantification, Semi-quantitative scoring system, Stereology.

INTRODUCTION

Water quality issues and its effects on living organisms in both freshwater and marine ecosystems remain the hot topic in the 21st century. 65% of the aquatic habitats throughout the world are in moderate to high threat concerning biodiversity [1]. One of the most important factors in this threat is pollution of anthropogenic origin, such as: nutrient, pesticide and organic loading, metal deposition, acidification, thermal change of waters, and agriculture [1]. Due to the complex interactions of these substances in water, it is literally impossible to follow all the effects on ecosystem integrity. Typical method in monitoring ecosystem pollution is to classify water in different categories, according to its physical and chemical parameters [2].

This approach is legitimate, but inadequate, in the way that monitoring of aquatic organisms should also be taken into account, since not all chemicals could be detected using analytical methods, or the price for this kind of research would be

* **Corresponding author Božidar Rašković:** University of Belgrade - Faculty of Agriculture, Institute of Animal Science, Nemanjina 6, 11080 Belgrade - Zemun, Serbia; Tel: +381114413289; E-mail: raskovic@agrif.bg.ac.rs

Panagiotis Berillis (Ed)

enormous. Fish are regarded as valuable organisms in monitoring of aquatic ecosystems and today they are established as indicator organisms and widely used in both laboratory studies and ecotoxicology. One of the most important fish species in laboratory research, and therefore the most studied is the zebrafish (*Danio rerio*), a famous model organism [3]. Apart from it, a large number of fish species are used as indicators in assessment of aquatic ecosystems [4]. Being on the top or near to the top of the food chain in the aquatic ecosystems, the environmental health of fish is of utmost importance. Variation in fish community abundance, ultimately a complete lack of this part of the food web, as well as variation in the diversity of fish communities, could represent a serious threat for the whole ecosystem. In order to monitor early fish response in the polluted environments, a number of different biomarkers were established. They include biochemical, physiological and pathological biomarkers [5]. The capacity of fish organisms to adapt to changed environment and maintain homeostasis is limited. Responses at different levels of organisation include: molecular, subcellular, cellular, tissue, organ changes, and the organism as a whole [5, 6]. Consequently, different biomarkers are available to monitor changes on each of these levels. With regard to morphological changes of cells and tissues of different fish organs exposed to environmental pollution, we focus on the histopathological approach.

Practical overview of different markers used in fish ecotoxicological research is given by van der Oost *et al.* [6]. Authors compiled a list of evaluated biomarkers according to the six different criteria, giving them overall grade from 1 to 5. In that list histopathology was graded with overall value of 3.5, with high toxicological significance. They concluded that histopathological method is not very sensitive to pollution, in the sense that histopathological alterations can persist even when stressor or xenobiotic is not present in the environment any more. Additionally, histopathology is highly valuable when chronic or sublethal pollutants are present in the water [7 - 9].

Histopathology was widely used in scientific research with model animals from the early 19th century. In aquatic toxicology histopathology was employed first for documenting tumor appearance in aquatic animals, in chemically polluted water bodies [10]. Especially from the 1970s, when the concern about environment started to gain importance due to the increasing pollution and environment degradation, histopathological approach became an increasingly used biomarker [8]. Later, from the early 1990s histopathology has developed into a recognized biomarker for environmental and fish health as "a higher level response following chemical and cellular interaction" [5]. Undoubtedly histopathology has its important place as biomarker of fish health status in both toxicological and ecosystem management perspective. However, there are several aspects in histopathological studies that should be considered while using histopathology for ecosystem and fish management. In this chapter we will discuss about histopathological method in assessing general fish fitness.

DEFINITION OF THE BIOMARKER

A number of different definitions for biomarker were given during the past years [5, 8, 11, 12]. Maybe the most accepted definition of biomarker is that it is "a biochemical, physiological and morphological response" as stated by Melancon [13]. This is actually the "common denominator" of all definition of biomarkers: that biochemical, physiological and morphological changes are induced. Definitions of biomarker differ according to the causing agent of changes.

Actually the cause of changes was extended over time: initially, biomarkers were considered, defined by NRC [14], as "indicators signaling events in biological systems or samples following chemical exposure". Nowadays biomarkers are defined as signals of early changes in an otherwise healthy organism, but also changes in community, or in the ecosystem. These changes are the consequence of stress induced by the environment itself, *i.e.* its changing conditions; chemical(s) and their metabolites; feed; and/or resource management. Also, NRC [14] recognizes use of three different types of biomarkers:

(1) Biomarkers of exposure: here, two types of exposure exist. First, there is environmental exposure defined as the quantity of xenobiotic in the environment, while internal exposure (or dose) includes amount of xenobiotic taken into the organism. They are characterized by half-life, circulating peak, or cumulative dose [14].

(2) Biomarkers of effect: any change that we can determine in the organism, including: health impairment, changes in organs, tissues or body fluids.

(3) Biomarkers of susceptibility: each organism, species or population has different capacity to respond to stressor or xenobiotic. This capacity includes genetic factors or background diseases that fish might have prior to exposure to xenobiotics.

GENERAL GUIDELINES FOR HISTOPATHOLOGY

Before the start of the histopathology study, some considerations have to be taken into account. First of all, target organs have to be chosen. Any organ of fish could be of interest, but due to the fact that for histopathological studies fish has to be sacrificed, here we suggest taking multiple organs, since it is not unusual that xenobiotics have an effect on one organ and do not affect another one [15, 16]. The most common choice in ecotoxicological studies are: gills, liver, skin, and kidney, but researchers also use gonads, brain, spleen, heart, eye, intestine, or muscles, depending on the nature of xenobiotic, level of pollution, state of the environment, or purpose of the analysis. Baseline studies of normal histology are required or at least a laboratory control group or a control site in order to compare lesions found to the morphology of non-affected organs. Guidelines for sampling have been summarized by Bernet *et al.* [17] and recommendations were given for: sampling size, species investigated, age, sex, time of the season, and migrations.

Susceptibility to pollutants of fish species living in the same environment is different [18, 19]. Age and sex of the fish are also important parameters but not always reported in articles. The morphology of organs is changing during the life span of fish, as documented in the literature [20], while there is morphological disparity in some organs depending of fish sex. The best example is the vitellogenin in liver, a yolk precursor protein in liver in females: due to the presence of oestrogenic compounds in the water it could also be found in immature or male fish [21, 22]. Temperature is one of the major factors influencing organ morphology. Organ histology is changing according to the time of the season [23 - 25]. Therefore, seasonal variations also have to be taken into account before the start of the study. Number of animals sampled per site or study is one of the issues in histopathology, since representative number has to be attained. It depends from study to study and there is definitely no preferred number, even though some researchers are aiming for a number of 30 animals per site, due to statistical methods employed. This number always depends on variations of histopathological alterations and methods used for quantifying alterations. Of course, conservation status of fish species has to be taken into account when field study is planned, and also a fact that in highly polluted sites it is not always possible to catch the desired number of fish. A shortcoming of working with fish is that fish are able to leave the studied area, contrary to some other, sedentary, indicator species, such as some shellfish or macrophytes.

Beside these requirements for fish sampling, there are also a number of other considerations when assessing histological slides. The biggest shortcoming of histopathological method is subjectivity. Even when properly trained and experienced, scientists can generate different results. Excellent guidelines for proper assessment of slides were given by Crissman *et al.* [26], and Wolf *et al.* [27]. They highlighted important steps in histopathological assessment, such as: blind evaluation of the slides due to the possible bias of researchers when slides are not blinded; two or more trained histopathologists assessing same slides and consensual grading of the slides; possible artifacts in fixation or processing of slides; background diseases; *etc.* Only if all these prerequisites are fulfilled, relevant results would be obtained from the study. Also, it has to be stressed out that tissue response is limited and that there is no specific tissue response to xenobiotics. Majority of xenobiotics are causing similar alterations in tissues, their severity depending only on xenobiotic's amount (concentration). Some slight statistical trends could be established when large number of different studies was taken into account [28].

QUANTIFYING ALTERATIONS

Histology is a quite old discipline, as the methods are dating from the 19th century, and at that time, results were based on description of cellular and histological features. After the 2nd World war, scientists were developing new techniques, which will help in quantification of the alterations. With the introduction of cheap personal computers, projectors and cameras, this field of research started to develop rapidly and basic histomorphometrical and stereological methods were emerging. Today, descriptive methods are not used very often, since in modern science, in most of the time, researcher needs to show statistical analysis of the data in order to publish the paper. It is impossible to do statistical analysis with descriptive histology, but it is helpful of course, when basic differences between cellular or histological structure in terms of treatments or doses in toxicology are established. In order to perform statistical methodology on histopathology data, researchers use one of the three available techniques today: semi-quantitative scoring system, histomorphometry or stereology.

Semi-Quantitative Scoring System

This is the most popular system for assessment of histopathological changes in fish today. The reason for this popularity is in time needed for obtaining scores. It is the least time consuming of three abovementioned systems, but the negative side is that it is subjective and depends on the skills and experience of the histopathologist. Typically, a histopathologist grades the tissue by giving certain score value, which range from 0, when no alteration is present, to maximum value, usually: 3, 5 or 10, when tissue is affected with alteration.

The score is usually given for a particular alteration, and rarely accounts for a larger number of alterations in the tissue, since that type of scoring is more error-prone. Number of grades is the first issue in semi-quantitative scoring system. In general, histopathologists are using three grading levels (mild, moderate and severe; Fig. **1**), but in some cases, that number is rising up to 10 grading levels. Gibson-Corley *et al.* [29] argue that 4 to 5 grading levels are optimal in histopathological studies in order to "to maximize detection and repeatability". When researcher applies more than five grading levels in the study, it usually lacks clear definitions between levels and the system faces a risk to become flawed. This also means that decimals scores should be avoided, for the same reason mentioned above.

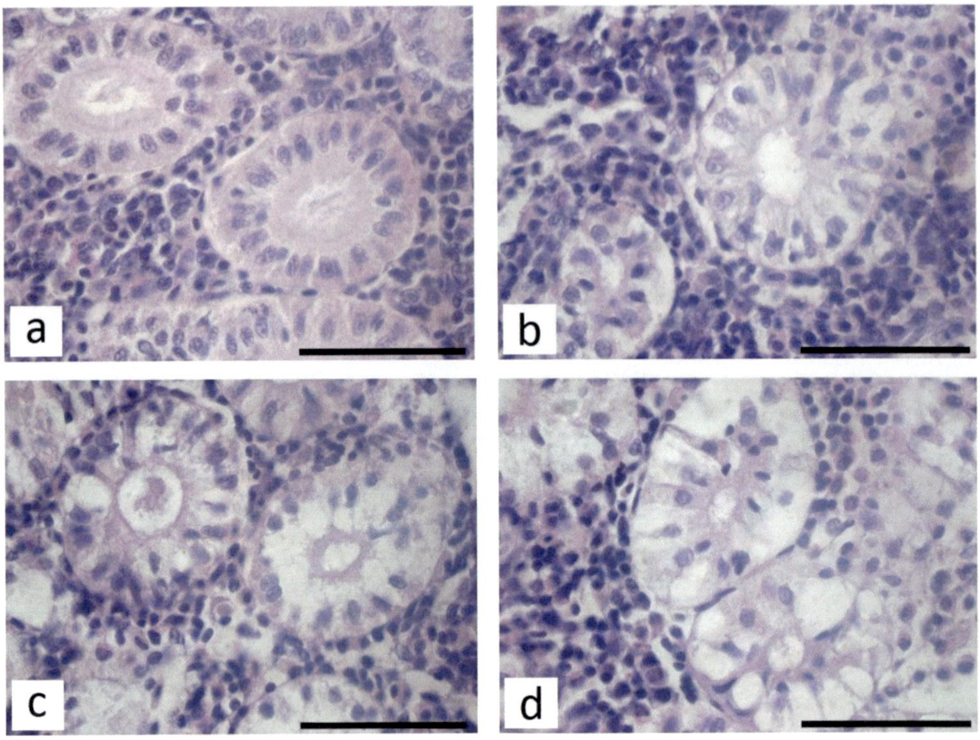

Fig. (1). Difference between grading levels - vacuolation of tubular epithelium in the kidneys of common carp: (**a**) normal; (**b**) mild; (**c**) moderate; (**d**) severe; micrographs previously unpublished; taken from study [129]; hematoxylin-eosin staining, x1000, scale bar = 50 μm.

Currently, there are 3 semi-quantitative scoring systems in use for fish histopathology. Combined, these three semi-quantitative scoring systems account for 928 citations. Poleksić and Mitrović-Tutundžić [30] is cited for 191 time, Bernet *et al.* [17] for 498 times and Schwaiger *et al.* [31] for 239 times (Google Scholar; accessed 13/11/2016). A brief overview of these three systems is given in the following part of the manuscript.

First system is published by Poleksić and Mitrović-Tutundžić [30] and it is used for assessment of histopathology of fish gills only. In this system, histopathological alterations of fish gills are classified in five groups, based on the type and location of alteration:

(1) hypertophy and hyperplasia of gill epithelia and related changes;
(2) changes in the mucous and/or chloride cells
(3) blood vessel changes;
(4) gill parasites;

(5) terminal stages.

Under these 5 groups, a detailed description of specific alteration is provided (Table **1**) and classified under three different stages:

Stage I - majority of histopathological alterations are classified in this group. They are easily reversible when environmental conditions improve, but in the case of long-term exposure they will progress to alterations grouped in stage II;

Stage II - more severe changes which are generally recoverable, but in the case of occupying large area of tissue or prolonged exposure to the toxicant, they will gradually progress to severe alterations in group III;

Stage III - tissue recovery is not possible, even when toxicant is not present in the water any more. These changes result in permanent damage to fish gills and impaired function of the gill apparatus.

In order to quantify changes, authors proposed adding increased exponential values for each stage:

Stage I - 10^0; stage II - 10^1; stage III - 10^2

Finally, number of observed histopathological changes would be multiplied by values for each stage, given in the following formula:

$$I = \sum_{i=1}^{n} a_i + 10\sum_{i=1}^{n} b_i + 100\sum_{i=1}^{n} c_i \quad \textbf{(1)}$$

where a, b and c are first, second and third stage alterations, respectively. Value should be calculated for each individual fish (I), based on the histopathological changes on their entire gill apparatus.

Grading of the final value (I) is also proposed in the same article, where: values up to 10 equal functionally normal gills, from 11 to 20 "slightly to moderately damaged gills"; from 21 to 50 "moderately to heavily damaged gills" and values above 100 "irreparably damaged gills" [30]. This scoring system is later adapted for use in other fish organs, like kidney and liver [23] and is particularly frequently used in the histopathological assessment of fish in polluted ecosystems of South America.

Second scoring system is published by Bernet *et al.* [17]. It is designed for assessing histopathological changes in gills, liver, skin and kidney. The system is based on the similar principles like the one described above, but with important difference of introducing grading levels. Bernet *et al.* [17] are proposing three grading levels (originally termed - score value), with scores of: 2 - mild occurrence; 4 - moderate occurrence; 6 - severe occurrence, but the use of intermediate values is allowed. Other difference is based in the way the lesions are reported, and instead of concentrating on particular lesion, the system is scoring general changes within cells and tissues (Table **2**).

Table 1. List of histopathological changes in fish gills [30].

Alteration	**Stage**
Hypertrophy and hyperplasia of gill epithelia and related changes	
Hypertrophy of respiratory epithelium	I
Lifting of respiratory epithelial cells	I
Leukocyte infiltration of gill epithelium	I
Thinning of respiratory epithelium	I
Rupture and peeling of lamellar epithelium	II
Focal hyperplasia of epithelial cells	I
Hyperplasia from the base to approximately half the length of the secondary lamella	I
Irregular (`chaotic`) hyperplasia of epithelial cells	I
Fusion of the tips of secondary lamellae	I
Hyperplasia of sponge-like eosinophilic cells	I
Fusion of the primary lamellae tips	I
Uncontrolled thickening of proliferated tissue	II
Fusion of several secondary lamellae	I
Shortening of secondary lamellae	I
Complete fusion of all the secondary lamellae	II
Changes in the mucous and/or chloride cells	
Hypertrophy and hyperplasia of mucous cells	I
Empty mucous cells or their disappearance	I
Hypertrophy and hyperplasia of chloride cells	I
Chloride cells present in secondary lamellae	I
Blood vessel changes	
Lamellar telangiectasis	I
Filament blood vessel enlargement	I
Haemorrhages with rupture of epithelium	II
Stasis	II
Gill parasites	I

(Table 1) contd.....

Alteration	Stage
Terminal stages	
Scar tissue - fibrosis	III
Necrosis	III

Moreover, Bernet *et al.* [17] are introducing five patterns of changes (reaction pattern (rp)) in fish organs: circulatory disturbances, progressive changes, regressive changes, inflammation and tumors. Due to the fact that alterations are not specified for the particular tissue, it is possible to combine them in four ways and calculation of four different histopathological indices is possible, namely: (1) organ index, (2) reaction index of an organ, (3) total index, and (4) total reaction index. What was designated as "stages" in Poleksić and Mitrović-Tutundžić [30] is termed "importance factor" (w) and is defined in similar way as in the first scoring system described: 1 - change is easily reversible if the stressor is neutralized; 2 - change is reversible in most cases when the stressor is neutralized; 3 - irreversible lesions.

Table 2. List of histopathological changes in fish skin [17].

Reaction pattern	Functional unit of the tissue	Alteration	Importance factor
Circulatory disturbances		Haemorrhage/hyperaemia/aneurysm	1
		Intercellular oedema	1
Regressive changes	Epidermis	Architectural and structural alterations	1
		Plasma alterations	1
		Deposits	1
		Nuclear alterations	2
		Atrophy	2
		Necrosis	3
	Basement membrane	Defect	2
	Dermis	Architectural and structural alterations	1
		Plasma alterations	1
		Deposits	1
		Nuclear alterations	2
		Atrophy	2
		Necrosis	3
Progressive changes	Epidermis	Hypertrophy	1
		Hyperplasia	2

(Table 2) contd.....

Reaction pattern	Functional unit of the tissue	Alteration	Importance factor
	Dermis	Hypertrophy	1
		Hyperplasia	2
Inflammation		Exudate	1
		Activation of RES	1
		Infiltration	2
Tumour		Benign tumour	2
		Malignant tumour	3

Distinctive and important difference between those two systems is in calculation of organ index, since Bernet *et al.* [17] proposed formula with grading levels, which makes this system more realistic than the previous one:

$$I = \sum_{rp} \sum_{alt} \left(a_{org\,rp\,alt} \times w_{org\,rp\,alt} \right) \qquad (2)$$

where: org = organ (constant); rp = reaction pattern; alt = alteration; a = score value; w = importance factor. This formula result in lower values obtained for assessment of a specific organ compared to previous scoring system, since it does not use exponential values. Also, using this formula, it is not possible to characterize functional anatomy of gill apparatus by placing them into categories, contrary to the previous scoring system. Poleksić and Mitrović-Tutundžić [30] in their article gave interesting hypothesis: that development of histopathological changes has exponential nature, and that explain their use of exponential factors for each stage of histopathological alterations.

Other important improvements in Bernet *et al.* [17] system is introducing neoplasia in the list of histopathological changes and giving more importance to inflammation changes, but on the other hand, it does not take into account presence of parasites in the calculation of histopathological indices, in contrast to Poleksić and Mitrović-Tutundžić [30] system. Parasite infestation is common in fish living in polluted ecosystems and parasites species as well as their number or abundance can also be a valuable indicator of environmental stress [19, 32 - 34].

The third system is proposed by Schwaiger *et al.* [31] and it is based on the different idea than two semi-quantitative scoring systems previously described. Similar to others, grades also exist in the ranking of alterations (1 - no pathological alterations; 2 - mild to moderate alterations and 3 - extended severe pathological alterations), but instead of introducing importance factor, like the two semiquantitative scoring systems previously described, they opted to give the same importance to each histopathological alteration. Schwaiger *et al.* [31] proposed calculating mean values of all alterations per fish and termed it "mean assessment values". They would later calculate statistical difference between histopathological scores of fish living in different environments and thus assess environmental stress.

Apart from standardized systems described above, a different method to increase objectivity in semi-quantitative system is used by various researchers. Instead of applying score values based on their experience (biased by default), some researchers are counting numbers of secondary lamellae in gills affected by specific histological alteration [7, 35, 36]. It is feasible when some specific structure could be recognized in the tissue, likewise with primary or secondary lamellae in gills (Fig. **2**) or tubuli and glomeruli in fish kidney, but it is also used in liver, when random fields of view were taken and assessed [36]. This approach is recommended whenever possible, since it could be later incorporated in any of above mentioned semi-quantitative scoring systems as it decrease bias and subjectivity, but it could also be used independently of scoring systems, since statistical analysis of the data is possible.

Alternative way for overcoming the issue of researcher subjectivity is to organize a pathology working group (PWG). This group consists of 3 or more trained histopathologists and their goal is to achieve consensus in all emerging issues, usually as majority vote [37]. Three main types of issues are: disagreement in opinions, inconsistent terminology and not including important information to final decisions [37]. Shortcomings of PWGs is that they are very expensive to organize and difficult to manage. Therefore, they are seldom formed, even in histopathology of mammals, which is a more developed scientific area than histopathology of fish. In fish, two PWGs were formed: one to establish criteria of liver lesions in medaka (*Oryzias latipes*), especially between neoplastic and non-neoplastic lesions [38] while the other one was formed in order to assess variation between three different studies of effects of diclofenac in rainbow trout (*Oncorhynchus mykiss*) [39]. The latter study illustrated some of the shortcomings in histopathological approach as variations between reported findings in three similar experiments were considerable, while PWG even found some unreported findings on the original slides [39].

Histomorphometry

Even though semi-quantitative scoring systems are widely and frequently used, several drawbacks still exist, and the most serious one remains - the subjectivity of researcher. In order to avoid this shortcoming, histopathologists are measuring some cytological and histological features, thus quantifying certain alterations (Fig. **2**). Some of the examples are morphometry of gills, proposed by Lease *et al.* [40]. Apart from descriptive histopathological assessment, authors measured lamellar thickness and lamellar diffusion distance in gills of the Lost River sucker (*Deltistes luxatus*) exposed to higher levels of environmental pH and ammonia and confirmed that level of ammonia has an effect on the degree of epithelial lifting. Diffusion distance is defined as "minimum distance from a randomly selected point on the outer lamellar epithelium to the inside of the nearest blood lacuna" [40, 41], while lamellar thickness represents "perpendicular distance across the lamella, from the outermost epithelium on each side of the lamellar cross-section" [40, 41]. Same methodology was later used in the field study with common carp (*Cyprinus carpio*) [42] and histopathology of the fish *Corydoras paleatus* contaminated with sublethal levels of organophosphorus in the water [7]. Nero *et al.* [43] assessed gill histopathological changes in yellow perch (*Perca flavescens*) and goldfish (*Carassius auratus*) exposed to oil sands process-affected water by measuring secondary lamellar length and width, interlamellar distance, basal epithelium thickness, and proportion of the secondary lamellae available for gas exchange. This is a different approach, but the aim remains the same: to quantify available gill surface for gas exchange. It is later used to assess toxicity of naphthenic acid to same two fish species [44], as well as in field study of *Prochilodus lineatus* from the Salado River basin in Argentina [45].

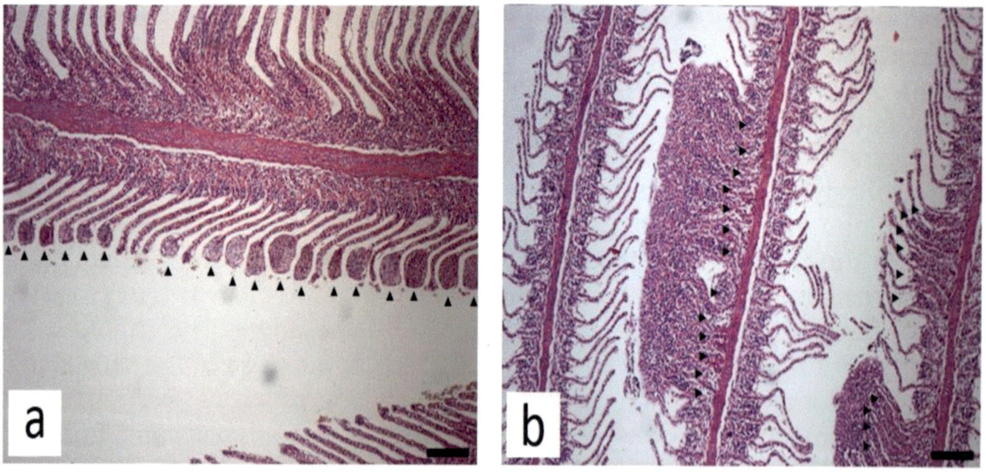

Fig. (2). Illustration of quantifying of histological alterations: (**a**) counting aneurysms of terminal capillaries (arrowheads) in secondary lamellae 18/41; (**b**) counting epithelial hyperplasia (arrowheads) between secondary lamellae 23/93; micrographs previously unpublished; taken from study [130]; hematoxylin-eosin staining, x100, scale bar = 50 μm.

Camargo and Martinez [23] measured nuclear and cellular profiles, as well as nuclear diameter of interrenal cells in kidney of *Astyanax altiparanae* exposed to municipal water. Apart from the kidney, this is a common approach in measuring hepatocytes in the liver, as those values can provide data about the metabolic activities of the cells, frequently used in starvation or fish nutrition experiments [46 - 48].

Stereology

Stereology represents the "gold standard" in histological and histopathological studies. This method is based on obtaining 3D structures from 2D histological sections and, if performed properly, is unbiased [49]. This methodology is one of the solutions for the issue of researcher subjectivity, although, substantial amount of training and experience is needed for mastering it. A combination of different strategies and techniques is employed at different steps of sample preparation protocol and each step is ensuring unbaisedness at different levels. This includes: (1) estimation of total volume; (2) systematic uniform random sampling, which is cascade sampling technique, enabling sampling of large organs; (3) estimation of volume, surface, length or number of different structures within organs using various probes. A large number of articles and books with theoretical background and detailed practical guidelines for preparation and assessment of slides is published in previous years [49 - 54] and therefore, no specific details will be given here.

As mentioned earlier, stereology is not excessively used in fish toxicopathology, but some researchers are successfully implementing these methods in their studies. One of the scholar examples of proper use of stereological methods is a study done by Da Costa *et al.* [55] where authors successfully estimated surface area and barrier thickness of *Arapaima gigas* gills. Estimating of gills surface area is one of the most difficult tasks in stereology, due to the anisotropy of this tissue. The same methodology is employed in studying effects of hypoxia and elevated temperature to gill remodeling of striped catfish (*Pangasianodon hypophthalmus*) [56]. The main advantage of stereology compared to histomorphometry is that it provides estimation of higher number of studied parameters in the same amount of time, thus making stereology less time consuming. The most frequently reported parameter in stereological studies in fish is volume density. It is estimated using a point-counting method provided by Weibel [57]. This technique is employed in immunohistochemical studies to assess volume density of catalase in liver [58] or PCNA in sagittal whole fish sections [59], exposed to ethanol and hypoxia, respectively. Similar strategy is used to estimate volume density of the respiratory epithelium, lamellae, central venous sinus, central axis, mucous cells, and chloride cells of the Nile tilapia (*Oreochromis niloticus*) exposed to copper sulfate [60]. An illustration of this method is provided in Fig. (**3**).

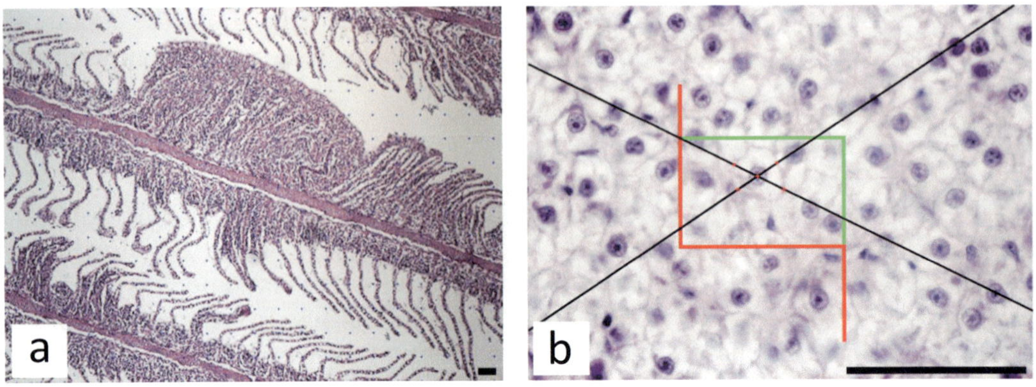

Fig. (3). Stereological techniques employed on histological slides of fish tissue: (**a**) grid superimposed on fish gills for estimating volume densities using point counting method (x100); (**b**) nucleator technique for estimating volume of hepatocyte in fish liver combined with unbiased counting frame (x1000); micrographs previously unpublished; taken from studies [47, 130]; hematoxylin-eosin staining, scale bar = 50 μm.

This method provides unbiased results and in only one effort, much more information is collected, than it could be done in histomorphometry, described earlier. The same methodology is used to assess effects of endocrine disruptors in gonads [61, 62] in order to estimate gonad developmental stages, in both ovaries and testis. Another commonly employed stereological method in estimation of the volume of hepatocyte nuclei is a nucleator [63]. Illustration of this method is shown in Fig. (**3b**) and used in ecotoxicological studies [64, 65]. The most complicated tissue for stereological research, apart from the gills, is the brain, but with the combination of different techniques, such as: optical (or physical) disector and unbiased counting frame, or a total volume and number of cells in different part of the brain could be estimated. This is done in several neurodegenerative studies in fish, assessing the effects of methylmercury [66] and dioxin [67].

ULTRASTRUCTURE STUDIES

Electron microscopy studies are seldom present in toxicological pathology. The most probable reason is the lack of expensive equipment needed and tedious sample preparation. Moreover, the research of cellular biomarkers in the past 20 years gives advantage to molecular methods, such as gene expression assays.

Ultrastructural studies can provide valuable insight in morphology of organelles and possible impaired cellular structure or function. Semi-quantitative scoring system is also used in ultrastructural studies, with clearly established end-points. Those systems are usually associated with alterations of organelles in the cell, and/or increase in their number. Useful example is a list of alterations suggested by Shramm *et al.* [68] in which increased number of several hepatocytes components (peroxisomes, lysosomes, mitochondria, macrophages, granulocytes and myelin-like bodies) is considered as pathology, as well as particular changes in nucleus, endoplasmic reticulum, Golgi apparatus, and mitochondria. In the same list cellular compartmentation and a ratio between storage products (lipid/glycogen) are given as important marker of cytopathology [68]. Using this methodology, authors could evaluate effects of pollutants on fish living in two streams in Germany [68], but it is used in many studies later. For ultrastructural studies, same organs are chosen like in light microscopy studies: gills, liver and kidney [69], even though liver is the most commonly assessed organ [70, 71]. Integrated approach of both light and electron microscopy methods is recommended [72], while in integrated study where liver ultrastructure, heat shock proteins and swimming behavior were studied, fish in highly polluted locations showed higher level of correlation of these biomarkers, comparing to fish living in locations with lower level of pollution [73]. Ultrastructural markers are especially important in studies of effects of carcinogenic substances on fish. Even though correlation between carcinogens and presence of neoplasms in fish is not yet clearly established in environmental assessment [6], a number of tumor induced studies in laboratory conditions are published [74, 75].

HISTOPATHOLOGY IN FIELD STUDIES

New formulations of pesticides and their degradation products, new pharmaceuticals, products from cosmetic industry, hormones and other endocrine disrupting compounds, industrial products, including nanoparticles are all emerging xenobiotics in aquatic ecosystems in the recent years [76 - 78]. Environmental monitoring and assessment today is provided with a difficult challenge of detecting these new xenobiotics in aquatic environment and research effects and complicated relationship between animals inhabiting it. Detailed analytical testing of water bodies for the presence of xenobiotics [79, 80] provides researchers with an idea of extent of pollution, but monitoring aquatic organisms using biomarkers is a recommended approach for assessment of ecosystem pollution [6, 8]. As reviewed by Cairns [81], when it comes to field studies of contaminated ecosystems - obtaining the biological response alone is not sufficient: the environmental concentration of a chemical(s) and the toxicological response(s) in different organisms, not necessarily the representative or the most sensitive ones, is the basis of hazard evaluation.

Histopathological approach is used in many field studies of different ecosystems, including research of environments burden with presence of metals [82 - 89], including uranium [90], organic pollutants [91 - 94], pulp and paper mill effluents [95], endocrine disruptors [96], plastic contamination [97], oil sands mining waters [98], waste water [99, 100], sewage water [23], cyanobacteria [101] or nanoparticles. Actually, one of the growing research fields in aquatic toxicology is pollution of nanomaterials. Histopathological evidence of toxicity of nanoparticles is already established in laboratory studies, but toxicity depends on nanoparticle source. Copper and silver nanoparticles are damaging branchial tissue in zebrafish, but that is not the case with titanium nanoparticles [102, 103]. Toxicity of copper and silver nanoparticles is confirmed in other fish species, such as: Siberian sturgeon (*Acipenser baerii*) [104], rainbow trout [105, 106], perch (*Perca fluviatilis*) [107] or orange-spotted grouper (*Epinephelus coioides*) [108].

Ideally, in field studies, histopathology needs to be combined with either other biomarkers for possible validation and comparison between different techniques and analytical methods, which could point to origin and/or mechanism of action of the stressor. Approach of using several biomarkers include: genotoxicity assays, such as comet assay [84, 109] and micronucleus test [110], fish antioxidant status [111, 112], blood plasma biochemistry [113], ethoxyresorufin O-deethylase (EROD) activity [114], expression of proliferating cell nuclear antigen (PCNA) and cytochrome P450 1A (CYP1A) [115], expression of heatshock proteins [116], or expression of metallothioneins in tissue [117]. Alternate approach is assessing fish status *via* histopathology and populations (and/or biomarkers) of other aquatic organisms living at the same site, such as zooplankton or zoobenthos [42], plants [118], shellfish [119] or mussels [120]. The best approach for monitoring polluted environments is to integrate two or more biomarkers in fish with analytical methods of testing water and sediment for the presence of metals, organic pollutants or other sources of contamination [114, 121 - 123].

Issues of fish mobility in field studies could be overcome in several different ways: by placing fish in cages [124], or by doing laboratory testing using mixture of chemicals corresponding to levels found in the water body [125]. The other possibility includes testing fish in laboratory conditions using contaminated sediments [126] or diverting contaminated water flow into experimental facility [73].

In addition to its "traditional" use in aquatic toxicology where histopathology is an important biomarker, pointing out organs affected, documenting first visible changes in fish cells and tissues, and the level of alterations, histopathology is also a valuable complementary method in assessment of consequences of the use of different rearing technologies in aquaculture such as feeding, stocking density, presence of the parasites and water quality [127 - 134].

CONCLUDING REMARKS

Histopathology as an important biomarker of fish fitness status is highly valuable especially when chronic or sublethal pollution of the aquatic environment occurs. Actually, all the stresses induced by changing environmental conditions; xenobiotics and their metabolites; rearing technologies in aquaculture; aquatic resource management, and even global climate changes, require adaptations that cause morphological changes in cells, tissues, organs, and organ systems and hence could be revealed by histopathology.

Some general guidelines for fish histopathological studies with the emphasis on methods available in quantifying alterations: semi-quantitative scoring system, histomorphometry and stereology are specified. The most important asset of modern fish histology is the fact that the histopathological response can be visualized, therefore documented, recorded, measured, analyzed, and compared with other environmental biomarkers.

Finally, a practical recommendation would be to use a "tailor-made" histology. Depending on the type of study and its goal, a proper approach should be selected: whether it will be a light microscopy or ultrastructural study, which are the organs to be studied, which is the method of quantification of alterations, and which additional research and/or monitoring method will be used besides histopathology? Those are some of the elements to be considered when choosing the appropriate research strategy.

CONFLICT OF INTEREST

The author (editor) declares no conflict of interest, financial or otherwise.

ACKNOWLEDGEMENT

We would like to acknowledge support of the Serbian Ministry of Education, Science and Technological Development (TR31075).

REFERENCES

[1] Vörösmarty CJ, McIntyre PB, Gessner MO, *et al.* Global threats to human water security and river biodiversity. Nature 2010; 467(7315): 555-61. [http://dx.doi.org/10.1038/nature09440] [PMID: 20882010]

[2] Xu Z, Boyd CE. Reducing the monitoring parameters of fish pond water quality. Aquaculture 2016; 465: 359-66. [http://dx.doi.org/10.1016/j.aquaculture.2016.09.031]

[3] Hill AJ, Teraoka H, Heideman W, Peterson RE. Zebrafish as a model vertebrate for investigating chemical toxicity. Toxicol Sci 2005; 86(1): 6-19. [http://dx.doi.org/10.1093/toxsci/kfi110] [PMID: 15703261]

[4] Lasne E, Bergerot B, Lek S, Laffaille P. Fish zonation and indicator species for the evaluation of the ecological status of rivers: example of the Loire basin (France). River Res Appl 2007; 23: 877-90. [http://dx.doi.org/10.1002/rra.1030]

[5] Schlenk D, Handy R, Steinert S, Depledge MH, Benson W. Biomarkers.The toxicology of fishes. Boca Raton: CRC Press 2008; pp. 683-733. [http://dx.doi.org/10.1201/9780203647295.ch16]

[6] van der Oost R, Beyer J, Vermeulen NP. Fish bioaccumulation and biomarkers in environmental risk assessment: a review. Environ Toxicol Pharmacol 2003; 13(2): 57-149. [http://dx.doi.org/10.1016/S1382-6689(02)00126-6] [PMID: 21782649]

[7] Fanta E, Rios FS, Romão S, Vianna AC, Freiberger S. Histopathology of the fish *Corydoras paleatus* contaminated with sublethal levels of organophosphorus in water and food. Ecotoxicol Environ Saf 2003; 54(2): 119-30. [http://dx.doi.org/10.1016/S0147-6513(02)00044-1] [PMID: 12550089]

[8] Handy RD, Galloway TS, Depledge MH. A proposal for the use of biomarkers for the assessment of chronic pollution and in regulatory toxicology. Ecotoxicology 2003; 12(1-4): 331-43. [http://dx.doi.org/10.1023/A:1022527432252] [PMID: 12739879]

[9] Wester PW, Roghair CJ. Monitoring of sublethal chronic effects in fish: the pathomorphological approach.Sublethal and chronic toxic effects of pollutants on freshwater fish. Oxford: Blackwell Scientific Publications 1994; pp. 7-14.

[10] Di Guglio RT, Hinton DE. The toxicology of fishes. Boca Raton: CRC Press 2008; p. 1071. [http://dx.doi.org/10.1201/9780203647295]

[11] Goodsaid FM, Frueh FW, Mattes W. Strategic paths for biomarker qualification. Toxicology 2008; 245(3): 219-23. [http://dx.doi.org/10.1016/j.tox.2007.12.023] [PMID: 18280028]

[12] McCarty LS, Munkittrick KR. Environmental biomarkers in aquatic toxicology: Fiction, fantasy, or functional? Hum Ecol Risk Assess 1996; 2: 268-74. [http://dx.doi.org/10.1080/10807039609383607]

[13] Melancon MJ. Bioindicators of Contaminant Exposure and Effect in Aquatic and Terrestrial Monitoring.Handbook of Ecotoxicology. Boca Raton: CRC Press 2003; pp. 257-78.

[14] Biological markers in environmental health research. Environ Health Perspect 1987; 74: 3-9. [PMID: 3691432]

[15] Molina R, Moreno I, Pichardo S, *et al.* Acid and alkaline phosphatase activities and pathological changes induced in Tilapia fish (*Oreochromis* sp.) exposed subchronically to microcystins from toxic cyanobacterial blooms under laboratory conditions. Toxicon 2005; 46(7): 725-35. [http://dx.doi.org/10.1016/j.toxicon.2005.07.012] [PMID: 16185737]

[16] Velisek J, Stara A, Machova J, Dvorak P, Zuskova E, Prokes M, *et al.* Effect of terbutryn at environmental concentrations on early life stages of common carp (*Cyprinus carpio* L.). Pestic Biochem Physiol 2012; 102: 102-8. [http://dx.doi.org/10.1016/j.pestbp.2011.11.005]

[17] Bernet D, Schmidt H, Meier W, Burkhardt-Holm P, Wahli T. Histopathology in fish: proposal for a protocol to assess aquatic pollution. J Fish Dis 1999; 22: 25-34. [http://dx.doi.org/10.1046/j.1365-2761.1999.00134.x]

[18] Rašković B, Poleksić V, Višnjić-Jeftić Ž, *et al.* Use of histopathology and elemental accumulation in different organs of two benthophagous fish species as indicators of river pollution. Environ Toxicol

2015; 30(10): 1153-61.
[http://dx.doi.org/10.1002/tox.21988] [PMID: 24677754]

[19] Stentiford GD, Longshaw M, Lyons BP, Jones G, Green M, Feist SW. Histopathological biomarkers in estuarine fish species for the assessment of biological effects of contaminants. Mar Environ Res 2003; 55(2): 137-59.
[http://dx.doi.org/10.1016/S0141-1136(02)00212-X] [PMID: 12502035]

[20] Fishelson L, Becker K. Development and aging of the liver and pancreas in the domestic carp, *Cyprinus carpio*: from embryogenesis to 15-year-old fish. Environ Biol Fishes 2001; 61: 85-97.
[http://dx.doi.org/10.1023/A:1011076722505]

[21] Wahli T, Meier W, Segner H, Burkhardt-Holm P. Immunohistochemical detection of vitellogenin in male brown trout from Swiss rivers. Histochem J 1998; 30(10): 753-8.
[http://dx.doi.org/10.1023/A:1003434629405] [PMID: 9874002]

[22] Zha J, Wang Z, Wang N, Ingersoll C. Histological alternation and vitellogenin induction in adult rare minnow (Gobiocypris rarus) after exposure to ethynylestradiol and nonylphenol. Chemosphere 2007; 66(3): 488-95.
[http://dx.doi.org/10.1016/j.chemosphere.2006.05.071] [PMID: 16844192]

[23] Camargo MM, Martinez CB. Histopathology of gills, kidney and liver of a Neotropical fish caged in an urban stream. Neotrop Ichthyol 2007; 5: 327-36.
[http://dx.doi.org/10.1590/S1679-62252007000300013]

[24] Haaparanta A, Valtonen ET, Hoffmann R, Holmes J. Do macrophage centres in freshwater fishes reflect the differences in water quality? Aquat Toxicol 1996; 34: 253-72.
[http://dx.doi.org/10.1016/0166-445X(95)00042-3]

[25] Haaparanta A, Valtonen ET, Hoffmann RW. Gill anomalies of perch and roach from four lakes differing in water quality. J Fish Biol 1997; 50: 575-91.
[http://dx.doi.org/10.1111/j.1095-8649.1997.tb01951.x]

[26] Crissman JW, Goodman DG, Hildebrandt PK, *et al.* Best practices guideline: toxicologic histopathology. Toxicol Pathol 2004; 32(1): 126-31.
[http://dx.doi.org/10.1080/01926230490268756] [PMID: 14713558]

[27] Wolf JC, Baumgartner WA, Blazer VS, Camus AC, Engelhardt JA, Fournie JW, *et al.* Nonlesions, misdiagnoses, missed diagnoses, and other interpretive challenges in fish histopathology studies: a guide for investigators, authors, reviewers, and readers. Toxicol Pathol 2014.
[PMID: 25112278]

[28] Mallatt J. Fish gill structural changes induced by toxicants and other irritants: a statistical review. Can J Fish Aquat Sci 1985; 42: 630-48.
[http://dx.doi.org/10.1139/f85-083]

[29] Gibson-Corley KN, Olivier AK, Meyerholz DK. Principles for valid histopathologic scoring in research. Vet Pathol 2013; 50(6): 1007-15.
[http://dx.doi.org/10.1177/0300985813485099] [PMID: 23558974]

[30] Poleksić V, Mitrović-Tutundžić V. Fish gills as a monitor of sublethal and chronic effects of pollution.Sublethal and chronic toxic effects of pollutants on freshwater fish. Oxford: Blackwell Scientific Publications 1994; pp. 339-52.

[31] Schwaiger J, Wanke R, Adam S, Pawert M, Honnen W, Triebskorn R. The use of histopathological indicators to evaluate contaminant-related stress in fish. J Aquat Ecosyst Stress Recovery 1997; 6: 75-86.
[http://dx.doi.org/10.1023/A:1008212000208]

[32] Landsberg JH, Blakesley BA, Reese RO, Mcrae G, Forstchen PR. Parasites of Fish as Indicators of Environmental Stress. Environ Monit Assess 1998; 51: 211-32.
[http://dx.doi.org/10.1023/A:1005991420265]

[33] Mackenzie K. Parasites as Pollution Indicators in Marine Ecosystems: a Proposed Early Warning System. Mar Pollut Bull 1999; 38: 955-9.
[http://dx.doi.org/10.1016/S0025-326X(99)00100-9]

[34] Schwaiger J. Histopathological alterations and parasite infection in fish: indicators of multiple stress factors. J Aquat Ecosyst Stress Recovery 2001; 8: 231-40.
[http://dx.doi.org/10.1023/A:1012954608541]

[35] De Boeck G, Van der Ven K, Meeus W, Blust R. Sublethal copper exposure induces respiratory stress in common and gibel carp but not in rainbow trout. Comp Biochem Physiol C Toxicol Pharmacol 2007; 144(4): 380-90.
[http://dx.doi.org/10.1016/j.cbpc.2006.11.008] [PMID: 17196885]

[36] Handy RD, Runnalls T, Russell PM. Histopathologic biomarkers in three spined sticklebacks, *Gasterosteus aculeatus*, from several rivers in Southern England that meet the freshwater fisheries directive. Ecotoxicology 2002; 11(6): 467-79.
[http://dx.doi.org/10.1023/A:1021061402491] [PMID: 12521142]

[37] Ward JM, Hardisty JF, Hailey JR, Streett CS. Peer review in toxicologic pathology. Toxicol Pathol 1995; 23(2): 226-34.
[http://dx.doi.org/10.1177/019262339502300218] [PMID: 7569679]

[38] Boorman GA, Botts S, Bunton TE, *et al.* Diagnostic criteria for degenerative, inflammatory, proliferative nonneoplastic and neoplastic liver lesions in medaka (*Oryzias latipes*): consensus of a National Toxicology Program Pathology Working Group. Toxicol Pathol 1997; 25(2): 202-10.
[http://dx.doi.org/10.1177/019262339702500210] [PMID: 9125779]

[39] Wolf JC, Ruehl-Fehlert C, Segner HE, Weber K, Hardisty JF. Pathology working group review of histopathologic specimens from three laboratory studies of diclofenac in trout. Aquat Toxicol 2014; 146: 127-36.
[http://dx.doi.org/10.1016/j.aquatox.2013.10.033] [PMID: 24292026]

[40] Lease HM, Hansen JA, Bergman HL, Meyer JS. Structural changes in gills of Lost River suckers exposed to elevated pH and ammonia concentrations. Comp Biochem Physiol C Toxicol Pharmacol 2003; 134(4): 491-500.
[http://dx.doi.org/10.1016/S1532-0456(03)00044-9] [PMID: 12727299]

[41] Tietge JE, Johnson RD, Bergman HL. Morphometric Changes in Gill Secondary Lamellae of Brook Trou (*Salvelinus fontinalis*) After Long-Term Exposure to Acid and Aluminum. Can J Fish Aquat Sci 1988; 45: 1643-8.
[http://dx.doi.org/10.1139/f88-194]

[42] Dulić Z, Poleksić V, Rašković B, Lakić N, Marković Z, Živić I, *et al.* Assessment of the water quality of aquatic resources using biological methods. Desalination Water Treat 2009; 11: 264-74.
[http://dx.doi.org/10.5004/dwt.2009.861]

[43] Nero V, Farwell A, Lister A, *et al.* Gill and liver histopathological changes in yellow perch (*Perca flavescens*) and goldfish (*Carassius auratus*) exposed to oil sands process-affected water. Ecotoxicol Environ Saf 2006; 63(3): 365-77.
[http://dx.doi.org/10.1016/j.ecoenv.2005.04.014] [PMID: 15964628]

[44] Nero V, Farwell A, Lee LE, Van Meer T, MacKinnon MD, Dixon DG. The effects of salinity on naphthenic acid toxicity to yellow perch: gill and liver histopathology. Ecotoxicol Environ Saf 2006; 65(2): 252-64.
[http://dx.doi.org/10.1016/j.ecoenv.2005.07.009] [PMID: 16129489]

[45] Troncoso IC, Cazenave J, Bacchetta C, Bistoni MdeL. Histopathological changes in the gills and liver of *Prochilodus lineatus* from the Salado River basin (Santa Fe, Argentina). Fish Physiol Biochem 2012; 38(3): 693-702.
[http://dx.doi.org/10.1007/s10695-011-9551-7] [PMID: 21850399]

[46] Ostaszewska T, Dabrowski K, Palacios ME, Olejniczak M, Wieczorek M. Growth and morphological changes in the digestive tract of rainbow trout (*Oncorhynchus mykiss*) and pacu (*Piaractus mesopotamicus*) due to casein replacement with soybean proteins. Aquaculture 2005; 245: 273-86. [http://dx.doi.org/10.1016/j.aquaculture.2004.12.005]

[47] Poleksić V, Stanković M, Marković Z, Relić R, Lakić N, Dulić Z, *et al.* Morphological and physiological evaluation of common carp (*Cyprinus carpio* L., 1758) fed extruded compound feeds containing different fat levels. Aquacult Int 2014; 22: 289-98. [http://dx.doi.org/10.1007/s10499-013-9654-5]

[48] Power D, Melo J, Santos CR. The effect of food deprivation and refeeding on the liver, thyroid hormones and transthyretin in sea bream. J Fish Biol 2000; 56: 374-87. [http://dx.doi.org/10.1111/j.1095-8649.2000.tb02112.x]

[49] Gundersen HJ, Mirabile R, Brown D, Boyce RW. Stereological Principles and Sampling Procedures for Toxicologic Pathologists.Haschek and Rousseaux's Handbook of Toxicologic Pathology. Boston: Academic Press 2013; pp. 215-86. [http://dx.doi.org/10.1016/B978-0-12-415759-0.00008-X]

[50] Gundersen HJ, Bagger P, Bendtsen TF, *et al.* The new stereological tools: disector, fractionator, nucleator and point sampled intercepts and their use in pathological research and diagnosis. APMIS 1988; 96(10): 857-81. [http://dx.doi.org/10.1111/j.1699-0463.1988.tb00954.x] [PMID: 3056461]

[51] Gundersen HJ, Bendtsen TF, Korbo L, *et al.* Some new, simple and efficient stereological methods and their use in pathological research and diagnosis. APMIS 1988; 96(5): 379-94. [http://dx.doi.org/10.1111/j.1699-0463.1988.tb05320.x] [PMID: 3288247]

[52] Marcos R, Monteiro RA, Rocha E. The use of design-based stereology to evaluate volumes and numbers in the liver: a review with practical guidelines. J Anat 2012; 220(4): 303-17. [http://dx.doi.org/10.1111/j.1469-7580.2012.01475.x] [PMID: 22296163]

[53] Mayhew TM. The new stereological methods for interpreting functional morphology from slices of cells and organs. Exp Physiol 1991; 76(5): 639-65. [http://dx.doi.org/10.1113/expphysiol.1991.sp003533] [PMID: 1742008]

[54] Mouton PR. Principles and practices of unbiased stereology: an introduction for bioscientists. Baltimore: Johns Hopkins University Press 2002; p. 214.

[55] Da Costa OT, Pedretti AC, Schmitz A, Perry SF, Fernandes MN. Stereological estimation of surface area and barrier thickness of fish gills in vertical sections. J Microsc 2007; 225(Pt 1): 1-9. [http://dx.doi.org/10.1111/j.1365-2818.2007.01710.x] [PMID: 17286690]

[56] Phuong LM, Huong DT, Nyengaard JR, Bayley M. Gill remodelling and growth rate of striped catfish *Pangasianodon hypophthalmus* under impacts of hypoxia and temperature. Comp Biochem Physiol A Mol Integr Physiol 2017; 203: 288-96. [http://dx.doi.org/10.1016/j.cbpa.2016.10.006] [PMID: 27768904]

[57] Weibel ER, Kistler GS, Scherle WF. Practical stereological methods for morphometric cytology. J Cell Biol 1966; 30(1): 23-38. [http://dx.doi.org/10.1083/jcb.30.1.23] [PMID: 5338131]

[58] Madureira TV, Lopes C, Malhão F, Rocha E. Estimation of volume densities of hepatocytic peroxisomes in a model fish: catalase conventional immunofluorescence versus cytochemistry for electron microscopy. Microsc Res Tech 2015; 78(2): 134-9. [http://dx.doi.org/10.1002/jemt.22455] [PMID: 25431324]

[59] Kong RY, Giesy JP, Wu RS, *et al.* Development of a marine fish model for studying *in vivo* molecular responses in ecotoxicology. Aquat Toxicol 2008; 86(2): 131-41. [http://dx.doi.org/10.1016/j.aquatox.2007.10.011] [PMID: 18055030]

[60] Monteiro SM, Rocha E, Mancera JM, Fontaínhas-Fernandes A, Sousa M. A stereological study of

copper toxicity in gills of *Oreochromis niloticus*. Ecotoxicol Environ Saf 2009; 72(1): 213-23. [http://dx.doi.org/10.1016/j.ecoenv.2008.02.008] [PMID: 18405973]

[61] Ortiz-Zarragoitia M, Cajaraville MP. Biomarkers of exposure and reproduction-related effects in mussels exposed to endocrine disruptors. Arch Environ Contam Toxicol 2006; 50(3): 361-9. [http://dx.doi.org/10.1007/s00244-005-1082-8] [PMID: 16328616]

[62] Silva P, Rocha MJ, Cruzeiro C, *et al.* Testing the effects of ethinylestradiol and of an environmentally relevant mixture of xenoestrogens as found in the Douro River (Portugal) on the maturation of fish gonads--a stereological study using the zebrafish (*Danio rerio*) as model. Aquat Toxicol 2012; 124-125: 1-10. [http://dx.doi.org/10.1016/j.aquatox.2012.07.002] [PMID: 22885794]

[63] Gundersen HJ. The nucleator. J Microsc 1988; 151(Pt 1): 3-21. [http://dx.doi.org/10.1111/j.1365-2818.1988.tb04609.x] [PMID: 3193456]

[64] Madureira TV, Rocha MJ, Cruzeiro C, Rodrigues I, Monteiro RA, Rocha E. The toxicity potential of pharmaceuticals found in the Douro River estuary (Portugal): evaluation of impacts on fish liver, by histopathology, stereology, vitellogenin and CYP1A immunohistochemistry, after sub-acute exposures of the zebrafish model. Environ Toxicol Pharmacol 2012; 34(1): 34-45. [http://dx.doi.org/10.1016/j.etap.2012.02.007] [PMID: 22445869]

[65] Santos M, Monteiro RA, Rocha E. A stereological study of the volume-weighted volume and of the relative volume of the nucleus of normal and preneoplastic hepatocytes in a trout model of hepatocarcinogenesis. Exp Toxicol Pathol 2013; 65(5): 623-30. [http://dx.doi.org/10.1016/j.etp.2012.07.001] [PMID: 22862953]

[66] Puga S, Pereira P, Pinto-Ribeiro F, *et al.* Unveiling the neurotoxicity of methylmercury in fish (*Diplodus sargus*) through a regional morphometric analysis of brain and swimming behavior assessment. Aquat Toxicol 2016; 180: 320-33. [http://dx.doi.org/10.1016/j.aquatox.2016.10.014] [PMID: 27780124]

[67] Hill A, Howard CV, Strahle U, Cossins A. Neurodevelopmental defects in zebrafish (*Danio rerio*) at environmentally relevant dioxin (TCDD) concentrations. Toxicol Sci 2003; 76(2): 392-9. [http://dx.doi.org/10.1093/toxsci/kfg241] [PMID: 14600291]

[68] Schramm M, Muller E, Triebskorn R. Brown trout *Salmo trutta f. Fario* liver ultrastructure as a biomarker for assessment of small stream pollution. Biomarkers 1998; 3(2): 93-108. [http://dx.doi.org/10.1080/135475098231264] [PMID: 23899294]

[69] Gernhöfer M, Pawert M, Schramm M, Müller E, Triebskorn R. Ultrastructural biomarkers as tools to characterize the health status of fish in contaminated streams. J Aquat Ecosyst Stress Recovery 2001; 8: 241-60. [http://dx.doi.org/10.1023/A:1012958804442]

[70] Grund S, Keiter S, Böttcher M, *et al.* Assessment of fish health status in the Upper Danube River by investigation of ultrastructural alterations in the liver of barbel *Barbus barbus*. Dis Aquat Organ 2010; 88(3): 235-48. [http://dx.doi.org/10.3354/dao02159] [PMID: 20377013]

[71] Wu WZ, Zhou BS, Xu Y, Schramm K, Kettrup A. Cytological and biochemical alterations in *Carassius auratus* hepatocytes from exposure to sediment containing dioxins and related compounds. Arch Environ Contam Toxicol 1999; 37(3): 358-63. [http://dx.doi.org/10.1007/s002449900525] [PMID: 10473792]

[72] Arockia Vasanthi L, Revathi P, Mini J, Munuswamy N. Integrated use of histological and ultrastructural biomarkers in *Mugil cephalus* for assessing heavy metal pollution in Ennore estuary, Chennai. Chemosphere 2013; 91(8): 1156-64. [http://dx.doi.org/10.1016/j.chemosphere.2013.01.021] [PMID: 23415490]

[73] Triebskorn R, Köhler H-R, Honnen W, Schramm M, Adams SM, Müller EF. Induction of heat shock proteins, changes in liver ultrastructure, and alterations of fish behavior: are these biomarkers related

and are they useful to reflect the state of pollution in the field? J Aquat Ecosyst Stress Recovery 1997; 6: 57-73.
[http://dx.doi.org/10.1023/A:1008224301117]

[74] Harada T, Hatanaka J, Enomoto M. Liver cell carcinomas in the medaka (*Oryzias latipes*) induced by methylazoxymethanol-acetate. J Comp Pathol 1988; 98(4): 441-52.
[http://dx.doi.org/10.1016/0021-9975(88)90092-8] [PMID: 3417912]

[75] Hinton DE, Lauren DJ, Teh SJ, Giam CS. Cellular composition and ultrastructure of hepatic neoplasms induced by diethylnitrosamine in Oryzias latipes. Mar Environ Res 1988; 24: 307-10.
[http://dx.doi.org/10.1016/0141-1136(88)90326-1]

[76] Gavrilescu M, Demnerová K, Aamand J, Agathos S, Fava F. Emerging pollutants in the environment: present and future challenges in biomonitoring, ecological risks and bioremediation. N Biotechnol 2015; 32(1): 147-56.
[http://dx.doi.org/10.1016/j.nbt.2014.01.001] [PMID: 24462777]

[77] Richardson SD, Kimura SY. Water Analysis: Emerging Contaminants and Current Issues. Anal Chem 2016; 88(1): 546-82.
[http://dx.doi.org/10.1021/acs.analchem.5b04493] [PMID: 26616267]

[78] Richardson SD, Ternes TA. Water analysis: emerging contaminants and current issues. Anal Chem 2014; 86(6): 2813-48.
[http://dx.doi.org/10.1021/ac500508t] [PMID: 24502364]

[79] Stipaničev D, Dragun Z, Repec S, Rebok K, Jordanova M. Broad spectrum screening of 463 organic contaminants in rivers in Macedonia. Ecotoxicol Environ Saf 2017; 135: 48-59.
[http://dx.doi.org/10.1016/j.ecoenv.2016.09.004] [PMID: 27685670]

[80] Triebskorn R, Hetzenauer H. Micropollutants in three tributaries of Lake Constance, Argen, Schussen and Seefelder Aach: a literature review. Environ Sci Eur 2012; 24: 8.
[http://dx.doi.org/10.1186/2190-4715-24-8]

[81] Cairns J. Ecotoxicology: An Opportunity for the Experimental Sciences. Asian J Exp Sci 2005; 19: 1-17.

[82] Abalaka SE. Heavy metals bioaccumulation and histopathological changes in Auchenoglanis occidentalis fish from Tiga dam, Nigeria. J Environ Health Sci Eng 2015; 13: 67.
[http://dx.doi.org/10.1186/s40201-015-0222-y] [PMID: 26448864]

[83] Abdel-Moneim AM, Al-Kahtani MA, Elmenshawy OM. Histopathological biomarkers in gills and liver of *Oreochromis niloticus* from polluted wetland environments, Saudi Arabia. Chemosphere 2012; 88(8): 1028-35.
[http://dx.doi.org/10.1016/j.chemosphere.2012.04.001] [PMID: 22546634]

[84] Javed M, Ahmad I, Usmani N, Ahmad M. Studies on biomarkers of oxidative stress and associated genotoxicity and histopathology in Channa punctatus from heavy metal polluted canal. Chemosphere 2016; 151: 210-9.
[http://dx.doi.org/10.1016/j.chemosphere.2016.02.080] [PMID: 26943742]

[85] Lujić J, Matavulj M, Poleksić V, *et al.* Gill reaction to pollutants from the Tamiš River in three freshwater fish species, *Esox lucius* L. 1758, *Sander lucioperca* (L. 1758) and *Silurus glanis* L. 1758: a comparative study. Anat Histol Embryol 2015; 44(2): 128-37.
[http://dx.doi.org/10.1111/ahe.12119] [PMID: 24809962]

[86] Poleksić V, Lenhardt M, Jarić I, *et al.* Liver, gills, and skin histopathology and heavy metal content of the Danube sterlet (*Acipenser ruthenus* Linnaeus, 1758). Environ Toxicol Chem 2010; 29(3): 515-21.
[http://dx.doi.org/10.1002/etc.82] [PMID: 20821473]

[87] Triebskorn R, Telcean I, Casper H, *et al.* Monitoring pollution in River Mureş, Romania, part II: metal accumulation and histopathology in fish. Environ Monit Assess 2008; 141(1-3): 177-88.
[http://dx.doi.org/10.1007/s10661-007-9886-9] [PMID: 17876715]

[88] van Dyk JC, Cochrane MJ, Wagenaar GM. Liver histopathology of the sharptooth catfish *Clarias gariepinus* as a biomarker of aquatic pollution. Chemosphere 2012; 87(4): 301-11. [http://dx.doi.org/10.1016/j.chemosphere.2011.12.002] [PMID: 22205049]

[89] van Dyk JC, Marchand MJ, Smit NJ, Pieterse GM. A histology-based fish health assessment of four commercially and ecologically important species from the Okavango Delta panhandle, Botswana. Afr J Aquat Sci 2009; 34: 273-82. [http://dx.doi.org/10.2989/AJAS.2009.34.3.9.985]

[90] Kelly JM, Janz DM. Assessment of oxidative stress and histopathology in juvenile northern pike (*Esox lucius*) inhabiting lakes downstream of a uranium mill. Aquat Toxicol 2009; 92(4): 240-9. [http://dx.doi.org/10.1016/j.aquatox.2009.02.007] [PMID: 19304330]

[91] Ayas Z, Ekmekci G, Ozmen M, Yerli SV. Histopathological changes in the livers and kidneys of fish in Sariyar Reservoir, Turkey. Environ Toxicol Pharmacol 2007; 23(2): 242-9. [http://dx.doi.org/10.1016/j.etap.2006.11.003] [PMID: 21783764]

[92] Koponen K, Myers MS, Ritola O, Huuskonen SE, Lindström-Seppä P. Histopathology of feral fish from a PCB-contaminated freshwater lake. Ambio 2001; 30(3): 122-6. [http://dx.doi.org/10.1579/0044-7447-30.3.122] [PMID: 11436658]

[93] Miranda AL, Roche H, Randi MA, Menezes ML, Ribeiro CA. Bioaccumulation of chlorinated pesticides and PCBs in the tropical freshwater fish Hoplias malabaricus: histopathological, physiological, and immunological findings. Environ Int 2008; 34(7): 939-49. [http://dx.doi.org/10.1016/j.envint.2008.02.004] [PMID: 18400298]

[94] Tuvikene A, Huuskonen S, Koponen K, Ritola O, Mauer U, Lindström-Seppä P. Oil shale processing as a source of aquatic pollution: monitoring of the biologic effects in caged and feral freshwater fish. Environ Health Perspect 1999; 107(9): 745-52. [http://dx.doi.org/10.1289/ehp.99107745] [PMID: 10464075]

[95] Orrego R, Burgos A, Moraga-Cid G, *et al.* Effects of pulp and paper mill discharges on caged rainbow trout (*Oncorhynchus mykiss*): biomarker responses along a pollution gradient in the Biobio River, Chile. Environ Toxicol Chem 2006; 25(9): 2280-7. [http://dx.doi.org/10.1897/05-385R.1] [PMID: 16986781]

[96] Rochman CM, Kurobe T, Flores I, Teh SJ. Early warning signs of endocrine disruption in adult fish from the ingestion of polyethylene with and without sorbed chemical pollutants from the marine environment. Sci Total Environ 2014; 493: 656-61. [http://dx.doi.org/10.1016/j.scitotenv.2014.06.051] [PMID: 24995635]

[97] Rochman CM, Hoh E, Kurobe T, Teh SJ. Ingested plastic transfers hazardous chemicals to fish and induces hepatic stress. Sci Rep 2013; 3: 3263. [http://dx.doi.org/10.1038/srep03263] [PMID: 24263561]

[98] van den Heuvel MR, Power M, Richards J, MacKinnon M, Dixon DG. Disease and gill lesions in yellow perch (*Perca flavescens*) exposed to oil sands mining-associated waters. Ecotoxicol Environ Saf 2000; 46(3): 334-41. [http://dx.doi.org/10.1006/eesa.1999.1912] [PMID: 10903831]

[99] Bernet D, Schmidt-Posthaus H, Wahli T, Burkhardt-Holm P. Evaluation of two monitoring approaches to assess effects of waste water disposal on histological alterations in fish. Hydrobiologia 2004; 524: 53-66. [http://dx.doi.org/10.1023/B:HYDR.0000036196.84682.27]

[100] Schlacher TA, Mondon JA, Connolly RM. Estuarine fish health assessment: evidence of wastewater impacts based on nitrogen isotopes and histopathology. Mar Pollut Bull 2007; 54(11): 1762-76. [http://dx.doi.org/10.1016/j.marpolbul.2007.07.014] [PMID: 17888457]

[101] Svirčev Z, Lujić J, Marinović Z, *et al.* Toxicopathology induced by microcystins and nodularin: a histopathological review. J Environ Sci Health C Environ Carcinog Ecotoxicol Rev 2015; 33(2):

125-67.
[http://dx.doi.org/10.1080/10590501.2015.1003000] [PMID: 26023756]

[102] Griffitt RJ, Hyndman K, Denslow ND, Barber DS. Comparison of molecular and histological changes in zebrafish gills exposed to metallic nanoparticles. Toxicol Sci 2009; 107(2): 404-15. [http://dx.doi.org/10.1093/toxsci/kfn256] [PMID: 19073994]

[103] Griffitt RJ, Weil R, Hyndman KA, *et al.* Exposure to copper nanoparticles causes gill injury and acute lethality in zebrafish (*Danio rerio*). Environ Sci Technol 2007; 41(23): 8178-86. [http://dx.doi.org/10.1021/es071235e] [PMID: 18186356]

[104] Ostaszewska T, Chojnacki M, Kamaszewski M, Sawosz-Chwalibóg E. Histopathological effects of silver and copper nanoparticles on the epidermis, gills, and liver of Siberian sturgeon. Environ Sci Pollut Res Int 2016; 23(2): 1621-33. [http://dx.doi.org/10.1007/s11356-015-5391-9] [PMID: 26381783]

[105] Federici G, Shaw BJ, Handy RD. Toxicity of titanium dioxide nanoparticles to rainbow trout (*Oncorhynchus mykiss*): gill injury, oxidative stress, and other physiological effects. Aquat Toxicol 2007; 84(4): 415-30. [http://dx.doi.org/10.1016/j.aquatox.2007.07.009] [PMID: 17727975]

[106] Shaw BJ, Al-Bairuty G, Handy RD. Effects of waterborne copper nanoparticles and copper sulphate on rainbow trout, (*Oncorhynchus mykiss*): physiology and accumulation. Aquat Toxicol 2012; 116-117: 90-101. [http://dx.doi.org/10.1016/j.aquatox.2012.02.032] [PMID: 22480992]

[107] Bilberg K, Malte H, Wang T, Baatrup E. Silver nanoparticles and silver nitrate cause respiratory stress in Eurasian perch (*Perca fluviatilis*). Aquat Toxicol 2010; 96(2): 159-65. [http://dx.doi.org/10.1016/j.aquatox.2009.10.019] [PMID: 19923013]

[108] Wang T, Long X, Cheng Y, Liu Z, Yan S. The potential toxicity of copper nanoparticles and copper sulphate on juvenile *Epinephelus coioides*. Aquat Toxicol 2014; 152: 96-104. [http://dx.doi.org/10.1016/j.aquatox.2014.03.023] [PMID: 24742820]

[109] Fatima M, Usmani N, Firdaus F, *et al. In vivo* induction of antioxidant response and oxidative stress associated with genotoxicity and histopathological alteration in two commercial fish species due to heavy metals exposure in northern India (Kali) river. Comp Biochem Physiol C Toxicol Pharmacol 2015; 176-177: 17-30. [http://dx.doi.org/10.1016/j.cbpc.2015.07.004] [PMID: 26191657]

[110] Koca S, Koca YB, Yildiz S, Gürcü B. Genotoxic and histopathological effects of water pollution on two fish species, *Barbus capito pectoralis* and *Chondrostoma nasus* in the Büyük Menderes River, Turkey. Biol Trace Elem Res 2008; 122(3): 276-91. [http://dx.doi.org/10.1007/s12011-007-8078-3] [PMID: 18214389]

[111] Gül S, Belge-Kurutaş E, Yildiz E, Şahan A, Doran F. Pollution correlated modifications of liver antioxidant systems and histopathology of fish (Cyprinidae) living in Seyhan Dam Lake, Turkey. Environ Int 2004; 30(5): 605-9. [http://dx.doi.org/10.1016/S0160-4120(03)00059-X] [PMID: 15051236]

[112] Pereira S, Pinto AL, Cortes R, Fontaínhas-Fernandes A, Coimbra AM, Monteiro SM. Gill histopathological and oxidative stress evaluation in native fish captured in Portuguese northwestern rivers. Ecotoxicol Environ Saf 2013; 90: 157-66. [http://dx.doi.org/10.1016/j.ecoenv.2012.12.023] [PMID: 23352272]

[113] Adams DH, Sonne C, Basu N, *et al.* Mercury contamination in spotted seatrout, *Cynoscion nebulosus*: an assessment of liver, kidney, blood, and nervous system health. Sci Total Environ 2010; 408(23): 5808-16. [http://dx.doi.org/10.1016/j.scitotenv.2010.08.019] [PMID: 20850170]

[114] Hinck JE, Blazer VS, Denslow ND, *et al.* Chemical contaminants, health indicators, and reproductive biomarker responses in fish from the Colorado River and its tributaries. Sci Total Environ 2007;

378(3): 376-402.
[http://dx.doi.org/10.1016/j.scitotenv.2007.02.032] [PMID: 17418376]

[115] Dabrowska H, Ostaszewska T, Kamaszewski M, *et al.* Histopathological, histomorphometrical, and immunohistochemical biomarkers in flounder (*Platichthys flesus*) from the southern Baltic Sea. Ecotoxicol Environ Saf 2012; 78: 14-21.
[http://dx.doi.org/10.1016/j.ecoenv.2011.10.025] [PMID: 22118815]

[116] Rajeshkumar S, Munuswamy N. Impact of metals on histopathology and expression of HSP 70 in different tissues of Milk fish (*Chanos chanos*) of Kaattuppalli Island, South East Coast, India. Chemosphere 2011; 83(4): 415-21.
[http://dx.doi.org/10.1016/j.chemosphere.2010.12.086] [PMID: 21257190]

[117] Montaser M, Mahfouz ME, El-Shazly SA, Abdel-Rahman GH, Bakry S. Toxicity of heavy metals on fish at Jeddah coast KSA: Metallothionein expression as a biomarker and histopathological study on liver and gills. WJ Fish Marine Sci 2010; 2: 174-85.

[118] Radić S, Gregorović G, Stipaničev D, *et al.* Assessment of surface water in the vicinity of fertilizer factory using fish and plants. Ecotoxicol Environ Saf 2013; 96: 32-40.
[http://dx.doi.org/10.1016/j.ecoenv.2013.06.023] [PMID: 23871567]

[119] Stentiford GD, Massoud MS, Al-Mudhhi S, Al-Sarawi MA, Al-Enezi M, Lyons BP. Histopathological survey of potential biomarkers for the assessment of contaminant related biological effects in species of fish and shellfish collected from Kuwait Bay, Arabian Gulf. Mar Environ Res 2014; 98: 60-7.
[http://dx.doi.org/10.1016/j.marenvres.2014.03.005] [PMID: 24680107]

[120] Marigómez I, Soto M, Cancio I, Orbea A, Garmendia L, Cajaraville MP. Cell and tissue biomarkers in mussel, and histopathology in hake and anchovy from Bay of Biscay after the Prestige oil spill (Monitoring Campaign 2003). Mar Pollut Bull 2006; 53(5-7): 287-304.
[http://dx.doi.org/10.1016/j.marpolbul.2005.09.026] [PMID: 16271373]

[121] Ameur WB, El Megdiche Y, de Lapuente J, *et al.* Oxidative stress, genotoxicity and histopathology biomarker responses in *Mugil cephalus* and *Dicentrarchus labrax* gill exposed to persistent pollutants. A field study in the Bizerte Lagoon: Tunisia. Chemosphere 2015; 135: 67-74.
[http://dx.doi.org/10.1016/j.chemosphere.2015.02.050] [PMID: 25912422]

[122] Lenhardt M, Poleksić V, Vuković-Gačić B, Rašković B, Sunjog K, Kolarević S, *et al.* Integrated use of different fish related parameters to assess the status of water bodies. Slovenian Veterinary Research 2015; 52: 5-13.

[123] Zimmerli S, Bernet D, Burkhardt-Holm P, Schmidt-Posthaus H, Vonlanthen P, Wahli T, *et al.* Assessment of fish health status in four Swiss rivers showing a decline of brown trout catches. Aquat Sci 2007; 69: 11-25.
[http://dx.doi.org/10.1007/s00027-006-0844-3]

[124] Vincze K, Scheil V, Kuch B, Köhler HR, Triebskorn R. Impact of wastewater on fish health: a case study at the Neckar River (Southern Germany) using biomarkers in caged brown trout as assessment tools. Environ Sci Pollut Res Int 2015; 22(15): 11822-39.
[http://dx.doi.org/10.1007/s11356-015-4398-6] [PMID: 25860546]

[125] Madureira TV, Rocha MJ, Cruzeiro C, Galante MH, Monteiro RA, Rocha E. The toxicity potential of pharmaceuticals found in the Douro River estuary (Portugal): assessing impacts on gonadal maturation with a histopathological and stereological study of zebrafish ovary and testis after sub-acute exposures. Aquat Toxicol 2011; 105(3-4): 292-9.
[http://dx.doi.org/10.1016/j.aquatox.2011.06.017] [PMID: 21781654]

[126] Costa PM, Caeiro S, Vale C, Delvalls TÀ, Costa MH. Can the integration of multiple biomarkers and sediment geochemistry aid solving the complexity of sediment risk assessment? A case study with a benthic fish. Environ Pollut 2012; 161: 107-20.
[http://dx.doi.org/10.1016/j.envpol.2011.10.010] [PMID: 22230075]

[127] Drobac D, Tokodi N, Lujić J, *et al.* Cyanobacteria and cyanotoxins in fishponds and their effects on

fish tissue. Harmful Algae 2016; 55: 66-76.
[http://dx.doi.org/10.1016/j.hal.2016.02.007] [PMID: 28073548]

[128] Rašković B, Čičovački S, Ćirić M, Marković Z, Poleksić V. Integrative approach of histopathology and histomorphometry of common carp (*Cyprinus carpio* L.) organs as a marker of general fish health state in pond culture. Aquacult Res 2016; 47: 3455-63.
[http://dx.doi.org/10.1111/are.12795]

[129] Rašković B, Ćirić M, Koko V, Stanković M, Živić I, Marković Z, *et al.* Effect of supplemental feeds on liver and intestine of common carp (*Cyprinus carpio*, L. 1758) in semi-intensive rearing system: histological implications. Biologia 2016; 71: 212-9.
[http://dx.doi.org/10.1515/biolog-2016-0017]

[130] Rašković B, Jarić I, Koko V, Spasić M, Dulić Z, Marković Z, *et al.* Histopathological indicators: a useful fish health monitoring tool in common carp (*Cyprinus carpio* Linnaeus, 1758) culture. Cent Eur J Biol 2013; 8: 975-85.
[http://dx.doi.org/10.2478/s11535-013-0220-y]

[131] Rašković B, Poleksić V, Živić I, Spasić M. Histology of carp (*Cyprinus carpio*, L.) gills and pond water quality in semiintensive production. Bulg J Agric Sci 2010; 16: 253-62.

[132] Saraiva A, Costa J, Eiras JC, Cruz C. Histological study as indicator of juveniles farmed turbot, *Scophthalmus maximus* L. health status. Aquaculture 2016; 459: 210-5.
[http://dx.doi.org/10.1016/j.aquaculture.2016.03.048]

[133] Saraiva A, Costa J, Serrão J, Cruz C, Eiras JC. A histology-based fish health assessment of farmed seabass (*Dicentrarchus labrax* L.). Aquaculture 2015; 448: 375-81.
[http://dx.doi.org/10.1016/j.aquaculture.2015.06.028]

[134] Saraiva A, Costa J, Serrão J, Eiras JC, Cruz C. Study of the gill health status of farmed sea bass (*Dicentrarchus labrax* L., 1758) using different tools. Aquaculture 2015; 441: 16-20.
[http://dx.doi.org/10.1016/j.aquaculture.2015.02.004]

CHAPTER 8

The Effect of Cyanobacteria and Their Toxins on Fish

Angeles Jos[1], **Ana M. Cameán**[1], **Reyhan Akcaalan**[2,*] and **Meriç Albay**[2]

[1] *Area of Toxicology, Faculty of Pharmacy, University of Sevilla, Spain*

[2] *Istanbul University, Faculty of Aquatic Sciences, Turkey*

Abstract: Cyanobacteria are grouped together with phytoplankton as primary producers in aquatic environments. These organisms are very diverse and under favorable conditions, their biomass increases dramatically, leading to blooms. Cyanobacterial blooms in aquatic environments are a well known global phenomenon, largely as a result of anthropogenic pressures, such as increas-ing nutrient inputs from catchment areas or climate change. Especially, planktonic cyanobacteria can thrive in surface waters and cause several adverse effects on aquatic organisms due to variations in pH, blocking light from entering the water and oxygen deficiency as a result of respiration at night. Moreover, they also produce toxins, called cyanotoxins, which can pose harmful effects on organisms in every trophic level; phytoplankton, zooplankton, invertebrates, fish, birds and also mammals. Cyanotoxins comprise of very diverse chemical compounds with several adverse effects on organisms. Considering the diversity of cyanotoxins, we chose two cyanotoxins, microcystin and cylindrospermopsin, since there are relatively more data about their toxic effects on fish, and also the former has a worldwide distribution and the latter is an emerging toxin with an extending geographical range. Together with the other adverse effects on fish, microcystins and cylindrospermopsin could cause oxidative stress and histopathological changes which we focus on and review in detail in this chapter.

Keywords: Cyanobacteria, Cylindrospermopsin, Fish, Histopathology, Microcystin, Oxidative stress.

INTRODUCTION

Cyanobacteria and Cyanotoxins

Cyanobacteria are prokaryotic organisms with a capacity for photosynthesis. The pigments; chlorophyll-a and phycobilins are responsible for the blue-green color, and the name blue-green algae has been used for years, however, the scientific

* **Corresponding author Reyhan Akcaalan:** Istanbul University, Fisheries Faculty, Turkey; Tel: +90 212 4555700; E-mail: rakcaalan@gmail.com

Panagiotis Berillis (Ed)

community now refer to them as cyanobacteria, due to them being Gram negative bacteria. Cyanobacteria are an ancient group with a long evolutionary history, ca. 3500 Ma [1] resulting from good adaptation strategies with changing geochemical and climatic conditions, and also anthropogenic pressures, such as nutrient enrichment, over-exploitation of water resources for drinking or irrigation purposes [2]. It is also a very diverse group with different morphologies [single cell, filamentous or colonial], variable size, ranging from 0.2 – 100 micron, and thriving in several habitats from freshwater to extreme saline environments. They have key roles in global primary production and nitrogen-fixation [3].

The successful growth of an organism is related to its ability to optimize resource utilization and when environmental conditions are in favor of one species in relation to its resource requirements, its population will thrive and reach high biomass [4]. Under normal circumstances, cyanobacteria contribute to the phytoplankton community of aquatic habitats, however, when environmental conditions are suitable for their growth, they form blooms. As nutrient concentrations have increased due to cultural eutrophication, cyanobacterial bloom frequency and severity have also risen. The main actor is phosphorus and there is a shift from a diverse phytoplankton community towards dominancy of cyanobacteria with increasing phosphorus concentrations [5]. Historical descriptions of cyanobacterial blooms have been recorded as early as 1188 in Wales and also in the 1800's in Australia [6, 7] and studies has dramatically increased worldwide after the 1970's. The proliferation of cyanobacteria has several unintended consequences in water bodies used for recreation or for the preparation of potable drinking water. These vary from economical [*e.g.* adverse economic effects for tourist attractions, increased costs of disinfection and treatment in drinking water plants] to environmental impacts *e.g.* suppression of growth of other organisms, depleting oxygen during bloom senescense [2, 4]. Besides these negative effects, cyanobacteria can produce a variety of toxic secondary metabolites, cyanotoxins, which cause serious intoxication to aquatic organisms and also mammals, including humans [2, 8, 9]. There are many reports of wild and farmed fish kills related to cyanobacterial blooms and their toxins [8].

Cyanotoxins are secondary metabolites synthesized within the cells showing great diversity [10]. It is possible to classify cyanotoxins depending on their chemical structures into three main groups; Cyclic peptides including microcystins and nodularins, alkaloids comprising a larger group including cylindrospermopsin, anatoxins, saxitoxins, applysiatoxins and lyngbyatoxins and the third group being lipopolysaccharides, which can be found in the cell membrane of many cyanobacteria. On the other hand, their toxicity mechanisms are more diverse and five groups can be identified in relation to their functions upon organisms, such as being hepatotoxins, cytotoxins, neurotoxins, dermatotoxins and irritant toxins [11

- 13]. Hepatotoxins include primarily microcystins and nodularins, although cylindrospermopsin also has hepatotoxic effects together with cytotoxic, genotoxic and dermatotoxic modes of action [14 - 16]. Neurotoxins interfere with the functioning of the neuro-muscular systems and can cause death very rapidly after the consumption of toxic cyanobacteria [12, 17]. Anatoxin-a, Homonatoxin-a, Anatoxin-a(S), Saxitoxins and BMAA are the potent neurotoxins and are produced by many cyanobacterial species including *Dolichospermum, Aphanizomenon, Oscillatoria, Phormidium* and *Planktothrix*. The only cytotoxin produced by cyanobacteria is cylindrospermopsin and mainly targets the liver and kidney, along with several other organs [7, 18]. Dermatotoxins are produced by marine benthic cyanobacteria and comprise aplysiatoxins, debromoaplysiatoxins and lyngbyatoxins. Irritant toxins are lipopolysaccharides [LPS] found in the cell wall of cyanobacteria as well as in all Gram-negative bacteria [12].

Effect of Cyanotoxins on Aquatic Organisms

Cyanobacterial blooms and their associated cyanotoxin risk are well-known around the world because of increasing anthropogenic pressures on aquatic environments. Cyanotoxins could affect all organisms in aquatic environments and can be transferred to higher trophic levels in aquatic food webs [18]. Organisms can be directly exposed to cyanobacteria or toxins *via* consumption, with some species of fish directly feeding on algae and cyanobacteria. On the other hand, zooplankton species, like Daphnia, can consume the cyanobacteria and pass the toxins up to the zooplanktivores and act as a vectors of cyanotoxins for fish species [19]. Similarly, shellfish can accumulate toxins which could be transferred to higher trophic levels [18, 20]. The examples above are related to intracellular toxins. However, the other possibility of exposure to cyanotoxins is *via* direct contact to extracellular toxins, for example toxins could pass through the gills of fish during breathing [21, 22]. Cyanotoxins could also bioaccumulate in the tissue of organisms at higher concentrations than in the environment and biomagnification could possibly occur through trophic levels [18, 23].

Cyanobacterial blooms are mainly associated with animal poisonings and deaths, especially fish and birds [20, 24]. Fishes are the main organisms affected adversely from cyanobacteria and cyanotoxins in aquatic environments. Wood [25] reported a total of 101 incidents concerning affected fish and aquatic animals related to cyanotoxins since 1800. Despite the fact that the number of incidents are relatively low, the total number of fish that died or were seriously affected can be expressed as thousands. Cyanotoxins are likely not the sole reason for fish kills since the exact cause of mortalities is not often determined. There are multiple stress factors that co-occur with blooms of cyanobacteria other than cyanotoxins to affect fish, *e.g.* oxygen depletion, high pH and ammonia and gill irritation [7,

11, 25]. On the other hand, sub-lethal effects of cyanotoxins could be observed in fish *e.g.* impaired development and survival of fish larvae [7, 23, 26].

In this chapter, we have focused on microcystins (MCs) and cylindrospermopsin (CYN) taking into consideration cyanotoxin diversity. From a toxicological point of view, the deleterious effects on a wide range of organisms, including fish is also relevant. Moreover, they have a frequent occurrence with a worldwide distribution [16].

Cyanotoxins have been reported to induce very different toxicological effects on fish, with oxidative stress and histopathological damage among the most extensively investigated. Thus, a thorough revision of the scientific data available on the effects of cyanotoxins on fish is provided, although the studies concerning MCs are considerably higher in comparison to CYN.

OXIDATIVE STRESS

Oxidative stress is classically defined as a disturbance in the prooxidant/antioxidant balance, leading to potential molecular damage [27]. Other authors suggest that oxidative stress may be better defined as a disruption of redox signaling and control [28, 29].

Cyanobacterial blooms can stimulate oxidative stress by generating hyperoxia/anoxia cycles as a consequence of photosynthetic and respiratory processes [30]. However, cyanotoxins *per se* can also induce oxidative stress.

Microcystins

Regarding MCs, apart from their well-known toxicological mechanism through inactivation of phosphatases, oxidative stress (Table **1**) is also a result of the exposure to MCs in different organisms from a toxicological point of view [31]. Moreover, Amado and Montserrat [29] published a review dealing with why and how these cyanotoxins generate oxidative stress in aquatic animals. Thus, they reported that MCs have been shown to induce a wide range of oxidative stress-related effects such as: increased production of reactive oxygen species [ROS] that can cause increased lipid peroxidation (LPO), DNA damage, DNA-protein crosslinking, mitochondrial damage (leading eventually to the release of apoptotic factors), and alteration of the antioxidant defense system. Moreover, it is known that conjugation with glutathione (GSH) is the first step in the detoxication of this cyanobacterial toxin in aquatic organisms [32]. Thus, GSH levels, the main intracellular antioxidant compound are depleted. Oxidative effects of MCs on fish have been widely investigated as the high number of reports in the scientific literature shows. In Table **1** a review of the latest studies (from 2010 to the present) is presented. From the analysis of these data it can be confirmed that oxidative stress plays a role in the pathogenicity of MCs in fish, as previous reports indicated.

Among MCs, MC-LR is the congener most frequently investigated (pure or extracted from cultures or blooms). There are also studies dealing with MC-RR [33 - 36] and with cyanobacterial cells [37, 38]. In all cases it has been proved oxidative-mediated toxicity, but in particular MC-RR did not induce increased LPO values in the liver of fish, although it did it in other organs. Some authors explain this finding due to the lower toxicity of MC-RR and/or the high level of antioxidant defenses in this organ [34]. There are no comparative oxidative stress studies, to the extent of our knowledge, between cyanobacterial cells and pure MCs on fish but it is known that cyanobacterial cells contain other compounds such as lipopolysaccharides and other bioactive peptides that can contribute to the toxic effects observed [39].

Different exposure routes have been investigated [immersion, oral through gavage or with the diet, or intraperitoneal]. From 2010, most of the studies have been performed under laboratory conditions, with immersion/balneation studies being the most frequent, and only a limited number are field studies [40 - 43]. It is difficult to elucidate the influence of the exposure route on MCs toxicity, as studies are independent and use, moreover, different doses/concentrations and exposure times. However, in general, immersion in water containing MCs appears to be less harmful than gavaging [31].

Regarding exposure times, their variability is very high, from acute exposures to subchronic and chronic ones. In any case all exposure periods can be considered relevant from an environmental point of view, as bloom lysis can result in a high concentration of MCs in the water (similarly to an acute exposure) and MCs can persist in aquatic systems for long periods of time due to their chemical stability and resistance to breakdown [31].

Apart from those important factors already mentioned such as dose/concentration, exposure time and route or MC congener that play a role in toxicity, the fish type and species would also have an impact on the kind of toxicological results observed. Sensitive fish species would show a higher degree of damage following exposure, as compared to species which are resistant [31, 44]. Toxicity studies have been performed in a wide variety of fish species (*Oreochromis niloticus, Oryzias latipes, Danio rerio, Cyprinus carpio, Oncorhynchus mykiss, etc.*).

Table 1. Oxidative stress studies conducted on fish exposed to MCs or MC-producing cyanobacterial cells from 2010 to the present time.

Fish species	Toxin / Cyanobacteria	Exposure conditions Concentration range	Main results	Reference
Tilapia (*Oreochromis niloticus*)	Cultures of a MCs-producing *Microcystis aeruginosa* strain (PCC7806) and nontoxic *M. aeruginosa* strain (PCC7005)	Oral exposure. 3 experimental groups: Control, MCs-producing cyanobacterial cells (60 μg MC-LR/fish/day), and MCs non-producing cyanobacterial cells. 21 days of exposure.	MCs containing cells induced an increase of LPO products in liver and kidney, and MCs containing and MCs non-containing cyanobacterial cells altered the activity, gene expression and protein abundance of the enzymes, indicating the importance of GPx and GST in MCs detoxification. Moreover, the liver experienced an adaptative response to the toxic insult.	Puerto *et al.*, 2011a
Medaka fish (*Oryzias latipes*)	MC (MC-RR, -HtyrR and -LR in decreasing order) from a *P. agardhii* natural bloom. Pure MC-LR	Gavage. Exposure to 5 μL of bloom extract containing 2.5 μg of MC-LR equiv. or 5 μg of MC-LR in 5 μL (control +). Sacrifice 2h after the exposure.	304 proteins were identified (147 with a high degree of identification confidence) in the fish livers with proteomic approach. Fifteen of these proteins were statistically significantly different from those of controls. Protein regulation discrepancies indicate oxidative stress and lipid regulation in the livers of the exposed medaka fish.	Marie *et al.*, 2012
Zebrafish (*Danio rerio*)	MC-LR	Balneation/immersion study. 30 days exposure to 2 or 20 μg/L MCLR.	The abundance of 22 proteins, measured by two-dimensional electrophoresis, was remarkably altered in the liver of fish in response to toxin exposure. These proteins were involved in cytoskeleton assembly, macromolecule metabolism, oxidative stress, and signal transduction.	Wang *et al.*, 2010a
Zebrafish (*Danio rerio*)	MC-LR	Balneation/immersion study. 30 days exposure to 2 or 20 μg/L MCLR.	The abundance of 30 protein spots was remarkably altered in the brain of fish in response to MCLR exposure. These proteins are involved in cytoskeleton assembly, macromolecule metabolism, oxidative stress, signal transduction, and other functions (*e.g.* transporting, protein degradation, apoptosis and translation).	Wang *et al.*, 2010b

(Table 1) contd.....

Fish species	Toxin / Cyanobacteria	Exposure conditions Concentration range	Main results	Reference
Silver carp (*Hypophthal- michthys molitrix*)	Purified MC-LR + NH_3–N	Balneation/immersion study. Newly-hatched larvae and 21-da- -old larvae. Samples of newly-hatched larvae were taken after 3 and 7 days, whereas samples of 21-day-olds were taken at 24, 48, and 96 h after fish were transferred to test solutions. MC-LR concentrations: 0, 10, 100 µg/L. NH_3–N concentrations: 0, 0.06, 0.264 mg/L.	Results showed a significant interaction between NH_3–N and MC-LR on GSH and MDA levels in newly-hatched larvae. In the 21-day-old larvae, MDA varied significantly in two toxin concentrations with time, and significant interactions were observed among any two or three factors (MC-LR, time, NH_3–N,); the SOD and GSH changed markedly with time and toxin concentrations and they were time or dose-dependent on NH_3–N and MC-LR exposure.	Sun *et al.*, 2011 [112]
Carp (*Cyprinus carpio*)	Pure MC-LR	Balneation/immersion study. 10 µg/L MC-LR for 0.5, 1, 2, 4, 6 and 14 days.	Hydroxyl radical (•OH) was significantly triggered at 0.5 day and then tended to decrease through the exposure period. As a response of antioxidant, CAT activity arose slightly at the beginning and decreased with exposure period. A distinct increase of GST pointed out that the conjugation reaction of MC-LR and GSH occurred. A time-dependent decrease of GSH with concurrent increase of GSSG level refer to the involving of GSH in detoxification of MC-LR in the liver. Oxidative damage was clear by the significant rise of MDA level at 2–6 days.	Jiang *et al.*, 2011
Zebrafish (*Danio rerio*)	Pure MC-LR	Immersion study. Fish embryos exposed to 0, 0.2, 0.5, 2, and 5 mg/L MC-LR until 96 hours post-fertilization	75 spots in the 0.5 mg/L MCLR condition indicated a significant increase or decrease in abundance in comparison to the control. Identified proteins were mainly a part of processes related to oxidative stress, energetic metabolism, and also the cytoskeleton assembly.	Li *et al.*, 2011 [113]

(Table 1) contd.....

Fish species	Toxin / Cyanobacteria	Exposure conditions Concentration range	Main results	Reference
Bighead carp (*Hypophthal-mythys nobilis*)	Purified MC-LR + NH_3–N	Balneation/immersion study. 14-day-old larvae were exposed to solutions with different combined concentrations of ammonia (0, 0.06, 0.264 mg/L) and MC-LR (0, 2, 10, 30 µg/L) for 10 days.	Antioxidant parameters were measured in whole-body homogenates. SOD, CAT, and MDA significantly changed with MC-LR concentration, whereas GSH was not affected by MC. Ammonia significantly affected the antioxidant system. There were significant interactions between ammonia and MC on SOD and MDA.	Sun *et al.*, 2012
Rainbow trout (*Oncorhynchus mykiss*)	Crude extracts of *Microcystis aeruginosa* (CPCC299) containing mainly MC-LR and <0.67% MC-RR	Balneation/immersion study. 4 days at 15 °C exposure to a single dose of crude extract from *M. aeruginosa* corresponding to MC-LR equivalent of 0, 0.75, 1.8 and 5 µg/L.	PP in the liver increased by at least 3-fold. Decreased phosphate content in proteins in the liver and brain. No increase in GST activity was observed and lipid peroxidation was unaffected in both liver and brain tissue. AChE activity in brain tissue was decreased after exposure to the highest concentration of crude extract.	Gélinas *et al.*, 2012 [114]
Carp *(Cyprinus carpio* L.)	Pure MC-LR	Balneation/immersion study. 0, 0.1, 1.0, 5.0, 10.0 µg/L MC-LR for 14 days.	SOD and CAT activities did not change at any concentration. GST increased at 1, 5 and 10 µg/L. GPx increased at 1 and decreased at 5 and 10 µg/L. GSH decreased in all groups and GSSG increased at 0.1 and 1 µg/L. MDA only increased at 5 µg/L. PP activity was inhibited in all treated groups. The overall toxicity of environmental MC-LR seems to be initiated in the liver *via* both the ROS pathway and the PP inhibition pathway, and the latter might be more important when ambient MC-LR concentration is greater than 0.1 µg/L.	Jiang *et al.*, 2012

(Table 1) contd.....

Fish species	Toxin / Cyanobacteria	Exposure conditions Concentration range	Main results	Reference
Zebrafish (*Danio rerio*)	Pure MC-LR and MC-RR	Balneation/immersion study. 0.1, 0.5, 1.0, 5.0 and 10 µg/L MC-LR or MC-RR. Fish were sampled out at 4, 7 and 15 days.	Responses varied (i) between MC-LR and MC-RR (for certain groups), (ii) for different enzymes (GST, GR, GPx, SOD), at all time points, and (iii) for different tissues (gills, liver, intestine and brain). Enzyme activities in general increased at lower concentrations (≤5.0 µg/L) and decreased at higher concentrations (≥5.0 µg/L). An abrupt change in enzyme activities was observed at a particular concentration in all the tissue enzymes. For GPx and GR, there was a differential response in the case of fish exposed to MC-LR and MC-RR.	Pavagadhi *et al.*, 2012
Zebrafish (*Danio rerio*)	Pure MC-LR	Balneation/immersion study. Adult zebrafish were continuously exposed to MC-LR (1, 5 and 20 µg/L) for 30 days. Afterwards fertilized eggs were collected and the following F1 generation was reared in water containing no MC-LR until 60 days post fertilization.	The activities of SOD, CAT, and GPx significantly dropped in 5 and 20 µg/L MC-LR groups, and the MDA levels markedly increased in all the three treatment groups. These biomarkers were measured in whole body fish samples of the F1 generation.	Liu *et al.*, 2014
Zebrafish (*Danio rerio*)	Pure MC-LR and nodularin	Immersion study. Exposure of hatched embryos to 0.1 and 1 µmol/L for 24, 48 and 96h.	No alteration in transcriptional expression of oxidative stress related genes *cat* and gst occurred.	Faltermann *et al.*, 2016

(Table 1) contd.....

Fish species	Toxin / Cyanobacteria	Exposure conditions Concentration range	Main results	Reference
Silver carp (*Hypophthalmichthys molitrix*) Bighead carp (*Aristichthys nobilis*) Goldfish *(Carassius auratus)* *Culter ilishaeformis*	*Microcystis aeruginosa* biomass	Field study. Monthly during 2005 in Lake Taihu (China). The concentration of MC in the water was different throughout the year.	CAT and GST in the kidney were significantly higher during blooms than before and after blooms in all fish species. The fluctuations of the other parameters (SOD, GPx, GSH) did not differ markedly before, during and after the cyanobacterial bloom.	Qiu *et al.*, 2012
Bighead carp (*Hypophthalmichthys nobilis*)	Microcystin	Field study (*in situ* treatment in lake Taihu, China), semi-*in situ* treatment (filtrated lake water), and aerated tap water (control). 4 weeks. Microcystin concentration in lake water < 1.0 µg/L.	The lake water (*in situ* and semi-*in situ* treatment) significantly elevated the contents of MDA, the expression of HSP70, and the activity of SOD in liver of fish. Meanwhile lake water significantly decreased the expression of CAT and GST genes. CAT activity and GSH content was not modified in any of the treatments.	Sun *et al.*, 2013
Roach (*Rutilus rutilus*) White bream (*Blicca bjoerkna*) Prussian carp (*Carassius gibelio*)	*Aphanizomenon flos-aquae* bloom	Field study. Before and later of a bloom in the Gruza Reservoir (Serbia). No qualitative or quantitative data reported about the cyanotoxins involved. The density of the bloom was 420 000 individual/L.	*B. bjoerkna* exhibited the higher antioxidant response. Liver total SOD, CuZn SOD and GST activities increased during the bloom only in this species. Liver Mn SOD activity in all 3 cyprinids increased during the bloom. A decrease in liver CAT and GPx activities was detected in all 3 species. All biomarkers were also inhibited during the bloom in gills.	Gavrilovic *et al.*, 2014
Tilapia (*Oreochromis niloticus*)	Eutrophic lagoon with MC. No qualitative or quantitative data reported about the cyanotoxin/s involved.	Field study. Jacarepaguá lagoon (Brazil).	Liver GSH values of exposed fish were significantly higher compared to fish from a reference site, showing an adaptation to oxidative conditions.	Hauser-Davis *et al.*, 2015

(Table 1) contd.....

Fish species	Toxin / Cyanobacteria	Exposure conditions Concentration range	Main results	Reference
Bighead carp (*Aristichthys nobilis*)	Extracted MCs from a bloom (RR>LR>YR)	Intraperitoneal injection of doses of approximately 0.5 ml extracted solution of MCs, amounting to 400 and 1,000 µg MC-LR + MC-RR per kg bw, respectively. Sampling points were set during a period of 48 h (1, 3, 12, 24 and 48 h post-treatment)	Contents of ROS and activities of antioxidant enzymes CAT, SOD, GPx, and GR as well as GST in the liver in both dose groups showed a biphasic change with an increase at initial 3 h followed by a decrease after injection, owing to the roles of the antioxidant system in eliminating excessive ROS and regenerating GSH.	Li *et al.*, 2010 [115]
Whitefish (*Coregonus lavaretus*)	Pure MC-LR	Intraperitoneal injection of MC-LR 100 µg/kg bw. Sacrifice after 8, 24, and 48 h of the exposure.	The mean level of miRNA expression in the liver was increased rapidly and temporarily. Increased mRNA level of *Dicer* may reflect high oxidative stress in the challenged tissue. Early up-regulation of *ferritin H* mRNA level indicated a rapid response by liver cells to confer protection against oxidative stress caused by ROS species. At 24 and 48h, a reduction in the expression of genes coding for *ferritin H* (frih) was observed.	Brzuzan *et al.*, 2012 [116]
Goldfish (*Carassius auratus*)	MC-RR extracted from freeze-dried crude algae	Intraperitoneal injection of 50 and 200 µg/kg bw under normoxia, hypoxia and reoxygenation.	MC-RR significantly induced SOD activities in both organs (liver and kidney) and doses but the total antioxidant capacity as well as CAT activities were low in these organs during hypoxia and reoxygenation in comparison to normoxia and control.	Okogwu *et al.*, 2014 [117]
Zebrafish (*Danio rerio*)	Pure MC-LR	Intraperitoneal injection at doses of 50 and 200 µg MC-LR/ kg bw. Sampling points 1, 3, 12, 24, 48, and 168 hours post inyection.	In the liver, the low dose intensified enzymatic activities and also up-regulated gene expression of antioxidant enzymes. The high dose suppressed enzymatic activities and down-regulated gene expression of antioxidant enzymes. At both doses, MDA, a marker of lipid peroxidation, increased and GSH decreased.	Hou *et al.*, 2015 [118]

(Table 1) contd.....

Fish species	Toxin / Cyanobacteria	Exposure conditions Concentration range	Main results	Reference
Carp (*Cyprinus carpio* L.)	Pure MC-LR	Intraperitoneal injection of 50 or 120 µg/kg bw. Carps were sacrificed at 1, 5, 12, 24 and 48 h after exposure to MC-LR.	In the liver, xanthine oxidase did not changed significantly and it might not contribute to over-production of ROS. SOD was significantly induced in the 50 µg/kg bw group at 1–24 h and the strongest inhibition of CAT activity was detected at 48 h after injection of 120 µg/kg dose. In general, a significant depletion of GSH was found at 5–12 h at both doses. Oxidative damage was evidenced by the significant elevation in MDA levels.	Shi *et al*., 2015 [119]

Abbreviations: AChE: acetylcholinesterase; CAT: catalase; GCS: γ-Glutamyl-cysteine synthetase; F1: first generation of offsprings; GR: glutathione reductase; GSH: reduced glutathione; GPx: glutathione peroxidase; GST: glutathione-S transferase; GSSG: oxidized glutathione; HSP: heat shock protein; LPO: lipid peroxidation; MDA: malondialdehyde; i.p.: intraperitoneal; PP: protein phosphatase; ROS: reactive oxygen species; SOD: superoxide dismutase.

Qiu *et al.* [40, 45] studied the hepatotoxic and nephrotoxic effects induced by a cyanobacterial bloom in four fish from different trophic levels: the phytoplanktivorous *Hypophthalmichthys molitrix* and *Aristichthys nobilis*, the omnivorous *Carassius auratus*, and the carnivorous *Culter ilishaeformis*. These authors suggested that the phytoplanktivorous fish showed higher resistance to toxic cyanobacterial blooms in comparison to omnivorous and carnivorous ones with respect to liver damage. On the other hand, all four species showed a similar nephrotoxicity. They indicated that the kidney might be more susceptible to toxic cyanobacteria than the liver. Also, Gavrilovic *et al.* [42] studied several oxidative stress biomarkers in three different cyprinids and observed that *Blicca bjoerkna* exhibited higher antioxidant responses to a cyanobacterial bloom than *Rutilus rutilus* and *Carassius gibelio*.

The change in the activity of antioxidant enzymes is one of the biomarkers most frequently studied [36, 41, 46] along with LPO and the levels of the non-enzymatic antioxidant glutathione [47, 48]. Actually, some reports have also evaluated these enzymes from a molecular perspective, investigating their gene expression [49, 50], their protein abundance, and the changes in the protein expression patterns using a proteomic approach [51 - 53]. In this regard, Paskerová *et al.* [54] consider that detoxification biomarkers (glutathione and glutathione-S-transferase) show very high variability. However, the oxidative stress biomarkers (*e.g.*, superoxide dismutase, catalase, glutathione peroxidase, and glutathione reductase) present more consistent responses and among them, LPO seems to fulfill the criteria with a sufficiently long half-life and systematic response. Toxic cyanobacteria systematically elevate levels of LPO in most of the reports, which indicates the important role of oxidative damage in cyanobacterial toxicity.

Cylindrospermopsin

Regarding CYN or cyanobacterial material [blooms, cultures, *etc.*] containing CYN, the number of oxidative stress studies in the scientific literature is much lower than those dealing with MCs (Table **2**). This toxin results in the inhibition of protein synthesis [55, 56]. It has also been reported that CYN causes genotoxicity [57, 58] and inhibition of glutathione synthesis [56, 59]. Therefore, the depletion of GSH levels might increase ROS levels. These effects are mediated by CYP450-activated CYN metabolites [58]. The oxidative effects of CYN on fish have been demonstrated by our research group using different exposure routes [gavage, i.p. injection, immersion] and CYN has been shown to increase LPO, reduce GSH levels and alter the activity and expression of different antioxidant enzymes, among other effects [18, 60 - 65]. Table **2** shows a compilation of studies available in the scientific literature, from 2010 to the present moment, dealing with the oxidative damage induced by CYN on fish *in vivo*. In this case, from 2010, only laboratory studies have been performed and the only fish species used was tilapia [*Oreochromis niloticus*]. Similar effects [oxidative stress induction] have been reported by other authors *in vitro* with a common carp (*Cyprinus carpio* L.) leucocyte cell line [66]. Also, as in the case of MCs, it has been reported that cyanobacterial cultures producing CYN are more toxic than the pure toxin [67].

Finally, the oxidative stress damage induced by cyanotoxins [MCs and CYN] on fish seems to be reversible, as depuration periods have shown that the biomarkers studied returned to their control levels [65, 68]. Moreover, different antioxidant compounds have been suggested as useful chemoprotectants to prevent or reduce the oxidative insult induced by cyanotoxins [69 - 71].

HISTOPATHOLOGICAL EFFECTS

Microcystins

The risks of exposure to MCs on fish have been assessed and reviewed by several authors [21, 72, 73] and more recently by Pavagadhi and Balasubramanian [31]. In this chapter a summary of the most relevant histopathological changes induced by MCs in fish is shown in Table **3**, depending on the species, route of exposure and concentration ranges.

In general, fish are less sensitive to MCs in comparison to mammals but their sensitivity depends on several factors, *e.g*., the species, and the exposure route [21]. However, the gross histopathological and ultrastructural changes induced by acute exposure to microcystins (crude extracts or pure toxin) in several fish species are similar to those seen in mammals [72]. Nevertheless, several dissimilarities do exist. In addition to severe damage of the liver (characterized by loss of the cord-like parenchymal structure, general glucogenic degeneration, macrovesicular steatosis of hepatocytes, *etc*.), MCs also induced damage in the kidney, gills, spleen, and gastrointestinal tract (g.i.) [74 - 76] as well as cardiac alterations [77], skeletal muscle [78] and ovaries [79, 80]. Moreover, reproductive toxicity [81] and effects on embryonic development, larval growth and histopahological changes in several species of fish, including the immune function have been demonstrated [82, 83]. Regarding the scientific literature, the susceptibility to MC-LR among fish species show considerable differences [75, 84]. Thus, in some species, damage was restricted to the liver and no changes were found in the kidney, gills or gut [85 - 87]. *T. tinca* presented lower resistance to toxic cyanobacterial cells in comparison to tilapias (*O. niloticus*) [88].

Table 2. Oxidative stress studies conducted on fish exposed to CYN or CYN-producing cyanobacterial cells from 2010 to the present time.

Fish species	Toxin/Cyanobacteria	Exposure conditions Concentration range	Main results	Reference
Tilapia (*Oreochromis niloticus*)	Pure CYN	Exposure to 200 μg/kg pure CYN through two different routes-gavage and i.p. Sacrifice after 24 h and 5 days.	The results in liver and kidney in general showed an increase in NADPH oxidase activity, LPO and protein oxidation; no changes in DNA oxidation; and a reduction in GSH levels and GCS activity. The time of sacrifice had a bigger influence on the results than the exposure route because after 5 days some of the biomarkers assayed had recovered their pre-intoxication levels.	Gutiérrez-Praena *et al.*, 2011
Tilapia (*Oreochromis niloticus*)	Pure CYN	Gavage. 200 or 400 μg CYN/kg bw. Sacrifice 24h after the exposure.	Results showed an increase of LPO products in liver and kidney. Moreover, CYN altered the activity, gene expression and protein abundance of the enzymes, indicating the importance of GPx and GST in CYN pathogenicity.	Puerto *et al.*, 2011b

(Table 2) contd.....

Fish species	Toxin/Cyanobacteria	Exposure conditions Concentration range	Main results	Reference
Tilapia (*Oreochromis niloticus*)	Pure CYN	Exposure by gavage or by i.p. injection to a single dose of 200 µg/kg bw of pure CYN and after 24 h or 5 days they were sacrificed.	A differential response in regard to the organ (liver or kidney), the type of exposure and the time of sacrifice was evident. No significant changes in GPx activity were appreciable in liver and kidney of fish in any of the experiments. Hepatic GST activity increased in fish orally exposed and sacrificed 24 h later, but the enzyme showed its basal activity when fish were sacrificed 5 d after the exposure. By i.p. injection this enzymatic activity increased at both times. The genetic expression of GPx and GST were modified in both organs but mainly in liver. Regarding to GST protein expression, liver experienced a significant increase when fish were exposed by the oral route. However, with the i.p. exposure changes were only induced in fish sacrificed after 24 h. In the kidney only an increase in fish sacrificed after 5 days was observed.	Gutiérrez-Praena *et al.*, 2013 [120]
Tilapia (*Oreochromis niloticus*)	Pure CYN CYN from lyophilized cells of *Aphanizomenon ovalisporum*	Oral exposure. 7 days pretreatment with absence or presence of 0, 22, and 45 mg N-acetylcysteine/fish/d and on day seven, fish received a single dose of 200 mg/kg CYN and were killed after 24 h.	Exposure to crude extract of CYN containing cyanobacteria cells induced higher toxic effects than pure CYN. The increase of LPO and protein oxidation as well as the decrease in GSH/GSSG and the alteration of the enzymatic activities were assayed as a result of oxidative stress inducedby CYN. The toxicity decreased with N-acetylcysteine supplementation particularly at the highest dose employed.	Gutiérrez-Praena *et al.*, 2012a

(Table 2) contd.....

Fish species	Toxin/Cyanobacteria	Exposure conditions Concentration range	Main results	Reference
Tilapia (*Oreochromis niloticus*)	*A. ovalisporum* cells containing CYN and deoxy-CYN (LEGE X-001), or pure standard CYN	Oral exposure. 21 days pretreatment with absence or presence of 400 or 880 mg L-Carnitine/kg fish/day. Afterwards, fish received single doses of 400 µg CYN/kg of fish (either that isolated from lyophilised *A. ovalisporum* cells or pure CYN).	L-carnitine is a chemoprotectant that reduces hepatic and renal oxidative stress in tilapia. In general, oxidative biomarkers such as LPO, protein oxidation, DNA oxidation, NADPH oxidase, SOD, CAT, GCS, GSH/GSSG, were restored to its basal levels with the chemoprotectant.	Guzmán-Guillén *et al.*, 2013a
Tilapia (*Oreochromis niloticus*)	CYN producer *A. ovalisporum* culture	Balneation/immersion study. *A. ovalisporum* cells to reach 10 or 100 µg CYN/L during two different exposure times: 7 and 14 days.	Major changes were observed at 100 µg CYN/L concentration after 14 days of exposure., DNA oxidation and GCS in kidney, CAT and GSH/GSSG decrease in the liver and kidney was assayed at 10 µg CYN/L concentration after 7 day of exposure. The most affected organ was kidney .	Guzmán-Guillén *et al.*, 2013b
Tilapia (*Oreochromis niloticus*)	CYN producer *A. ovalisporum* culture	Balneation/immersion study. Addition of *A. ovalisporum* cells to reach 10 µg CYN/L every two days during 14 days to the water. Two distinct depuration periods: 3 or 7 days.	LPO and DNA oxidation increased by the toxin in liver and kidney, returned to normal values after 7 d of depuration, however, to restore the oxidatively damaged proteins a longer period was needed. 3 d of depuration was sufficient for the recovery of SOD and GCS activities, while 7 days was necessary for CAT activity to return to the control values. The ratio of GSH/GSSG returned to control levels in liver and kidney after adepuration period of 7 d. The depuration process is a highly effective for the detoxification of fish contaminated with CYN.	Guzmán-Guillén *et al.*, 2014a

(Table 2) contd.....

Fish species	Toxin/Cyanobacteria	Exposure conditions Concentration range	Main results	Reference
Tilapia (*Oreochromis niloticus*)	CYN-containing culture of *A. ovalisporum*	Balneation/immersion study. Exposure to repeated doses for 14 days. CYN concentration in the water of the aquaria varied from 11.2 ± 0.5 µg/L and increased up to 42.4 ± 0.2 µg/L after 14 days of exposure. 2 depuration periods (3 or 7 days).	Results showed marked alterations in GPx and GST activity, gene expression and relative abundance. Also, their recovery varied with time, 3 days for GPx activity, GST relative abundance and 7 days for GPx gene expression and GST activity.	Ríos *et al.*, 2014
Tilapia (*Oreochromis niloticus*)	Pure CYN	Gavage. 200 or 400 µg CYN/kg bw. Sacrifice 24h after the exposure.	Increased protein oxidation products in liver and kidney. NADPH oxidase activity was significantly increased with the kidney being the most affected organ. Decreased GSH contents were also detected in both organs, at the higher dose assayed.	Puerto *et al.*, 2014
Tilapia (*Oreochromis niloticus*)	Pure CYN	Oral study. Fish were pretreated with vitamin E (700 mg/kg of fish bw/day) for 7 days by oral route, and they received a single oral dose of 400 mg pure CYN/kg fish bw on day 7, and were sacrificied after 24 h.	The biomarkers (LPO, GST, GPx, SOD, CAT, GCS activities and GSH/GSSG) recovered mainly in liver as well as in kidney to basal levels. Result showed that for reducing the toxic effects of CYN on fish, vitamin E supplementation is highly effective	Guzmán-Guillén *et al.*, 2015a

Abbreviations: CAT: catalase; GCS: γ-Glutamyl-cysteine synthetase; GSH: reduced glutathione; GPx: glutathione peroxidase; GST: glutathione-S transferase; GSSG: oxidized glutathione; LPO: lipid peroxidation; i.p.: intraperitoneal; SOD: superoxide dismutase.

In relation to the exposure conditions, most of the pioneer studies have focused on the histopathological changes after acute administration of MCs by intraperitoneal injection [i.p.] (Table **3**), using pure standard MC-LR [89, 90] or from cyanobacterial cells (mainly *M. aeruginosa*) [76, 84, 91, 92]. As far as we know only one study was undertaken to determine the potential differences between the effects induced by MC-LR and MC-RR pure standards in fish [93]: both MC variants caused lesions in hepatic tissues with a more pronounced effect by MC-LR and MC-RR led to marked degenerative renal changes which may be due its greater hydrophilicity. Later, Li *et al.* [94] confirmed the severity and extension of ultrastructural lesions in the kidney with increasing accumulation of MC-RR in crucian carp exposed by i.p. injection to cyanobacterial extracts (containing mainly MC-LR and MC-RR). These authors suggested that the membranous structure is the main action site of MCs in the kidney, among which mitochondrial damage in the tubules is a critically important effect. As far as we are concerned, only one subchronic study has been carried out by repeated i.p. injections of MC-LR standard [10 or 100 μg toxin/kg fish], which demonstrated time-dependent effects in white fish [95] by this route. For a better characterization of the MC-LR-induced changes in the liver, a quantitative analysis of the liver cells morphology, immunodetection of MC-LR, and mRNA expression of genes involved in membrane transport, cell proliferation and apoptosis were carried out by these authors. The induction of severe liver injury by MC-LR in the early phase of exposure was reported, and active reparative processes and regeneration of the liver parenchyma in the late phase of exposure was followed; it was demonstrated that subsequent injections did not cause any further damage. These results indicated for the first time that whitefish liver must have compensatory mechanisms that not only regenerate tissue structure but also make the liver resistant to subsequent exposures [95]. In fact, the recovery of main target organs has been demonstrated. Thus, a quick histological recovery was observed in the liver of bighead carp at 48 hours post injection (hpi) (200 and 500 μg MC-LR/kg) [96], and renal ultrastructural recovery has been reported in crucian carp at 168 hpi [94]. *In vivo* studies on the reproductive system are limited, and by i.p. route, Hou *et al.* [80], have shown for the first time the structural recovery of zebrafish ovary which was evident in both dose groups (50 and 200 μg MC-LR/kg b.w fish) from 24 to 168 hpi.

With respect to oral (feeding, gavage) and balneation/immersion exposure, only a few studies have been performed with MC-LR pure standard (>95%), in medaka fish (*O. latipes*) [79, 97, 98], *Pognichthys macrolepidotus* [87] and more recently in zebrafish (*D. rerio*) [81, 83, 99]. The use of extracts of cyanobacterial cells in oral toxicology is likely a more effective replicate of natural poisoning considering exposure to a toxic scum, or persistent blooms with MCs in contaminated aquatic ecosystems [39].

Table 3. Histopathogical changes induced by MCs on fish, depending on the fish species, route of exposure, and concentration ranges.

Fish species	Toxin/Cyanobacteria	Exposure conditions Concentration ranges	Main results	Reference
Injection (intraperitoneal, dorsal aorta *etc.*)				
Atlantic Salmon	Pure MC-LR	Three i.p. injections of MC-LR at 555 µg/Kg/injection spaced 3 d apart. Livers of exposed fish were collected at 8 d, 15d, 27 d and 36 d post-injection.	A severe, diffuse necrosis, hydropic degeneration and hypertrophy of the hepatocytes was evident at 8 d, necrotic changes were reduced at 15 d, but livers lost the normal architecture and hepatic megalocytosis. Identical changes were detected at 27 d. At 36 d the livers did not exhibit necrotic changes, but some alterations were still evident.	Andersen *et al.*, 1993
Brown trout (*Salmo trutta* L) and rainbow trout (*Oncorhynchus mykiss)*	MC-LR from cyanobacterial culture.	Injection *via* the dorsal aorta. 25, 75 or 300 µg/Kg	Increased liver damage as doses increased. At 25 µg/Kg swelling of hepatocytes and increase in the frequency of necrotic cells was detected. At 75 µg/Kg areas of confluent necrosis and cellular degeneration, which were increased at the highest doses, with a higher number of pycnotic nuclei and parenchyma of the liver disintegrated. No kidney damage was seen.	Bury *et al.*, 1997
Bighead carp *(Aristichthys nobilis)*	MCs from cyanobacterial extract (containing MC-RR, and –LR)	i.p. injection at two doses, 50 and 200 µg MC-LR eq./kg b.w. Samples were studied at 1,3,12,24, and 48 hpi.	Two new findings were detected in fish: widening of intercellular spaces and ultrastructural recovery of hepatocytes at 48 hpi in both dose groups.	Li *et al.*, 2005b

(Table 3) contd.....

Fish species	Toxin/Cyanobacteria	Exposure conditions Concentration ranges	Main results	Reference
Injection (intraperitoneal, dorsal aorta *etc.*)				
Common carp (*Cyprinus carpio*)	MC-LR from *M. aeruginosa* culture	i.p. injection of 130, 150,250,300 and 550 µg/Kg bw	Disturbances in the swimming behavior at all doses. The lowest lethal dose was 550 µg/Kg. Severe liver damage with dissociation of the hepatocytes and hydropic degeneration in the liver was observed at sublethal doses, and the Bowman's capsules of the glomeruli were dilated in the kidney. At he highest dose, degeneration of the kidney tubuli and total loss of the liver parenchymal architecture were observed. Ultraestructural observations were also reported.	Rabergh *et al.*, 1991
Crucian carp *(Carassius auratus)*	MCs from *Microcystis* extracts (mainly MC-RR and MC-LR)	Injection of 150 and 600 µg/Kg bw	Irregular swimming behavior and changes in ventilation rates were observed after injection of MCs, at sublethal and lethal doses. Elevated cortisol levels and reduced plasma concentrations of thyroid hormones.	Li *et al.*, 2008
Crucian carp *(Carassius auratus)*	Extract *from M. aeruginosa* crude algae from blooms (mainly MC-RR and MC-LR)	i.p. injection at two doses, 50 and 200 µg MC-LR eq./kg b.w.	Ultrastructural lesions in the kidney (swollen mitochondria in proximal tubules (noticed at 1 h), fusion of foot processes and marked widening of glomerular basement membrane, *etc*) progressed in severity with accumulated MC-RR and exposure times within 48 hpi at two doses. Hydropic degeneration of mitochondria, proliferation of lysosomes in proximal tubules, extensive fusion of podocyte foot processes, disruption of mesangial cells in glomeruli was observed between 24-48 h. The severity of the injury in renal tubules was more pronounced than that in glomeruli. Renal ultrastructural recovery of fish in the low dose group was evident at 168 h phi.	Li *et al.*, 2013

(Table 3) contd.....

Fish species	Toxin/Cyanobacteria	Exposure conditions Concentration ranges	Main results	Reference
Injection (intraperitoneal, dorsal aorta *etc.*)				
Rainbow trout (*Oncorhynchus mykiss*)	MC-LR obtained by purification of a cyanobacterial extract	i.p. injection 400 μg/Kg and 1000 μg/Kg MC-LR	1000 μg/Kg MC-LR caused 100% mortality. 400 μg/Kg caused hepatocellular swelling and lysis of hepatocyte plasma membranes, resulting in liquifactive necrosis. Also kidney lesions were observed, such as coagulative tubular necrosis with a dilation of Bowman's space.	Kotak *et al.*, 1996
Silver carp (*Hypophthalmichthys molitrix* Val.)	MC-LR standard	I.p. injection of pure MCLR in dose of 250 μg/kg bw, or 400 μg/kg bw	Impairment of hepatopancreas (enlargement, yellow-brownish colour, *etc*) large dystrophic and necrobiotic alterations of hepatocytes, as well as of kidney tubuli.	Vajcová *et al.*, 1998
Tilapias (*Oreochromis* sp.)	MC-LR and MC-RR standards	Single i.p. injection of the pure toxins at a dose of 500 μg/Kg biotoxin	Both MCs, particularly MC-LR, caused megalocytosis, necrotic process, and microvesicular steatosis in liver, Degenerative renal changes and glomerulopathy, were more severe after application of MC-RR. Both MCs led to a significant myopathy in the heart, whereas gills did not show histopathological changes.	Atencio *et al.*, 2008b
White fish (juvenile)	MC-LR standard	I.p. injection of 10 and 100 μg/kg bw and different treatment periods: 1/3, 1, 2, 7 14 and 28 days. The injections were repeated every week	The higher dose (100 μg/kg) posed severe liver injury at the beginning of the exposure period, and the effects were time-dependent. The lower dose (10 μg/kg) caused less injury and its effects began to be observed later in the exposure period.	Wozny *et al.*, 2016
Zebrafish *(Danio rerio)*	MC-LR standard	i.p. injection at doses of 50 and 200 μg MC-LR/kg bw. Ovarian samples were studied at 1,3,12,24,48 and 168 h post-injection	Pathological lesions progressed in severity with the increasing exposure time and dose within 12 hpi (vacuolation of gonadosomatic tissue). However, lesions were not aggravated but gradually alleviated from 24 to 48 hpi, and at 168 hpi ovarian tissue showed great recovery.	Hou *et al.*, 2014

(Table 3) contd.....

Fish species	Toxin/Cyanobacteria	Exposure conditions Concentration ranges	Main results	Reference
Injection (intraperitoneal, dorsal aorta *etc.*)				
Oral (feeding, gavage) and Balneation/immersion				
Brown trout alevins (*Salmo trutta*)	Extracts from *M. aeruginos*a strains PCC 7813 and MC-LR purified	5,50, 500 µg MC-LR equivalents/L.	Significant increases of heart rate after exposure to aqueous extracts but no to purified MC-LR. Cardiac output increased.	Best *et al.*, 2001
Feral carp (*Cyprinus carpio*)	Natural exposure in a lake with significant growth of *M. aeruginosa*	Microcystin content in *M. aeruginosa* blooms oscillated between 2.2-4.0 mg/g of lyophilized scum	Impaired hepatocyte function, some degenerative changes in the branchial epithelium were observed, and no mortalities were reported.	Carbis *et al.*, 1997
Common carp *(Cyprinus carpio)*	Lyophilized *M. aeruginosa* strain PCC 7806	Gavaging with a single dose of *Microcystis* PCC806 equivalent to 400 µg MC-LR/kg bw.	Early alterations in hepatopancreas characterized by a dissociation of hepatocytes, apoptotic cell death and delayed cell lysis. In kidney, increased vacuolization of tubular epithelial cells, apoptosis, cell shedding and proteinaceous casts at the cortico-medullary junction. Changes in the GI tract were observed at 12 hpi.(pyknotic nuclei and apoptotic cells) MC-immunopositive staining in the organs.	Fischer and Dietrich, 2000
Common carp *(Cyprinus carpio)*	*Microcystis* bloom scum mixed with the food powder	50 µg MC/kg b.w., 4 weeks of exposure	Inhibition the growth of carp. Severe damage in hepatocytes: swollen endomembrane system and RER, extensive vacuolization of the cytoplasm, and lipid droplets	Li *et al*, 2004 [121]
Common carp *(Cyprinus carpio)*	Cyanobacterial extract containing MC-LR, -RR and –YR.	Carp embryos and larvae exposed for a short term (8 d) or a long-term (30 d). The cumulative doses were: 130 µg/L (high concentration), 13 µg/ L (medium) and 1.3 µg/L (low).	Vacuolar dystrophy of hepatocytes with damage of nuclei in the group exposed to low concentrations. Focal necrosis and dystrophic changes of hepatocytes with vacuolisation and nucleic damage at medium concentrations. No changes were found in the kidney.	Palikova *et al.*, 2004

(Table 3) contd.....

Fish species	Toxin/Cyanobacteria	Exposure conditions Concentration ranges	Main results	Reference
Injection (intraperitoneal, dorsal aorta *etc.*)				
Silver carp (*Hypophthalmichthys molitrix*, Hm), bighead carp (*Aristichthys nobilis, An*), *Carassius auratus* (Ca), *Culter ilishaeformis* (Ci)	Cyanobacterial blooms in a Lake (Lake Taihu), containing MCs	Immersion in natural conditions, monthly in 2005	Kidney impairment from chronic exposure of toxic cyanobacterial blooms might be the first step, and then followed by hepatic failure. At ultrastructural level, pathological changes of mitochondria in the proximal tubules of fish kidney were observed.	Qiu *et al.*, 2012
Silver carp (*Hypophthalmichthys molitrix)*	Cells from a culture of *M. aeruginosa* NPLJ4 containing [D-Leu[1]]Microcystin-LR	Immersion (10^5 cells/mL) for 15 d, and fish were removed at 5-day intervals. The remaining fish were removed and sacrificed after 15 and 17 d to evaluate tissue regeneration	Liver, skeletal muscle and intestinal tract were similarly affected, such as dissociation of cells, necrosis and haemorrhage. The recovery period of 17 days was insufficient time for regeneration of organs affected.	Ferreira *et al.*, 2010
Loach (*Misgurnus mizolepis* Gunthe)	MC-LR from *Microcystis aeruginosa*	Embryos and juveniles loach were exposed to different concentrations (1-1000 μg/L)	The developmental toxicity was concentration-dependent. Liver and heart were the main targets of MC-LR toxicity. MC-LR was lethal to loach juvenile after 7 d of exposure.	Liu *et al.*, 2002
Medaka fish (*Oryzias latipes*)	MC-LR standard	Exposure by gavage (5 μg MC-LR/g bw). Fishes were collected at different time intervals: 3h, 24h, 72h.	Liver changes characterized by cellular dissociation and loss of its normal architecture. Hepatocytes smaller, round, hypereosinophilic cytoplasm, pyknotic nucleus, and a mitosis blockage able to produce binucleated cells. Then, process of apoptonecrosis after 24 h of exposure occurred. Studies on MC-LR distribution and proteomic approaches were also carried out.	Mezhoud *et al.*, 2008

(Table 3) contd.....

Fish species	Toxin/Cyanobacteria	Exposure conditions Concentration ranges	Main results	Reference
Injection (intraperitoneal, dorsal aorta *etc.*)				
Medaka fish (*Oryzias latipes*)	MC-LR standard	Fish were fed dietary MC-LR (0, 0.46, 0.85, 2.01 and 3.93 µg MC-LR/ g dry diet) for 8 weeks	Significant single cell necrosis were observed in female and male fish with 0.85 and 3.93 µg MC-LR/ g in their diet, respectively.	Deng *et al.*, 2010
Medaka fish (*Oryzias latipes*)	MC-LR standard	Exposure by balneation (5 µg MC-LR/L bw) during 30 days	Cell disjunction and abundant cellular lysis were observed in the liver, spleen, and intestine. Cell disjunctions together with a decreasing functional activitywere detected in gonads. The effects on ovaries are reported for the first time.	Trinchet *et al.*, 2011
Medaka fish (*Oryzias latipes*)	Cyanobacterial bloom dominated by *P. agardhii,* containing MCs. MC-LR standard	Oral gavage with 5 µL cyanobacterial extracts containing 2.5 µg equiv. MC-LR, and also MC-LR standard	Hepatocytes presented glycogen storage loss, and disorganization of the hepatic parenchyma, and cellular damage. Immunological localization of MCs.	Marie *et al.*, 2012
Rainbow trout (*Oncorhynchus mykiss)*	MCs from cells of *M aeruginosa* PCC 7806,	Exposure by gavage of 5700 µg MC-LR/bw) Samples were taken after 1,3,12,24,48 and 72 h after exposure	Massive necrosis of the liver. The typical chord structure of liver disappeared and the cytoplasm of the hepatocytes was condensed after 1 h of exposure. The structure damage increased, and limited punctual microhemorrhagia *via* ruptured vessels was detected between 3-12 h. The whole liver damaged, with lysed hepatocyte membranes and pyknotic nuclei after 24-48 h,. No damage was observed in kidney, spleen, heart.	Tencalla and Dietrich, 1997
Rainbow trout (*Oncorhynchus mykiss)*	MCs from cells of *M aeruginosa* PCC 7806	Acute exposure by gavage of 5700 µg MC-LR/bw. Samples were taken at 1,3,12,24,48 and 72 h after bolus dosing	Changes in the cord-like organization of hepatocytes were seen al 1 h postdosing. After 3 h, marked MC-immunopositivity was observed in cytoplasm and nuclei of hepatocytes. Apoptotic cells were detected after 48 h.	Fischer *et al.*, 2000

(Table 3) contd.....

Fish species	Toxin/Cyanobacteria	Exposure conditions Concentration ranges	Main results	Reference
Injection (intraperitoneal, dorsal aorta *etc.*)				
Sacramento splittail (*Pogonichthys macrolepidotus)*	Two sources: MC-LR standard and MC-LR from *Microcystis* harvested in the upper S. Francisco Estuary (SFE)	Exposure to 5 diets for 28 d of MCs from both sources: Three diets containing 3.55 (D5), 9.14 (D0) and 17.13 (D20) mg MC-LR/kg, and the other 2 diets contained purified MC-LR at 3.89 mg/kg (D5R) or no MC-LR (D0)	Liver, gill, kidney and gut were examined, and only severe liver toxicity were observed. Higher MCs content caused cytoplasmic inclusion bodies and single cell necrosis with a greater severity in the fish.	Acuña *et al.*, 2012
Southern catfish (*Silurus meridionalis)*	Blooms being *M aeruginosa* the predominant species	Fertilized eggs were incubated with MCs (0, 1,10,100 µg MC-LR equiv. /L)	Egg development and larval growth retarded, hatching rate reduced, and damage in hepatocytes characterized by disorganization of cell structure and a loss of adherence between cells. Cellular degeneration with vacuolar hepatocytes and marginal nuclei, even hepatocellular necrosis were observed.	Zhang *et al.*, 2008
Tencas *(Tinca tinca)*	MCs from *Microcystis* waterblooms	Fish were exposed to cyanobacterial cells dosing 0,5,11,25 and 55 µg MC-LR/fish mixed with the food. Fish were sacrifized at 96 h post-exposure	Cord-like parenchymal structure of the liver was lost, primarily in the perilobular regions. Glucogenic degeneration and macrovesicular steatosis were observed. Glomerulopathy with dilated Bowman's space and decrease in the glomerular components, tubulonephrosis and microhemorrhages were observed in the kidney. Loss of miofibrils in the heart and vacuolated enterocytes in the GI tract were the main changes observed.	Atencio *et al.*, 2008a

(Table 3) contd.....

Fish species	Toxin/Cyanobacteria	Exposure conditions Concentration ranges	Main results	Reference
Injection (intraperitoneal, dorsal aorta *etc.*)				
Tilapia (*Oreochromis* sp.)	MCs from lyophilized cells from a *Microcystis* bloom	Exposure to 60.0 µg MC-LR/fish/day by two types or oral administration: mixed with commercial fish food o crushed into this fish food. Two different exposure times: 14 or 21 days	The liver parenchymal architecture was disassociated, round hepatocytes with the appearance of pyknotic nuclei were detected. Dilation of Bowman's space and necrotic epithelial cells with pyknotic nuclei in renal tubules. Gills and intestines were also affected.	Molina *et al.*, 2005
European White fish (*Coregonus lavaretus)*	MCs from *Planktothirix rubescen*s culture	0.6-0.9 µg MCLR equiv/Kg b.w. by gavage	Behavioral changes, increased opercular beat rates. Histopathological alterations in the liver (focal disintegration of the parenchymal liver architecture, infiltration of leucocytes, coagulative necrosis, *etc*), the GI tract (epithelial degeneration and cell lysis, loss of the gut architecture, *etc*) and in the kidney (focal and coagulative interstitial necrosis, degeneration of the tubular lining, *etc*) were observed.	Ernst *et al.*, 2006 [122]
European White fish (*Coregonus lavaretus)*	MCs from *Planktothirix rubescen*s culture	Exposure by immersion at low (1500), medium (15,000) and high (55,000) *P. rubescens* cells/ml for up to 28 days	Alterations in liver (disintegration of the parenchymal architecture, necrosis, apoptosis), GI tract (loss of the mucosas structure, extensive epithelial degeneration) and kidney (tubular degeneration resulted in hyalinization of tubuli epithelia and interstitial cell lyses), which were also immunopositive for MCs. The range and severity of patholological changes were exposure time and *P. rubescens* cell density dependent. Gill pathology appeared to result from mechanical abrasion and irritation.	Ernst *et al.*, 2007

(Table 3) contd.....

Fish species	Toxin/Cyanobacteria	Exposure conditions Concentration ranges	Main results	Reference
Injection (intraperitoneal, dorsal aorta *etc.*)				
Yellow catfish (*Pelteobagrus fulvidraco* Richardson)	MCs from fresh cyanobacteria (*M. aeruginosa)*	Fish were exposed by diet: low –cyanobacteria diet (LCD, 32.3 µg MCs/g), and high diet (HCD, 71.96 µg MCs/g), during 60 d and then the fish were free for a further 60 days	Liver tissues in both doses exhibited dose –dependent histopathological modifications, such as non-uniform distribution of hepatocytes, dilation of intercellular space of hepatocytes and vacuolization. After 60-d recovery, fish with the LCD showed no pathological alterations, whereas fish fed with HCD showed a significant alleviation of liver injury but they did not recover completely.	Dong *et al.*, 2012
Zebrafish *(Danio rerio)*	MC-LR standard	Male and female fish were exposed by subchronic immersion in 1, 5,20 µg /L of MC-LR for 30 days	MC-LR exposure exerts diverse reproductive toxic effects with females exhibiting more sensitivity than males. Marked histological lesions were observed in the livers, ovaries and testes in MC-LR treated fish. In the liver of treated fish, a considerable disruption of the spatial architecture and intercellular spaces were observed, similar between females and males. In the testis, cell density seemed to reduce, cellular deteriorations and reduced proportion of matures sperm were visible. In ovary, loss of contact between the oocyte cell membranes and the follicular cells layer were observed; degenerating vitellogenic oocytes were also found.	Qiao *et al.*, 2013
Zebrafish *(Danio rerio)*	MC-LR standard	Adult fish were continuously exposed to MC-LR, with dose of 1,5,20 µg /L for 30 days	Juveniles of treated parents were subject to a considerable disruption of the spatial architecture and widened intercellular spaces in their liver, and pycnotic nuclei were observed. Moreover, MCs influence the growth and immune function in F1 offspring	Liu *et al.*, 2014

(Table 3) contd.....

Fish species	Toxin/Cyanobacteria	Exposure conditions Concentration ranges	Main results	Reference
Injection (intraperitoneal, dorsal aorta *etc.*)				
Zebrafish *(Danio rerio)*	MC-LR standard	Female fish were exposed to 2,10,50 µg /L of MC-LR for 21 days	MC-LR affected the growth of follicles, having smaller vitellogenic follicles of exposed fish (10 and 50 µg /L). MC-LR exposure of 10 µg /L shifted pre-vitellogenic oocytes towards vitellogenic induction, while 50 µg /L inhibited pre-vitellogenic oocytes growth and maduration. Decreases in fertilization and hatching success were observed in 50 µg /L MC-LR group.	Zhao *et al.*, 2015

Abbreviations: b.w.: body weight; G.I.: gastrointestinal; hpi: hours post injection; RER: rough endoplasmic reticulum

In this sense, many study were conducted using cyanobacterial blooms to determine the toxic effects of MCs on fish (see Table **3**) especially under laboratory conditions, whereas natural exposure in lakes by monitoring the exposure to M. aeruginosa cells are very scarce [44, 74]. It is important to consider the metabolites produced by cyanobacteria other than toxins while evaluating the results obtained from extracts or lyophilized cells [39].

Acute toxic episodes are quite rare concerning oral and immersion exposure of fish to MCs [72]. However, subchronic oral toxicity of MCs demonstrated that low concentrations cause alterations in different organs, mainly in the liver [disintegration of the parenchymal architecture, necrosis, apoptosis] and kidney [tubular degeneration, dilation of Bowman's space and necrotic epithelial cells]. Gills, intestinal tract, heart [88] and skeletal muscle [78] were also affected. The range and severity of these pathological changes were exposure time and cell density dependent [78, 87, 88, 100 - 102], but in some cases fish can adapt to the toxic environment over time and survive [100]. The recovery of some histopathological modifications after a depuration period was demonstrated to be a function of the time the fish were free of MCs, and depending on the doses of MCs administered in the diet: liver of yellow catfish after 60 days of depuration did not show pathological alterations on the low cyanobacteria diet, whereas with a high diet of MCs a significant elevation of liver injury was detected but fish did not recover completely [102].

Studies indicated the effects of MCs on the reproductive system of fish. Trinchet *et al.* were the first to report the effects of MCs on reproductive function [79] using medaka fish exposed for 30 d to MC-LR. The reduction of the vitellus storage and lysis of the gonadosomatic tissue was observed in female gonads. The spermatogenesis appeared to be disrupted in males. These authors pointed out that MC-LR may have toxic effects on reproduction in the long term and at low concentration, which may not involve conventional effects on estrogen or androgen receptors but result from indirect interactions. It is not clear whether MC-LR toxicity is caused *via* entering the gonad cells or has a secondary effect due to the damage of the hypothalamic-pituitary-gonad axis. Later, Qiao *et al.* [99] demonstrated that females are more sensitive than males to the toxicity of MC-LR in zebrafish exposed by subchronic immersion. Marked histological lesions were observed in the livers, ovaries and testes and this study confirmed that MC-LR did not cause any estrogenic effects in adult zebrafish. General reproductive parameters and pathological observations in gonads were considered after MCs exposure, but these have not been well characterized. Recently, Zhao *et al.* [81], showed a link among follicular development, transcriptional regulations in the endocrine system, sex hormones and the reproductive performance of female zebrafish after subchronic exposure to MC-LR, and the data demonstrate that the toxin deteriorates the follicular development with an influence of intra- and extra-ovarian factors.

Pronounced toxic effects of MCs and also cyanobacterial extracts have been reported in the development of embryos and fish larvae histopathologically [31, 83]. Among these effects, an enlarged opaque yolk sac, small head, curved body and tail and hepatobiliary abnormalities, including hepatocyte damage (hepatocellular necrosis) were particularly noteworthy [82, 83, 86, 103].

Cylindrospermopsin

Data on CYN toxicity in fish are scarce compared to MCs, especially those relative to the more relevant histopathological changes observed, which have been summarized in Table **4**. In fact, the effects have only been reported in one fish species, Tilapias (*Oreochromis niloticus*) which have been exposed under different laboratory conditions, mainly by the oral route (feeding) or by immersion to CYN or cyanobacterial extracts containing CYN. Only in two studies were histopathological changes observed in fish after acute i.p. injection of pure CYN.

Exposure of tilapias by gavage to pure CYN (200 or 400 μg CYN/L) caused dose-dependent pathological lesions in different organs (liver, kidney, heart, intestines, gills) [61], which were more pronounced in fish euthanized after 5 days post exposure [104], confirming the delayed toxicity of CYN reported in mice [105]. The observed effects of CYN on fish liver and kidney are comparable to those previously published in rodents, revealing the same selectivity for those tissues. Similar to MCs, the liver and kidney were the main targets of CYN [106], although the hepatodegeneration process characterized by a loss of the parenchymal architecture, general glucogenic degeneration and lipid droplets inducing steatosis were less severe in the case of CYN than MCs. Degenerative changes in glomeruli, tubuli, and the intersticial tissue of kidney were observed after exposure to CYN. In this organ, vascular processes were more pronounced in comparison to MC-RR [93], with the presence of large haemorrhagic areas. CYN is highly hydrophilic, which could explain the greater damage to the kidney, although the toxic mechanisms remain to be explored. Histological changes in the heart (myopathy, fibrolysis, generalized edema), intestines (necrotic enteritis, partial desquamation of intestinal villi, edema), and gills (haemorrhage, desquamation areas, tumefaction) of tilapias exposed to CYN showed similarities to those reported by MCs in tilapias [100, 107] or were even more pronounced. The direct action of CYN on the heart could be be concerned with cardiac muscular fibers and be indirectly coupled with the effects on the kidney [61].

Table 4. Histopathological changes induced by CYN on fish, depending on route of exposure and concentration ranges.

Fish species	**Toxin/ Cyanobacteria**	**Exposure conditions Concentration ranges**	**Main results**	**Reference**
Injection (intraperitoneal, i.p. dorsal aorta *etc.*).				
Tilapias (*Oreochromis niloticus*)	CYN standard	Acute i.p. injection of 200 µg/kg fish, which were sacrificed after 24 h or 5 days after administration	Histopathological findings were more pronounced in the liver, kidney, heart and gills after i.p. administration, in comparison to administration by gavage, with the exception of the GI tract (necrotic enterocytes, increased caliciform cells). Histological damage in all organs was more severe after 5 d compared to 24 h. The morphological changes in diameter of hepatocytes and renal tubules were also more severe by i.p. route, and with the time of sacrifice.	Gutiérrez-Praena *et al.*, 2012b
Tilapias (*Oreochromis niloticus*)	CYN standard	Fish i.p. exposed to 200 µg/kg fish and euthanized after 23 h or 5 days	The i.p. administration of CYN caused the appearance of more CYN-immunoreactive cells in the liver, kidney, intestines and gills in comparison to the oral exposure. In some cases only were evident in animals euthanized at 5 days post-i.p. In the kidney by i.p. route CYN was mainly present in glomeruli and erytrocytes.	Guzmán-Guillén *et al.*, 2014b
Oral (feeding) and Balneation/immersion				
Tilapias (*Oreochromis niloticus*)	CYN pure standard	Acute exposure by gavage (200 µg/kg fish), and fish were sacrificed after 24 h or 5 days after administration	Light degenerative changes in the liver, with glycogen accumulation and lipid droplets. In kidney glomerulopathy with glomerular capillary atrophy, hyperemia and alterations in tubules was observed. The GI tract showed a catarrhal enteritis process, with necrotic changes. In gills, tumefaction processes, hyperemia and certain inflammatory foci. All these alterations were more pronounced at 5 days.	Gutiérrez-Praena *et al.*, 2012b

(Table 4) contd.....

Fish species	Toxin/ Cyanobacteria	Exposure conditions Concentration ranges	Main results	Reference
Tilapias (*Oreochromis niloticus*)	CYN pure standard	Acute exposure by gavage of 200 or 400 µg/kg fish, which were sacrificed after 24 h of exposure	Disorganized parenchymal architecture in the liver, dilated Bowman's space in the kidney, fibrolysis in the heart, necrotic enteritis in the intestines and hemorrhages in the gills appeared more severe in the highest dose of 400 µg CYN /kg fish.	Puerto *et al.*, 2014
Tilapias (*Oreochromis niloticus*)	CYN pure standard and CYN from lyophilized cells from an *Aphanizomenon ovalisporum* CYN-producer	Acute exposure of a single dose of 200 µg/kg fish by feeding fish	The pathological changes were more pronounced with CYN from the cyanobacterial strain, being the liver and kidney the main targets. The IHC analysis showed that CYN-antigen appeared mainly in the liver and GI tract.	Gutiérrez-Praena *et al.*, 2014
Tilapias (*Oreochromis niloticus*)	CYN pure standard and CYN from lyophilized cells from an *Aphanizomenon ovalisporum*	Fish were exposed by gavage to 200 µg CYN/kg bw and euthanized after 24 h or 5 days. Moreover, fish were exposed subchronically by immersion to cells using 10 or 100 µg/L for two different exposure times: 7 or 14 days	The IHC analysis revealed that the organ which presented more immunopositive results was the liver, followed by the kidney, intestine, and gills. The delayed and dose dependent toxicity of CYN was confirmed.	Guzmán-Guillén *et al.*, 2014b
Tilapias (*Oreochromis niloticus*)	CYN from lyophilized cells from an *Aphanizomenon ovalisporum*	Subchronic exposure by immersion to cells using 10 or 100 µg/L for two different exposure times: 7 or 14 days	Subchronic exposure to two concentrations induced degenerative processes and steatosis in the liver, membranous glomerulopathy in the kidney, myofibrolysis and edema in the heart, necrotic enteritis in the GI tract and hyperaemia in gill lamellae and microhaemorrhages.	Guzmán-Guillén *et al.*, 2015b

(Table 4) contd.....

Fish species	Toxin/ Cyanobacteria	Exposure conditions Concentration ranges	Main results	Reference
Tilapias (*Oreochromis niloticus*)	CYN from lyophilized cells from an *Aphanizomenon ovalisporum*	Subchronic exposure by immersion to cells using 10 µg CYN/L during 14 d followed by no depuration or by 3, or 7 d of depuration.	The brains of exposed and no depurated fish irregular showed degenerative process signs of necrosis, and vascular processes. Small necrotic and basophilic neurons with irregular borders were observed, hyperaemia and haemorrhage. After 7 d of depuration, the brains recovered their normal structure.	Guzmán-Guillén *et al.*, 2015c
Tilapias (*Oreochromis niloticus*)	CYN from lyophilized cells from an *Aphanizomenon ovalisporum*	Immersion to cells using 10 µg CYN/L during 14 d followed by no depuration or by 3, or 7 d of depuration.	Gills were totally recovered after 3 d of depuration, 7 d is effective for the liver, kidney and GI tract. A longer depuration might be necessary for a full recovery of the heart.	Guzmán-Guillén *et al.*, 2016

Abbreviations: b.w.: body weight; G.I.: gastrointestinal; IHC: Immunohistochemical analysis.

After acute exposure to CYN by i.p. injection, all the pathological changes mentioned above were more pronounced, with the exception of the g.i. tract [104], and the immunohistochemical analysis demonstrated more CYN-immunoreactive cells by this route in comparison to oral exposure [108]. Subchronic exposure of tilapias by immersion to lyophilized cells of CYN-producing *A. ovalisporum* (10-100 μg CYN/L for 7 or 14 days) demonstrated damage in all organs at 10 μg CYN/L [109], including neurotoxic alterations in brains of exposed fish [110]. This could pose a high risk for fish. The extent of the damage according to histopathological findings is related to the CYN concentration and to the length of exposure. These changes were more intense in subchronic immersion exposure than with acute dose of CYN (200 μg CYN/kg bw fish) [104].

Recent studies demonstrated the usefulness of some depuration processes (3 or 7 days) in reversing the above-mentioned damage in gills (3 d), in liver, kidney, g.i. tract and brain (7d), whereas longer depuration periods are needed for a full recovery of the heart [110, 111]. The scarce morphometric studies found in the scientific literature indicated that the average nuclear diameter of hepatocytes, and cross sections of proximal and distal convoluted tubules may be useful for quantifying the extent or the severity of CYN in these target organs, namely the liver and kidney [104, 109, 111].

In summary, in tilapias (*O. niloticus*), the only fish species investigated, a variability in the intensity of effects to the toxic insult induced by CYN has been demonstrated, and it might be due to the different concentrations and exposure periods considered, such as single dose (200 or 400 μg CYN/kg bw fish) *vs* subchronic concentrations of CYN (10 or 100 μg CYN/L), the route assayed (oral, immersion, or i.p. injection), and the type of material administered (pure standard CYN vs CYN, deoxy-CYN and other bioactive substances from CYN-producers cyanobacterial cells).

CONCLUDING REMARKS

The current literature indicates that both cyanobacterial toxins, MCs and CYN, bioaccumulate in a wide range of fish species and induce toxic injuries. Studies dealing with different congeners of MCs, such as MC-RR or MC-YR, and also with CYN are still very rare in comparison to MC-LR. Therefore, further *in vivo* experiments should be performed in different fish species to study the potential toxicity differences induced by these MCs-congeners and CYN, depending on the experimental model, to confirm the influence of the route, dose and time of exposure, and to validate if the depuration periods are effective for detoxification of contaminated fish. Moreover, exposure studies of CYN by immersion of fish under natural conditions where cyanobacterial blooms with MCs or especially

CYN-producer species predominate would be of great interest, as well as studies exploring the potential synergistic effects of both cyanotoxins.

CONFLICT OF INTEREST

The author (editor) declares no conflict of interest, financial or otherwise.

ACKNOWLEDGEMENTS

The authors would like to acknowledge the Ministerio de Economía y Competitividad of Spain (AGL2015-64558-R, MINECO/FEDER, UE) for its financial support and the European Cooperation in Science and Technology, COST Action ES 1105 "CYANOCOST" for adding value to this work through networking and knowledge sharing with European researchers. We are also grateful to Dr. James S. Metcalf for the valuable comments and language correction.

REFERENCES

[1] Whitton BA, Potts M. Introduction to the cyanobacteria. 2012. [http://dx.doi.org/10.1007/978-94-007-3855-3_1]

[2] Paerl HW, Otten TG. Harmful cyanobacterial blooms: causes, consequences, and controls. Microb Ecol 2013; 65(4): 995-1010. [http://dx.doi.org/10.1007/s00248-012-0159-y] [PMID: 23314096]

[3] Chorus I, Bartram J. Toxic cyanobacteria in water: a guide to their public health consequences, monitoring and management: E&FN Spon. 2012.

[4] Oliver RL, Ganf GG. Freshwater blooms The ecology of cyanobacteria. Springer 2000.

[5] O'Neil JM, Davis TW, Burford MA, Gobler CJ. The rise of harmful cyanobac-teria blooms: The potential roles of eutrophication and climate change. Harmful Algae 2012; 14: 313-34. [http://dx.doi.org/10.1016/j.hal.2011.10.027]

[6] Codd G, Beattie K. Cyanobacteria (blue-green algae) and their toxins: aware-ness and action in the United Kingdom. PHLS Microbiol Dig 1991; 8: 82-6.

[7] Meriluoto J, Spoof L, Codd G. Handbook of Cyanobacterial Monitoring and Cyanotoxin Analysis. John Wiley & Sons 2016. [http://dx.doi.org/10.1002/9781119068761]

[8] Codd GA, Lindsay J, Young FM, Morrison LF, Metcalf JS. Harmful cyano-bacteria (From mass mortalities to management measures).Harmful cyanobacteria. Netherlands: Springer 2005; pp. 1-23. [http://dx.doi.org/10.1007/1-4020-3022-3_1]

[9] Akcaalan R, Young FM, Metcalf JS, Morrison LF, Albay M, Codd GA. Microcystin analysis in single filaments of Planktothrix spp. in laboratory cultures and environmental blooms. Water Res 2006; 40(8): 1583-90. [http://dx.doi.org/10.1016/j.watres.2006.02.020] [PMID: 16597454]

[10] Falconer IR. Cyanobacteria□Toxins in Drinking Water. Wiley Online Library 2003. [http://dx.doi.org/10.1002/0471263397.env025]

[11] Kaebernick M, Neilan BA. Ecological and molecular investigations of cyanotoxin production. FEMS Microbiol Ecol 2001; 35(1): 1-9. [http://dx.doi.org/10.1111/j.1574-6941.2001.tb00782.x] [PMID: 11248384]

[12] Mankiewicz J, Tarczynska M, Walter Z, Zalewski M. Natural toxins from cyanobacteria. Acta Biol Cracov Bot 2003; 45: 9-20.

[13] Rastogi RP, Sinha RP, Incharoensakdi A. The cyanotoxin-microcystins: cur-rent overview. Rev Environ Sci Biotechnol 2014; 13: 215-49.
[http://dx.doi.org/10.1007/s11157-014-9334-6]

[14] Akcaalan R, Mazur-Marzec H, Zalewska A, Albay M. Phenotypic and toxico-logical characterization of toxic Nodularia spumigena from a freshwater lake in Turkey. Harmful Algae 2009; 8: 273-8.
[http://dx.doi.org/10.1016/j.hal.2008.06.007]

[15] de la Cruz AA, Hiskia A, Kaloudis T, *et al.* A review on cylindrospermopsin: the global occurrence, detection, toxicity and degradation of a potent cyanotoxin. Environ Sci Process Impacts 2013; 15(11): 1979-2003.
[http://dx.doi.org/10.1039/c3em00353a] [PMID: 24056894]

[16] Akcaalan R, Köker L, Oğuz A, Spoof L, Meriluoto J, Albay M. First report of cylindrospermopsin production by two cyanobacteria (Dolichospermum mendotae and Chrysosporum ovalisporum) in Lake Iznik, Turkey. Toxins (Basel) 2014; 6(11): 3173-86.
[http://dx.doi.org/10.3390/toxins6113173] [PMID: 25412284]

[17] Bruno M, Ploux O, Metcalf JS, Mejean A, Pawlik-Skowronska B, Furey A. Anatoxin-a, Homoanatoxin-a, and Natural Analogues. Jussi Meriluoto LSGAC Handbook of Cyanobacterial Monitoring and Cyanotoxin Analy-sis. John Wiley & Sons 2016; pp. 138-44.

[18] Gutiérrez-Praena D, Jos Á, Pichardo S, Moreno IM, Cameán AM. Presence and bioaccumulation of microcystins and cylindrospermopsin in food and the effectiveness of some cooking techniques at decreasing their concentrations: a review. Food Chem Toxicol 2013; 53: 139-52.
[http://dx.doi.org/10.1016/j.fct.2012.10.062] [PMID: 23200893]

[19] Sotton B, Guillard J, Anneville O, Maréchal M, Savichtcheva O, Domaizon I. Trophic transfer of microcystins through the lake pelagic food web: evidence for the role of zooplankton as a vector in fish contamination. Sci Total Environ 2014; 466-467: 152-63.
[http://dx.doi.org/10.1016/j.scitotenv.2013.07.020] [PMID: 23906853]

[20] Landsberg JH. The effects of harmful algal blooms on aquatic organisms. Rev Fish Sci 2002; 10: 113-390.
[http://dx.doi.org/10.1080/20026491051695]

[21] Malbrouck C, Kestemont P. Effects of microcystins on fish. Environ Toxicol Chem 2006; 25(1): 72-86.
[http://dx.doi.org/10.1897/05-029R.1] [PMID: 16494227]

[22] Funari E, Testai E. Human health risk assessment related to cyanotoxins exposure. Crit Rev Toxicol 2008; 38(2): 97-125.
[http://dx.doi.org/10.1080/10408440701749454] [PMID: 18259982]

[23] Ibelings BW, Bruning K, de Jonge J, *et al.* Distribution of microcystins in a lake foodweb: no evidence for biomagnification. Microb Ecol 2005; 49(4): 487-500.
[http://dx.doi.org/10.1007/s00248-004-0014-x] [PMID: 16052377]

[24] Albay M, Akcaalan R, Tufekci H, Metcalf JM, Beattie KA, Codd GA. Depth profiles of cyanobacterial hepatotoxins (microcystins) in three Turkish freshwater lakes. Hydrobiologia 2003; 505: 89-95.
[http://dx.doi.org/10.1023/B:HYDR.0000007297.29998.5f]

[25] Wood R. Acute animal and human poisonings from cyanotoxin exposure - A review of the literature. Environ Int 2016; 91: 276-82.
[http://dx.doi.org/10.1016/j.envint.2016.02.026] [PMID: 26995270]

[26] Oberemm A, Becker J, Codd G, Steinberg C. Effects of cyanobacterial toxins and aqueous crude extracts of cyanobacteria on the development of fish and amphibians. Environ Toxicol 1999; 14: 77-

88.
[http://dx.doi.org/10.1002/(SICI)1522-7278(199902)14:1<77::AID-TOX11>3.0.CO;2-F]

[27] Halliwell B, Gutteridge JM. Free radicals in biology and medicine. 4th ed., New York: Oxford University Press 2007.

[28] Jones DP. Redefining oxidative stress. Antioxid Redox Signal 2006; 8(9-10): 1865-79. [http://dx.doi.org/10.1089/ars.2006.8.1865] [PMID: 16987039]

[29] Amado LL, Monserrat JM. Oxidative stress generation by microcystins in aquatic animals: why and how. Environ Int 2010; 36(2): 226-35. [http://dx.doi.org/10.1016/j.envint.2009.10.010] [PMID: 19962762]

[30] Seki H, Takahashi M, Hara Y, Ichimura S. Dynamics of dissolved oxygen during algal bloom in lake Kasumigaura. Japan. Water Res 1980; 14: 179-83. [http://dx.doi.org/10.1016/0043-1354(80)90235-3]

[31] Pavagadhi S, Balasubramanian R. Toxicological evaluation of microcystins in aquatic fish species: current knowledge and future directions. Aquat Toxicol 2013; 142-143: 1-16. [http://dx.doi.org/10.1016/j.aquatox.2013.07.010] [PMID: 23948073]

[32] Pflugmacher S, Wiegand C, Oberemm A, *et al.* Identification of an enzymatically formed glutathione conjugate of the cyanobacterial hepatotoxin microcystin-LR: the first step of detoxication. Biochim Biophys Acta 1998; 1425(3): 527-33. [http://dx.doi.org/10.1016/S0304-4165(98)00107-X] [PMID: 9838216]

[33] Li XY, Chung IK, Kim JI, Lee JA. Oral exposure to Microcystis increases activity-augmented antioxidant enzymes in the liver of loach (Misgurnus mizolepis) and has no effect on lipid peroxidation. Comp Biochem Physiol C Toxicol Pharmacol 2005; 141(3): 292-6. a [http://dx.doi.org/10.1016/j.cca.2005.07.004] [PMID: 16055386]

[34] Prieto AI, Jos A, Pichardo S, Moreno I, Cameán AM. Differential oxidative stress responses to microcystins LR and RR in intraperitoneally exposed tilapia fish (Oreochromis sp.). Aquat Toxicol 2006; 77(3): 314-21. [http://dx.doi.org/10.1016/j.aquatox.2005.12.012] [PMID: 16455146]

[35] Cazenave J, Bistoni MdeL, Pesce SF, Wunderlin DA. Differential detoxification and antioxidant response in diverse organs of Corydoras paleatus experimentally exposed to microcystin-RR. Aquat Toxicol 2006; 76(1): 1-12. [http://dx.doi.org/10.1016/j.aquatox.2005.08.011] [PMID: 16263184]

[36] Pavagadhi S, Gong Z, Hande MP, Dionysiou DD. Cruz AAdl, Balasubrama-nian R. Biochemical response of diverse organs in adult Danio rerio (zebrafish) exposed to sub-lethal concentrations of microcystin-LR and mi-crocystin-RR: A balneation study. Aquat Toxicol 2012; 109: 1-10. [http://dx.doi.org/10.1016/j.aquatox.2011.11.009] [PMID: 22207040]

[37] Jos A, Pichardo S, Prieto AI, *et al.* Toxic cyanobacterial cells containing microcystins induce oxidative stress in exposed tilapia fish (Oreochromis sp.) under laboratory conditions. Aquat Toxicol 2005; 72(3): 261-71. [http://dx.doi.org/10.1016/j.aquatox.2005.01.003] [PMID: 15820106]

[38] Prieto AI, Pichardo S, Jos A, Moreno I, Cameán AM. Time-dependent oxidative stress responses after acute exposure to toxic cyanobacterial cells containing microcystins in tilapia fish (Oreochromis niloticus) under laboratory conditions. Aquat Toxicol 2007; 84(3): 337-45. [http://dx.doi.org/10.1016/j.aquatox.2007.06.012] [PMID: 17669513]

[39] Falconer IR. Cyanobacterial toxins present in Microcystis aeruginosa extracts--more than microcystins! Toxicon 2007; 50(4): 585-8. [http://dx.doi.org/10.1016/j.toxicon.2007.03.023] [PMID: 17509636]

[40] Qiu T, Xie P, Li L, Guo L, Zhang D, Zhou Q. Nephrotoxic effects from chronic toxic cyanobacterial blooms in fishes with different trophic levels in a large Chinese lake. Environ Toxicol Pharmacol

2012; 33(2): 252-61.
[http://dx.doi.org/10.1016/j.etap.2011.12.004] [PMID: 22240185]

[41] Sun H, Wang W, Geng L, Chen Y, Yang Z. In situ studies on growth, oxidative stress responses, and gene expression of juvenile bighead carp (Hypophthalmichthys nobilis) to eutrophic lake water dominated by cyanobacterial blooms. Chemosphere 2013; 93(2): 421-7.
[http://dx.doi.org/10.1016/j.chemosphere.2013.05.022] [PMID: 23769464]

[42] Gavrilovic BR, Despotovic SG, Gavric JP, Borkovic-Mitic SS, Ognjanovic BI, Pavlovic SZ, *et al.* Changes in antioxidant enzyme activities in the livers and gills of three cyprinids after exposure to a cyanobacterial bloom in the Gruza Reservoir, Serbia. Ecol Indic 2014; 38: 141-8.
[http://dx.doi.org/10.1016/j.ecolind.2013.10.016]

[43] Hauser-Davis RA, Lavradas RT, Lavandier RC, Rojas EG, Guarino AW, Ziolli RL. Accumulation and toxic effects of microcystin in tilapia (Oreochromis niloticus) from an eutrophic Brazilian lagoon. Ecotoxicol Environ Saf 2015; 112: 132-6.
[http://dx.doi.org/10.1016/j.ecoenv.2014.10.036] [PMID: 25463863]

[44] Cairns J, Niederlehner BR. Problems associated with selecting the most sensi-tive species for toxicity testing. Hydrobiologia 1987; 153: 87-94.
[http://dx.doi.org/10.1007/BF00005507]

[45] Qiu T, Xie P, Ke Z, Li L, Guo L. In situ studies on physiological and biochemical responses of four fishes with different trophic levels to toxic cyanobacterial blooms in a large Chinese lake. Toxicon 2007; 50(3): 365-76.
[http://dx.doi.org/10.1016/j.toxicon.2007.04.006] [PMID: 17572463]

[46] Sun H, Lü K, Minter EJ, Chen Y, Yang Z, Montagnes DJ. Combined effects of ammonia and microcystin on survival, growth, antioxidant responses, and lipid peroxidation of bighead carp Hypophthalmythys nobilis larvae. J Hazard Mater 2012; 221-222: 213-9.
[http://dx.doi.org/10.1016/j.jhazmat.2012.04.036] [PMID: 22560242]

[47] Jiang J, Gu X, Song R, *et al.* Time-dependent oxidative stress and histopathological changes in Cyprinus carpio L. exposed to microcystin-LR. Ecotoxicology 2011; 20(5): 1000-9.
[http://dx.doi.org/10.1007/s10646-011-0646-9] [PMID: 21455607]

[48] Jiang J, Shi Y, Shan Z, Yang L, Wang X, Shi L. Bioaccumulation, oxidative stress and HSP70 expression in Cyprinus carpio L. exposed to microcystin-LR under laboratory conditions. Comp Biochem Physiol C Toxicol Pharmacol 2012; 155(3): 483-90.
[http://dx.doi.org/10.1016/j.cbpc.2011.12.008] [PMID: 22240489]

[49] Puerto M, Gutiérrez-Praena D, Prieto AI, *et al.* Subchronic effects of cyanobacterial cells on the transcription of antioxidant enzyme genes in tilapia (Oreochromis niloticus). Ecotoxicology 2011; 20(2): 479-90. a
[http://dx.doi.org/10.1007/s10646-011-0600-x] [PMID: 21279735]

[50] Faltermann S, Grundler V, Gademann K, Pernthaler J, Fent K. Comparative effects of nodularin and microcystin-LR in zebrafish: 2. Uptake and molecular effects in eleuthero-embryos and adult liver with focus on endoplasmic reticulum stress. Aquat Toxicol 2016; 171: 77-87.
[http://dx.doi.org/10.1016/j.aquatox.2015.12.001] [PMID: 26748408]

[51] Wang M, Chan LL, Si M, Hong H, Wang D. Proteomic analysis of hepatic tissue of zebrafish (Danio rerio) experimentally exposed to chronic microcystin-LR. Toxicol Sci 2010; 113(1): 60-9. a
[http://dx.doi.org/10.1093/toxsci/kfp248] [PMID: 19822598]

[52] Wang M, Wang D, Lin L, Hong H. Protein profiles in zebrafish (Danio rerio) brains exposed to chronic microcystin-LR. Chemosphere 2010; 81(6): 716-24. b
[http://dx.doi.org/10.1016/j.chemosphere.2010.07.061] [PMID: 20800265]

[53] Marie B, Huet H, Marie A, Djediat C, Puiseux-Dao S, Catherine A, *et al.* Effects of a toxic cyanobacterial bloom (Planktothrix agardhii) on fish: Insights from histopathological and quantitative proteomic assessments following the oral exposure of medak. 2012; 114-115: 39-48.

[54] Hilscherová K, Bláha L. Oxidative stress and detoxification biomarker responses in aquatic freshwater vertebrates exposed to microcystins and cy-anobacterial biomass. Environ Sci Pollut Res Int 2012; 19: 2024-37.
[http://dx.doi.org/10.1007/s11356-012-0960-7] [PMID: 22767295]

[55] Terao K, Ohmori S, Igarashi K, *et al.* Electron microscopic studies on experimental poisoning in mice induced by cylindrospermopsin isolated from blue-green alga Umezakia natans. Toxicon 1994; 32(7): 833-43.
[http://dx.doi.org/10.1016/0041-0101(94)90008-6] [PMID: 7940590]

[56] Runnegar MT, Kong SM, Zhong YZ, Lu SC. Inhibition of reduced glutathione synthesis by cyanobacterial alkaloid cylindrospermopsin in cultured rat hepatocytes. Biochem Pharmacol 1995; 49(2): 219-25.
[http://dx.doi.org/10.1016/S0006-2952(94)00466-8] [PMID: 7840799]

[57] Humpage AR, Fenech M, Thomas P, Falconer IR. Micronucleus induction and chromosome loss in transformed human white cells indicate clastogenic and aneugenic action of the cyanobacterial toxin, cylindrospermopsin. Mutat Res 2000; 472(1-2): 155-61.
[http://dx.doi.org/10.1016/S1383-5718(00)00144-3] [PMID: 11113708]

[58] Humpage AR, Fontaine F, Froscio S, Burcham P, Falconer IR. Cylindrospermopsin genotoxicity and cytotoxicity: role of cytochrome P-450 and oxidative stress. J Toxicol Environ Health A 2005; 68(9): 739-53.
[http://dx.doi.org/10.1080/15287390590925465] [PMID: 16020200]

[59] Runnegar MT, Kong SM, Zhong YZ, Ge JL, Lu SC. The role of glutathione in the toxicity of a novel cyanobacterial alkaloid cylindrospermopsin in cultured rat hepatocytes. Biochem Biophys Res Commun 1994; 201(1): 235-41.
[http://dx.doi.org/10.1006/bbrc.1994.1694] [PMID: 8198579]

[60] Puerto M, Jos A, Pichardo S, Gutiérrez-Praena D, Cameán AM. Acute effects of pure cylindrospermopsin on the activity and transcription of antioxidant enzymes in tilapia (Oreochromis niloticus) exposed by gavage. Ecotoxicology 2011; 20(8): 1852-60. b
[http://dx.doi.org/10.1007/s10646-011-0723-0] [PMID: 21695509]

[61] Puerto M, Jos A, Pichardo S, Moyano R, Blanco A, Cameán AM. Acute exposure to pure cylindrospermopsin results in oxidative stress and pathological alterations in tilapia (Oreochromis niloticus). Environ Toxicol 2014; 29(4): 371-85.
[http://dx.doi.org/10.1002/tox.21764] [PMID: 22331699]

[62] Gutiérrez-Praena D, Jos A, Pichardo S, Cameán AM. Oxidative stress responses in tilapia (Oreochromis niloticus) exposed to a single dose of pure cylindrospermopsin under laboratory conditions: influence of exposure route and time of sacrifice. Aquat Toxicol 2011; 105(1-2): 100-6.
[http://dx.doi.org/10.1016/j.aquatox.2011.05.015] [PMID: 21703978]

[63] Guzmán-Guillén R, Prieto AI, Vázquez CM, Vasconcelos V, Cameán AM. The protective role of l-carnitine against cylindrospermopsin-induced oxidative stress in tilapia (Oreochromis niloticus). Aquat Toxicol 2013a; 132– 133: 141-50.

[64] Guzmán-Guillén R, Prieto AI, Vasconcelos VM, Cameán AM. Cyanobacterium producing cylindrospermopsin cause oxidative stress at environmentally relevant concentrations in sub-chronically exposed tilapia (Oreochromis niloticus). Chemosphere 2013; 90(3): 1184-94. b
[http://dx.doi.org/10.1016/j.chemosphere.2012.09.027] [PMID: 23072785]

[65] Guzmán-Guillén R, Prieto AI, Moreno I, Ríos V, Vasconcelos VM, Cameán AM. Effects of depuration on oxidative biomarkers in tilapia (Oreochromis niloticus) after subchronic exposure to cyanobacterium producing cylindrospermopsin. Aquat Toxicol 2014; 149: 40-9. a
[http://dx.doi.org/10.1016/j.aquatox.2014.01.026] [PMID: 24561425]

[66] Sieroslawska A, Rymuszka A. Cylindrospermopsin induces oxidative stress and genotoxic effects in the fish CLC cell line. J Appl Toxicol 2015; 35(4): 426-33.

[http://dx.doi.org/10.1002/jat.3040] [PMID: 25219470]

[67] Gutiérrez-Praena D, Puerto M, Prieto AI, *et al.* Protective role of dietary N-acetylcysteine on the oxidative stress induced by cylindrospermopsin in tilapia (Oreochromis niloticus). Environ Toxicol Chem 2012; 31(7): 1548-55. a
[http://dx.doi.org/10.1002/etc.1838] [PMID: 22511408]

[68] Ríos V, Remedios Guzmán-Guillén IMM, Prieto AI, Puerto M, Jos A, Cameán AM. Influence of two depuration periods on the activity and transcription of antioxidant enzymes in tilapia exposed to repeated doses of cylindrospermopsin under laboratory conditions. 2014; 6: 1062-79.

[69] Prieto AI, Jos A, Pichardo S, *et al.* Time-dependent protective efficacy of Trolox (vitamin E analog) against microcystin-induced toxicity in tilapia (Oreochromis niloticus). Environ Toxicol 2009; 24(6): 563-79.
[http://dx.doi.org/10.1002/tox.20458] [PMID: 19051281]

[70] Puerto M, Prieto AI, Pichardo S, *et al.* Effects of dietary N-acetylcysteine on the oxidative stress induced in tilapia (Oreochromis niloticus) exposed to a microcystin-producing cyanobacterial water bloom. Environ Toxicol Chem 2009; 28(8): 1679-86.
[http://dx.doi.org/10.1897/08-520.1] [PMID: 19290682]

[71] Guzmán-Guillén R, Prieto Ortega AI, Martín-Caméan A, Cameán AM. Beneficial effects of vitamin E supplementation against the oxidative stress on Cylindrospermopsin-exposed tilapia (Oreochromis niloticus). Toxicon 2015; 104: 34-42. a
[http://dx.doi.org/10.1016/j.toxicon.2015.07.336] [PMID: 26235908]

[72] Zurawell RW, Chen H, Burke JM, Prepas EE. Hepatotoxic cyanobacteria: a review of the biological importance of microcystins in freshwater environments. J Toxicol Environ Health B Crit Rev 2005; 8(1): 1-37.
[http://dx.doi.org/10.1080/10937400590889412] [PMID: 15762553]

[73] Prieto AI, Atencio L, Puerto M, Pichardo S, Jos A, Moreno I, *et al.* Efectos tóxicos producidos por las microcistinas em peces. Rev Toxicol 2008; 25(1-3): 22-31.

[74] Carbis CR, Rawlin GT, Grant P, Mitchell GF, Anderson JW, McCauley I. A study of feral carp, Cyprinus carpio L.,exposed to Microcystis aeruginosa at lake Mokoan, Australia, and possible implications for fish health. J Fish Dis 1997; 20: 81-91.
[http://dx.doi.org/10.1046/j.1365-2761.1997.d01-111.x]

[75] Fischer WJ, Dietrich DR. Pathological and biochemical characterization of microcystin-induced hepatopancreas and kidney damage in carp (Cyprinus carpio). Toxicol Appl Pharmacol 2000; 164(1): 73-81.
[http://dx.doi.org/10.1006/taap.1999.8861] [PMID: 10739746]

[76] Rabergh CM, Bylund G, Eriksson JE. Histopathological effects of microcys-tin-LR, a cyclic peptide toxin from the cyanobacterium (blue-green algae) Microcystis aeruginosa, on common carp (Cyprinus carpio L). Aquat Toxicol 1991; 20: 131-46.
[http://dx.doi.org/10.1016/0166-445X(91)90012-X]

[77] Best JH, Eddy FB, Codd GA. Effects of purified microcystin-LR and cell extracts of Microcystis strains PCC 7813 and CYA 43 on cardiac function in brown trout (Salmo trutta) alevine. Fish Physiol Biochem 2011; 24: 171-8.
[http://dx.doi.org/10.1023/A:1014081827372]

[78] Ferreira M, Oliveira V, Oliveira R. Cunha Pd, Grisolia C, Junior OP. Histo-pathological effects of [D-Leu1]Microcystin-LR variants on liver, skeletal muscle and intestinal tract of Hypophthalmichthys molitrix. Toxicon 2010; 55: 1255-62.
[http://dx.doi.org/10.1016/j.toxicon.2010.01.016] [PMID: 20144637]

[79] Trinchet I, Djediat C, Huet H, Dao SP, Edery M. Pathological modifications following sub-chronic exposure of medaka fish (Oryzias latipes) to microcystin-LR. Reprod Toxicol 2011; 32(3): 329-40.
[http://dx.doi.org/10.1016/j.reprotox.2011.07.006] [PMID: 21839164]

[80] Hou J, Li L, Xue T, Long M, Su Y, Wu N. Damage and recovery of the ovary in female zebrafish i.p.-injected with MC-LR. Aquat Toxicol 2014; 155: 110-8. [http://dx.doi.org/10.1016/j.aquatox.2014.06.010] [PMID: 25005048]

[81] Zhao Y, Xie L, Yan Y. Microcystin-LR impairs zebrafish reproduction by affecting oogenesis and endocrine system. Chemosphere 2015; 120: 115-22. [http://dx.doi.org/10.1016/j.chemosphere.2014.06.028] [PMID: 25014902]

[82] Zhang X, Xie P, Wang W, *et al.* Dose-dependent effects of extracted microcystins on embryonic development, larval growth and histopathological changes of southern catfish (Silurus meridionalis). Toxicon 2008; 51(3): 449-56. [http://dx.doi.org/10.1016/j.toxicon.2007.11.005] [PMID: 18166207]

[83] Liu W, Qiao Q, Chen Y, Wu K, Zhang X. Microcystin-LR exposure to adult zebrafish (Danio rerio) leads to growth inhibition and immune dysfunction in F1 offspring, a parental transmission effect of toxicity. Aquat Toxicol 2014; 155: 360-7. [http://dx.doi.org/10.1016/j.aquatox.2014.07.011] [PMID: 25105566]

[84] Kotak BG, Semalulu S, Fritz DL, Prepas EE, Hrudey SE, Coppock RW. Hepatic and renal pathology of intraperitoneally administered microcystin-LR in rainbow trout (Oncorhynchus mykiss). Toxicon 1996; 34(5): 517-25. [http://dx.doi.org/10.1016/0041-0101(96)00009-8] [PMID: 8783446]

[85] Tencalla F, Dietrich D. Biochemical characterization of microcystin toxicity in rainbow trout (Oncorhynchus mykiss). Toxicon 1997; 35(4): 583-95. [http://dx.doi.org/10.1016/S0041-0101(96)00153-5] [PMID: 9133713]

[86] Palikova M, Navratil S, Tichy F, Sterba F, Marsalek B, Blaha L. Histopathol-ogy of carp (Cyprinus carpio L.) larvae exposed to cyanobacteria extract. Acta Vet Brno 2004; 73: 253-7. [http://dx.doi.org/10.2754/avb200473020253]

[87] Acuña S, Deng DF, Lehman P, Teh S. Sublethal dietary effects of Microcystis on Sacramento splittail, Pogonichthys macrolepidotus. Aquat Toxicol 2012; 110-111: 1-8. [http://dx.doi.org/10.1016/j.aquatox.2011.12.004] [PMID: 22245626]

[88] Atencio L, Moreno I, Pichardo S, Jos A, Moyano R, Blanco A, *et al.* Dose-dependent antioxidant responses and pathological changes in Tenca (Tinca tinca) after acute oral exposure to Microcystins from toxic cyanobacterial blooms. Toxicon 2008; 52: 1-12. a [http://dx.doi.org/10.1016/j.toxicon.2008.05.009] [PMID: 18588906]

[89] Andersen RJ, Luu HA, Chen DZ, *et al.* Chemical and biological evidence links microcystins to salmon 'netpen liver disease'. Toxicon 1993; 31(10): 1315-23. [http://dx.doi.org/10.1016/0041-0101(93)90404-7] [PMID: 8303725]

[90] Vajcová V, Mavrátil S, Palíková M. The effect of intraperitoneally applied pure microcystin LR on haematological biochemical and morphological indi-ces of silver carp (Hypophthalmichthys molitrix Val.). Acta Vet Brno 1998; 67: 281-7. [http://dx.doi.org/10.2754/avb199867040281]

[91] Bury N, McGeer J, Eddy F, Codd G. Liver damage in brown trout, Salmo trutta L., and rainbow trout, Oncorhynchus mykiss (Walbaum), following administration of the cyanobacterial hepatotoxin MC-LR *via* the dorsal aor-ta. J Fish Dis 1997; 20: 209-15. [http://dx.doi.org/10.1046/j.1365-2761.1997.00292.x]

[92] Li D, Xie P, Zhang X. Changes in plasma thyroid hormones and cortisol levels in crucian carp (Carassius auratus) exposed to the extracted microcystins. Chemosphere 2008; 74(1): 13-8. [http://dx.doi.org/10.1016/j.chemosphere.2008.09.065] [PMID: 18990424]

[93] Atencio L, Moreno I, Prieto AI, Moyano R, Molina AM, Cameán AM. Acute effects of microcystins MC-LR and MC-RR on acid and alkaline phosphatase activities and pathological changes in intraperitoneally exposed tilapia fish (Oreochromis sp.). Toxicol Pathol 2008; 36(3): 449-58. b

[http://dx.doi.org/10.1177/0192623308315356] [PMID: 18441257]

[94] Li L, Xie P, Lei H, Zhang X. Renal accumulation and effects of intraperitoneal injection of extracted microcystins in omnivorous crucian carp (Carassius auratus). Toxicon 2013; 70: 62-9. [http://dx.doi.org/10.1016/j.toxicon.2013.03.022] [PMID: 23608020]

[95] Woźny M, Lewczuk B, Ziółkowska N, *et al.* Intraperitoneal exposure of whitefish to microcystin-LR induces rapid liver injury followed by regeneration and resilience to subsequent exposures. Toxicol Appl Pharmacol 2016; 313: 68-87. [http://dx.doi.org/10.1016/j.taap.2016.10.014] [PMID: 27765657]

[96] Li L, Xie P, Chen J. In vivo studies on toxin accumulation in liver and ultrastructural changes of hepatocytes of the phytoplanktivorous bighead carp i.p.-injected with extracted microcystins. Toxicon 2005; 46(5): 533-45. b [http://dx.doi.org/10.1016/j.toxicon.2005.06.025] [PMID: 16084552]

[97] Mezhoud K, Bauchet AL, Château-Joubert S, *et al.* Proteomic and phosphoproteomic analysis of cellular responses in medaka fish (Oryzias latipes) following oral gavage with microcystin-LR. Toxicon 2008; 51(8): 1431-9. [http://dx.doi.org/10.1016/j.toxicon.2008.03.017] [PMID: 18471846]

[98] Deng DF, Zheng K, Teh FC, Lehman PW, Teh SJ. Toxic threshold of dietary microcystin (-LR) for quart medaka. Toxicon 2010; 55(4): 787-94. [http://dx.doi.org/10.1016/j.toxicon.2009.11.012] [PMID: 19941879]

[99] Qiao Q, Liu W, Wu K, *et al.* Female zebrafish (Danio rerio) are more vulnerable than males to microcystin-LR exposure, without exhibiting estrogenic effects. Aquat Toxicol 2013; 142-143: 272-82. [http://dx.doi.org/10.1016/j.aquatox.2013.07.002] [PMID: 24076620]

[100] Molina R, Moreno I, Pichardo S, *et al.* Acid and alkaline phosphatase activities and pathological changes induced in Tilapia fish (Oreochromis sp.) exposed subchronically to microcystins from toxic cyanobacterial blooms under laboratory conditions. Toxicon 2005; 46(7): 725-35. [http://dx.doi.org/10.1016/j.toxicon.2005.07.012] [PMID: 16185737]

[101] Ernst B, Hoeger S. OBrien E, Dietrich D. Physiological stress and pathology in European whitefish (Coregonus lavaretus) induced by subchronic expo-sure to environmentally relevant densities of Planktothrix rubescens. Aquat Toxicol 2007; 82: 15-26. [http://dx.doi.org/10.1016/j.aquatox.2007.01.007] [PMID: 17320197]

[102] Dong G, Xie S, Zhu X, *et al.* Responses of yellow catfish (Pelteobagrus fulvidraco Richardson) exposed to dietary cyanobacteria and subsequent recovery. Toxicon 2012; 60(7): 1298-306. [http://dx.doi.org/10.1016/j.toxicon.2012.08.013] [PMID: 22982458]

[103] Liu Y, Song L, Li X, Liu T. The toxic effects of microcystin-LR on embryo-larval and juvenile development of loach, Misguruns mizolepis Gunthe. Toxicon 2002; 40(4): 395-9. [http://dx.doi.org/10.1016/S0041-0101(01)00173-8] [PMID: 11738232]

[104] Gutiérrez-Praena D, Jos A, Pichardo S, *et al.* Time-dependent histopathological changes induced in Tilapia (Oreochromis niloticus) after acute exposure to pure cylindrospermopsin by oral and intraperitoneal route. Ecotoxicol Environ Saf 2012; 76(2): 102-13. b [http://dx.doi.org/10.1016/j.ecoenv.2011.10.008] [PMID: 22019311]

[105] van Apeldoorn ME, van Egmond HP, Speijers GJ, Bakker GJ. Toxins of cyanobacteria. Mol Nutr Food Res 2007; 51(1): 7-60. [http://dx.doi.org/10.1002/mnfr.200600185] [PMID: 17195276]

[106] Gutiérrez-Praena D, Risalde MA, Pichardo S, *et al.* Histopathological and immunohistochemical analysis of Tilapia (Oreochromis niloticus) exposed to cylindrospermopsin and the effectiveness of N-Acetylcysteine to prevent its toxic effects. Toxicon 2014; 78: 18-34. [http://dx.doi.org/10.1016/j.toxicon.2013.11.014] [PMID: 24291634]

[107] Puerto MA, Jos A, Moreno I, Moyano R, Blanco A, Cameán A. Dietary N-Acetylcysteine (NAC) prevents histopathological changes in tilapias (Oreochromis niloticus) exposed to a microcystin-producing cyanobacterial water bloom. Aquaculture 2010; 306: 35-48. [http://dx.doi.org/10.1016/j.aquaculture.2010.06.020]

[108] Guzmán-Guillén R, Gutiérrez-Praena D, Risalde MdeL, *et al.* Immunohistochemical approach to study cylindrospermopsin distribution in tilapia (Oreochromis niloticus) under different exposure conditions. Toxins (Basel) 2014; 6(1): 283-303. b [http://dx.doi.org/10.3390/toxins6010283] [PMID: 24406981]

[109] Guzmán-Guillén R, Prieto AI, Moreno I, *et al.* Cyanobacterium producing cylindrospermopsin cause histopathological changes at environmentally relevant concentrations in subchronically exposed tilapia (Oreochromis niloticus). Environ Toxicol 2015; 30(3): 261-77. b [http://dx.doi.org/10.1002/tox.21904] [PMID: 24000190]

[110] Guzmán-Guillén R, Manzano IL, Moreno IM, *et al.* Cylindrospermopsin induces neurotoxicity in tilapia fish (Oreochromis niloticus) exposed to Aphanizomenon ovalisporum. Aquat Toxicol 2015; 161: 17-24. c [http://dx.doi.org/10.1016/j.aquatox.2015.01.024] [PMID: 25661706]

[111] Guzmán-Guillén R, Prieto Ortega AI, Moreno IM, *et al.* Effects of depuration on histopathological changes in tilapia (Oreochromis niloticus) after exposure to cylindrospermopsin. Environ Toxicol 2017; 32(4): 1318-32. [http://dx.doi.org/10.1002/tox.22326] [PMID: 27463828]

[112] Sun H, Yang W, Chen Y, Yang Z. Effect of purified microcystin on oxidative stress of silver carp *Hypophthalmichthys molitrix* larvae under different ammonia concentrations. . Biochem Syst Ecol 2011; 39: 536-43. [http://dx.doi.org/10.1016/j.bse.2011.08.001]

[113] Li G, Chen J, Xie P, Jiang Y, Wu L, Zhang X. Protein expression profiling in the zebrafish (*Danio rerio*) embryos exposed to the microcystin-LR. Proteomics 2011; 11(10): 2003-18. [http://dx.doi.org/10.1002/pmic.201000442] [PMID: 21472853]

[114] Gélinas M, Juneau P, Gagné F. Early biochemical effects of Microcystis aeruginosa extracts on juvenile rainbow trout (*Oncorhynchus mykiss*). Comp Biochem Physiol B Biochem Mol Biol 2012; 161(3): 261-7. [http://dx.doi.org/10.1016/j.cbpb.2011.12.002] [PMID: 22178707]

[115] Li L, Xie P, Guo L. Antioxidant response in liver of the phytoplanktivorous bighead carp (*Aristichthys nobilis*) intraperitoneally-injected with extracted microcystins. Fish Physiol Biochem 2010; 36(2): 165-72. [http://dx.doi.org/10.1007/s10695-008-9228-z] [PMID: 20467858]

[116] Brzuzan P, Wozny M, Wolinska L, Piasecka A. Expression profiling *in vivo* demonstrates rapid changes in liver microRNA levels of whitefish (*Coregonus lavaretus*) following microcystin-LR exposure. Aquat Toxicol 2012; 122– 123: 188-96.

[117] Okogwu OI, Xie P, Zhao Y, Fan H. Organ-dependent response in antioxidants, myoglobin and neuroglobin in goldfish (*Carassius auratus*) exposed to MC-RR under varying oxygen level. Chemosphere 2014; 112: 427-34. [http://dx.doi.org/10.1016/j.chemosphere.2014.05.011] [PMID: 25048936]

[118] Hou J, Li L, Xue T, Long M, Su Y, Wu N. Hepatic positive and negative antioxidant responses in zebrafish after intraperitoneal administration of toxic microcystin-LR. Chemosphere 2015; 120: 729-36. [http://dx.doi.org/10.1016/j.chemosphere.2014.09.079] [PMID: 25462319]

[119] Shi Y, Jiang J, Shan Z, Bu Y, Deng Z, Cheng Y. Oxidative stress and histopathological alterations in liver of *Cyprinus carpio* L. induced by intraperitoneal injection of microcystin-LR. Ecotoxicology 2015; 24(3): 511-9.

[http://dx.doi.org/10.1007/s10646-014-1399-z] [PMID: 25586193]

[120] Gutiérrez-Praena D, Jos A, Pichardo S, Puerto M, Cameán AM. Influence of the exposure way and the time of sacrifice on the effects induced by a single dose of pure Cylindrospermopsin on the activity and transcription of glutathione peroxidase and glutathione-S-transferase enzymes in Tilapia (*Oreochromis niloticus*). Chemosphere 2013; 90(3): 986-92. [http://dx.doi.org/10.1016/j.chemosphere.2012.06.067] [PMID: 22850279]

[121] Li X-Y, Chung I-K, Kim J-I, Lee J-A. Subchronic oral toxicity of microcystin in common carp (*Cyprinus carpio* L.) exposed to *Microcystis* under laboratory conditions. Toxicon 2004; 44(8): 821-7. [http://dx.doi.org/10.1016/j.toxicon.2004.06.010] [PMID: 15530963]

[122] Ernst B, Hoeger SJ, O'Brien E, Dietrich DR. Oral toxicity of the microcystin-containing cyanobacterium *Planktothrix rubescens* in European whitefish (*Coregonus lavaretus*). Aquat Toxicol 2006; 79(1): 31-40. a [http://dx.doi.org/10.1016/j.aquatox.2006.04.013] [PMID: 16806524]

CHAPTER 9

Review on Rainbow Trout Desert Farming Using Underground Brackish Water

Morteza Alizadeh[1,*], **Shahram Dadgar**[1] and **Shohre Masaeli**[2]

[1] *Iranian Fisheries Science Research Institute, Agricultural Research, Education and Extension Organization, Tehran, Iran*

[2] *Fisheries Office of Isfahan Agriculture organization, Isfahan, Iran*

Abstract: Inland saline aquaculture in salt-affected areas presents an opportunity for income modification and a potentially productive use of land that cannot support traditional agriculture production and the investment levels are characteristically low. Such systems need to be developed in a manner that both prevents further degradation of agricultural land and provide opportunities for an alternative and sustained economic growth of rural communities. Most central areas of Iran are at high risk of salinization through surface water. Using saline groundwater for aquaculture production is a potential adaptive use of these otherwise degraded resources. Expansion of aquaculture in these areas is limited by some factors such as shortage of suitable sites and strict environmental regulations. These limitations, with an abundance of salt-affected land and water resources, have led to the logical progression of investigating the suitability of these resources for aquaculture. Rainbow trout, which could well adapt to rapid changes in salinity, has been promoted as a potential candidate for aquaculture in these areas. Through using suitable production systems well-set to climate conditions, trout desert farming using brackish water could be a profitable method to develop inland aquaculture in Iran.

Keywords: Aquaculture, Earth pond, Inland saline water, Rainbow trout, Underground brackish water.

INTRODUCTION

Secondary salinization is a major environmental problem in many arid and semi-arid lands of the world, adversely affecting up to 380 million hectares of land, including 100 million hectares of arable land [1, 2].

Inland saline aquaculture using the groundwater in the salty areas offers the possibility for positive social, economic, and environmental outcomes. The issues

* **Corresponding author Morteza Alizadeh:** Iranian Fisheries Science Research Institute, Agricultural Research, Education and Extension Organization, Tehran, Iran; Tel: +989135214228; E-mail: m_alizadeh47@yahoo.com

Panagiotis Berillis (Ed)

are closely interlinked. From a social perspective, declines in crop production on salt-affected farms reduce income and decrease the capital value, forcing many farmers to leave their homeland, thereby compromising the structure and capability of rural communities. Beresford *et al.* [3], described increased incidence of stress and depression in rural communities as a consequence of the economic impacts of salinization. New economically viable industries such as inland saline aquaculture could capitalize upon an established workforce experience in animal husbandry and agribusiness, thus helping to maintain the establishments of the rural communities. From an environmental perspective, inland saline aquaculture production is typically an adaptive approach to salinity and will therefore, indirectly remediate this environmental problem [4, 5]. Salt-affected area is meaningfully more economic to exploit than other types of land and is associated with an added benefit of being typically freehold. Where inland saline aquaculture can be effectively combined with salinity prevention schemes, the costs of pumping water can be decreased resulting in substantial investment reductions for both enterprises [6]. The isolation of inland farms from pathogens and parasites may provide additional economic benefits through reduction or elimination of costly disease outbreaks and the production of certified disease-free seed stock. Also, the incidence of certain diseases that occur in fresh water are sometimes reduced by salinity augmentation [7].

More than fifty percent of the world's aquaculture production currently arises in marine or brackish coastal waters [8]. Salinity can change the amount of available energy for growth of fish by altering the energetic cost of ionic and osmotic regulation; however, the relation between salinity and growth is complex and cannot be readily predictable [9]. The main standards for fish species selection suitable for a good inland saline aquaculture are the same for any aquaculture industry. The selected species must be robust, should have a rapid rate of growth, well-established hatchery techniques, and a good marketability. Rainbow trout (*O. mykiss*) appears to be well-adapted to rapid changes in salinity and is often directly transferred from fresh to oceanic waters in aquaculture grow-out situations. Finstad *et al.* [10] reported that *O. mykiss* (40 - 120 g) tolerated direct transfer from freshwater to saline water (26ppt) without any visible signs of stress. Johnsson *et al.* [11] also stated that *O. mykiss* was better able to acclimatize to seawater during winter when water temperatures were lower, although not extreme. Tsintsadze [12] reported that rainbow trout had the highest growth rate in 15–18% salinity (highest tested), slower growth in lower salinities, and had the lowest growth rate in fresh water.

Although some reports have been published over the last years on rainbow trout culture in sea water through different systems [13 - 16], to the best of our knowledge, there is no record on rainbow trout aquaculture by use of inland underground brackish water of particular characteristics especially high temperature and inconstant salt composition.

THE IDEA TO ACTION

Rainbow trout culture has been practiced in Iran since 1960. This activity experienced a slight development and only three farms were constructed till 1985. Then, the development of the rainbow trout aquaculture industry entered into a new phase, with the production of trout rising from about 500 tons in 1990 to more than 70,000 tons in 2009. This significant production was mostly achieved using freshwater resources but in areas with low-rain climate. The abundance of salt-affected land rich in underground saline water resources with different salinities (5-40g/L) brought about brought about the idea of investigating the appropriateness of these resources for aquaculture. In fact, saline underground water usage for aquaculture is a probable adaptive use of this resource.

The first experiment on trout aquaculture in inland underground saline water was conducted in Inland Brackish Water Fish Research Station in 1995 in the central area of Iran where the climate is dry-warm and extensive plains rich in underground brackish water exist. Considering the availability of relatively cheap land in salt-affected areas as well as low construction cost, earth ponds were selected as a culturing system. Since rainbow trout adapts very well to salinity changes and that its production success was frequently experienced in Iran, this species was considered as a good candidate for culture in these areas. Based on the experiments, semi-annual production system in earth ponds during fall and winter was suitable for rainbow trout culture in Iran. Definitely, the key principle in saline aquaculture is water quality requirements for optimal function of a species which must be adapted to the characteristics of a water source [6]. The aim of the current review is to summarize certain key issues in the success of inland saline aquaculture rainbow trout. Due to the importance of product quality and production in unit area, it is necessary to construct an appropriate trout inland brackish water aquaculture according to the behavior and needs of rainbow trout. As fish culture welfare is a growing area of research, this is by no means a comprehensive review covering all aquaculture fish species or indeed all situations where welfare may be a concern but can be expanded as general guidelines to develop rainbow trout aquaculture in inland waters.

STOCKING DENSITY

Stocking density is an essential factor affecting fish welfare in the aquaculture industry, particularly where high production is the target of culture. The stocking density is hard to measure in the field because at any point in time, it will increase as fish grow or decrease following sorting. Stocking density or the conception of space for a fish, is therefore, more complex than for terrestrial species as fish use a three-dimensional medium [17 - 19]. Because fish are dependent on this medium for both physiological and behavioral needs, the welfare concerns associated with stocking density should address both the carrying capacity of the holding environment and the spatial and behavioral needs of the species. Carrying capacity states the maximum number of fish that an environment can supply oxygen and remove metabolic wastes. It is determined by the oxygen consumption rate of the fish and their response to metabolic waste products such as CO_2 and ammonia [17]. Beside the physiological requirements, fish need enough space to show their best normal behavior with minimal distress [20]. Stocking density is, therefore, an area that demonstrates both the importance of species differences and the existence of a complex web of cooperating features that affect fish welfare.

Because of the particular conditions of arid areas, the stocking density of rainbow trout in brackish water earth pond system is an important factor which must be assessed to attain more production in selected breeding system leading to more favorable economic assessment. The best growth performance of rainbow trout was reportedly obtained in a density of 1.5 fish/m^2 in 0.3 ha earth ponds in inland brackish water earth pond which was stocked by rainbow trout juveniles with an initial weight of about 15g for 150 d by 5% water exchange rate [21].

According to some measured environmental factors such as dissolved oxygen, turbidity, Sechi disk depth, and total suspended particles in inland saline systems, it seems that by increasing density per earth pond area unit, water quality will be decreased leading to low growth performance and increased feed conversion ratios. Under such conditions, exchanging pond water more than 5% is inevitable, while the high temperature of well water could probably increase pond water temperature up to thermal tolerance threshold of rainbow trout during the culture period. Therefore, considering the type of used water source which is normally significantly affected by climatic conditions, low stocking density (1-1.5fish/m^2) could generally bring about better access to optimal growth performance. It should be considered by managers that this density should only be performed in cold months in semi-annual production.

PROTEIN AND ENERGY REQUIREMENTS

In the aquaculture industry, feeds represent 40–60% of the production costs. Formulated fish feeds based on fish nutrition requirements have progressed in recent years with an aim to stimulate optimal fish health and growth. Feeding programs should also be modified and developed according to environmental conditions (climate, geography, economic condition, *etc.*). Moreover, nutritional requirements for optimal growth of fish appear to be affected by several parameters such as temperature, salinity, fish age and size, *etc* [22]. Comparison of results from different studies is difficult because of the existence of confusing factors, such as different experimental designs, different diet compositions, feeding protocols, experimental conditions, *etc*. Furthermore, understanding the protein and energy requirements during the grow-out period is a significant factor in fish culture management. Realization of the optimum protein and energy level for cultured fish would help reduce the costs and maximize the feed conversion efficiency [23 - 25].

When introducing rainbow trout in brackish water as a new culturing medium, diet may play an important role in stress sensitivity [26], and therefore it is important to recognize some basic dietary requirements in order to prepare balanced feeds leading to optimum growth and feed utilization. Investigating various ratios of dietary protein to energy (P/E) in growing stage of rainbow trout in brackish water medium, it was shown that by increasing energy level from 370 to 430 Kcal/l00g, weight gain (WG %), average daily growth (ADG %), protein efficiency ratio (PER), apparent net protein utilization (ANPU %), specific growth rate (SGR), condition factor (CF) increased and feed conversion ratio (FCR) decreased [27]. In inland brackish water system, the best growth performance was achieved in 35% protein level and 430kcal/100g digestible energy level.

Most of the published results on rainbow trout in different culture systems mainly in fresh water underlined that the protein level of 35% was the optimum value in the growing stage in moderate energy diet. So, protein level can be decreased as low as minimum needs if the required energy and amino acids are supplied [28 - 30]. Lever *et al.* [31], reported specific growth rates of 1.09 and 1.21%/d from intensive and semi-intensive rainbow trout farms, respectively. Under small-scale experimental conditions, Okumus and Mazlum [32] attained an SGR of 2.0 and an FCR of 1.0 in rainbow trout grown for 22 weeks in water with 7 – 10 ppt salinity.

Altinok and Grizzle [7] showed that rainbow trout grew faster in 3 and 9% salinities with the highest SGR and lowest FCR than in 1% salinities. In 3 and 9% salinity, the digestibility of feed in energy terms was highest for rainbow trout in 9% salinities, so by increasing salinity, digestibility of energy in feed increased for this species. Therefore, the salinity of 9% can be used to increase growth and FCR in rainbow trout. Overall, there is no obvious difference in fresh and brackish water in growth performance.

UTILITY IMPROVEMENT IN CULTURAL EARTH PONDS

The culturing system enhancement for increasing production and economic benefits is an important goal in aquaculture systems. Some technical approaches for utility improvement of earth ponds according to climatic conditions are described as bellows:

Pond Aeration

Aeration, like feeds and feeding, is one of the most frequently discussed topics in aquaculture. This is true because, at any given temperature, the availability of oxygen plus food are considered as factors determining the extent to which oxygen consuming organisms such as finfish and shellfish grow, both in nature and in aquaculture systems. With no sufficient oxygen, such organisms will not grow, regardless of good nutrition and near-optimum environmental conditions. The oxygen levels required to support life, good health and growth vary depending on species, body size and water temperature among other factors. One basic principle of pond aquaculture is that natural aeration as well as biological and chemical processes affecting the concentration of dissolved oxygen and other gasses are normally far exceed anything that can be achieved by mechanical aeration. In fact, mechanical aeration can be used to provide emergency or supplementary oxygen, but such valuable uses can be readily overwhelmed by poor pond design or management practices.

Referring to the specific condition of farms in the desert area, higher growth performance and lower water utilization in inland earth pond could be obtained without aeration [33]. Soil condition in aquaculture should also be considered as importantly as water condition. As there is an equilibrium between water and air, there is also an equilibrium between water and soil. Water quality can be greatly affected by its interaction with soil. Many of the suspended particles found in water are derived from its contact with soil. According to Boyd [34] earth pond aeration caused soil disturbance of benthic sediment and consequently re-suspended them. It can be asserted that under such conditions particularly in the case of adapted water temperature, increasing plankton growth could affect food consumption by fish. It also exaggerates the exchange of water that can cause increases in pond water temperature beyond the tolerance threshold of rainbow trout. In fish culturing earth ponds, where the water is full of biological and chemical activity, the aerators will not be so efficient for rainbow trout farming.

In contrast, different aeration methods are extensively used in earth pond warm water aquaculture. Aeration in these systems provides economic advantages in aquaculture. Nightly aeration of channel catfish to increase production by Hollerman and boyd [35] showed that aerated ponds yielded an average of 5,307 kg/ha of channel catfish and had a net economic profit of $1,500/ha. The unaerated ponds produced an average of 1,400 kg/ha, *i.e.*, a noticeable economic failure. Hybrid carp, *Ctenopharyngodon idella* × *Hypophthalmichthys nobilis*, was stocked into 0·04 ha ponds with and without paddlewheel aeration, and grew at meaningfully greater rates in the paddlewheel ponds [36]. The average shrimp production was 2,852 kg/ha in the aerated ponds and only 2,061 kg/ha in unaerated ponds. The net value of the shrimp crop of the aerated ponds was 42% greater than that of the unaerated ponds [37]. One experiment in America compared aerated catfish ponds with unaerated ponds. The net returns to land, management, and equity capital was found to be 92% higher for the aerated ponds than they were for the unaerated ponds [38]. Nighttime aeration for 5 h improved the growth of tilapia in the treatment with 4cages/pond and increased the carrying capacity in the integrated cage-cum-pond system [39].

Therefore, it appears that the application of aerators such as force group, injecting air in water, is not appropriate for rainbow trout earth ponds where the depth is less than 2 m. Apparently, these devices do not have a major effect either on increasing production or decreasing required water in semi-intensive earth pond systems of rainbow trout farming by using warm-brackish well water. In such environments with this type soil bottom, it can be suggested that the aerator must work at a depth sufficient to oxygenate and the pond bottom soil, keeping it oxygenated and free of deadly toxins that can pollute the water and harm or kill aquatic animals, it must do so without eroding the pond bottom soil though.

Reuse of Earth Ponds Wastewater

Nutrient recycling efficiency and economic characteristics should be explicitly stated when reporting the efficiency of the aquaculture system. Reporting overall purification results (effluent/influent) is not enough. Sound nutrient budgeting is a significant step towards improving the perfrmance of aquaculture plants [40]. So, reuse of wastewater for aquaculture may be a suitable approach to consider multiple exploitation of water resources and development in aquaculture sector through using non-drinkable water to cultivate protein products.

Study on some limnological parameters of rainbow trout brackish water earth pond proved the reasonable quality of waste water as well as water in pond [41, 42]. Taking into consideration the importance of water in the semi-arid area and appropriate quality of the earth ponds wastewater, fish production in an aquaculture wastewater reuse system would be an environmental as well as economic occupation.

According to Alizadeh *et al.* [43], waste water reuse of rainbow trout brackish water earth ponds with the aim of utility improvement of farm led to an increased yield quantity and gross income of the farm up to 31% and 41%, respectively. This additional yield was obtained by constructing a 30 m^3 round cement tank, at the end of earth pond and by regulating the outlet flow water back to them using both pumping and gravity. During the culture period in the cement tank, none of the environmental parameters was critical for rainbow trout. Accordingly, it can be asserted that reusing wastewater of rainbow trout earth pond in the semi-intensive method is a good approach for farmers to better exploit the possibilities in order to achieve more yields and revenues.

Establishing Enclosure Medium (Net Pen)

Fish production from enclosures could be increased through the implementation of a number of strategies, all of which would result in a better utilization of many pressurized resources. Cages and pens have several advantages over other culture methods and they are popular with farmers, extension workers, and development programs. This is because they use existing water bodies, require comparatively low capital investments and use simple technologies. They can be used not only mainly as a method for producing high quality protein cheaply but also, as is happening in Malaysia and Singapore, to clean up eutrophicated waters through the culturing and harvesting of caged planktivorous species [44]. Thus, despite accounting for only 5–10% of current inland water aquaculture production, growth in this sector is rapid.

Establishing enclosure medium in earth pond of rainbow trout to increase the yield of farm is generally performed by separating 20% of earth pond's space *via* a fixed net wall with 15 mm in the mesh. Fish density in enclosure part is 7.5 fish/m^2 so the number of fish in this area is similar to an open pond. The percentage of harvested marketable size (larger than 250g), production net income and total yield have been reportedly increased significantly in the enclosure medium rather than open medium [46].

In semi-intensive rainbow trout farming in large earth pond, the low fish density may lead to ununiformed fish distribution throughout the pond which may result in unequal access to food and consequently dissimilarity in final weight. Culturing fish in enclosed parts of the pond is effective to remove or moderate these limitations in this system. Therefore, through pen culture of rainbow trout in 0.5 ha earth ponds and likely larger ones, utility and production of the farm can be improved considerately.

MATURATION

Growth and development of fish are affected by internal and external parameters such as genetic [47, 48], biologic [49] and environmental conditions. In spite of basic role of genetic and biological factors on growth rate and the correlation between high growth rates and underage maturation [50], water quality among the outer factors, is the most important and it appears that the effects of salinity on growth, gonad development and maturation of fishes are remarkable. There are many reports on gonadal growth and maturation in fresh water, while a few in brackish or saline water for rainbow trout.

Considering rapid growth and maturation of rainbow trout in semi-intensive brackish water earth pond system, undoubtedly seems that salinity plays an operational role in growth and gonad development in rainbow trout. The somatic and gonadal growth of one-year-old rainbow trout with the initial weight of 250g was compared in fresh and brackish water for 180 d. The somatic growth of rainbow trout from culture period until sexual ripening was about 20% higher in brackish water than fresh water. It was also demonstrated that the time required for passing any sexual stage of testis and ovary was different in brackish and fresh water, so the process of gonadal development of rainbow trout was faster in brackish water [51]. According to Albrektsen & Torrissen [52], brackish water environment was principally encouraging for pre-grown generating in waters with higher temperatures. It was also proved that in different light regimes, during the stage before making the yolk and yolk with the internal origin, it affected by environmental factors, but which with external origin was under internal biological phases [53]. Sexual ripening and artificial propagation of two year old rainbow trout (initial weight of 20g) reared in brackish water with a salinity of about 14g/l were investigated and it was shown that a maturity of 80% was achieved in which 55% were female and 45% were male [54]. Fertilized eggs were incubated in fresh water and led to favorable results of 92% hatching. The larvae weight of 250 mg were capable of directing move in brackish water without any strange behavior or mortality [54]. Therefore, it can be claimed that the salinity as a factor along with other environmental, genetic and biological factors, accelerate the growth of reproductive material in a variety of male and female rainbow trout.

CONCLUSION

Inland brackish and saline waters have high potentials for aquaculture. Rainbow trout is a euryhaline fish and considering the results of the studies performed, this is a very important species to develop inland saline aquaculture. Underground saline waters according to the geographic environments vary in salt composition and temperature at the time of pumping. Despite the resistance of trout to salinity, the water temperature has an influential and main role in exploiting such waters for aquaculture. Some investigations have been conducted on the potential of rainbow trout for inland saline aquaculture based on the use of underground saline and warm water commonly found in areas with low altitude and hot climate. The results suggest the utilization of these water sources for trout culture based on the construction of 0.25 ha earth ponds and culture in cold months in semi-annual production. Semi-intensive system (1.5 fish/m^2) and water exchange rate of at least 5% pond water volume in a day have been determined ideal for production and can produce 5-7 tons trout per hectare in each growing period.

To increase production in ponds, aeration is also a significant part of commercial aquaculture. Application of some techniques such as the reuse of earth ponds waste water and establishing enclosure medium (net pen) could also considerably increase productivity of trout farms.

Some of the essential nutritional requirements of trout, including protein, fat, and energy in brackish water are not specifically different from fresh water and the best growth performance has been reportedly attained with diets including 35% crude protein and 20.6% crude fat. In another word, common commercial feeds used for trout have a high efficiency in the brackish water conditions as well.

It was also showed that the somatic and gonadal growth in brackish water is greater than fresh water. Thus, the importance of these water resources for pre-grown generating and meat production while observing water temperature management is considerable. Moreover, inland saline aquaculture can be considered as the transitional environment for creating pre-grown generating herds of trout.

Overall the benefits of rainbow trout desert farming using underground brackish water can be summarized as follows:

1. Aquaculture development in inland saline waters, particularly in unsuitable areas for agriculture.
2. Economic utilization of brackish water as non-consumption water resources.
3. Meat production according to its growing market demand and high economic value.
4. Developing exports and foreign exchange income.
5. Job creation in less developed areas and prevent migration.
6. Increasing the per capita consumption of fish to help improve public health.

CONFLICT OF INTEREST

The author (editor) declares no conflict of interest, financial or otherwise.

ACKNOWLEDGEMENT

Declared none.

REFERENCES

[1] Ghassemi F, Jakeman AJ, Nix HA. Salinisation of land and water resources: human causes, extent, management and case studies. Cab International 1995.

[2] Lambers H. Introduction, dryland salinity: a key environmental issue in southern Australia. Plant Soil 2003; 257(2): 5-7. [http://dx.doi.org/10.1023/B:PLSO.0000003909.80658.d8]

[3] Beresford Q, Bekle H, Phillips H, Mulcock J. The salinity crisis: landscapes, communities and politics. ERA Trial 2004; 2009: 42.

[4] Braaten RO, Flaherty M. Salt balances of inland shrimp ponds in Thailand: implications for land and water salinization. Environ Conserv 2001; 28(04): 357-67. [http://dx.doi.org/10.1017/S0376892901000388]

[5] Starcevich MR, Lymbery AJ, Doupé RG. Potential environmental impacts from farming rainbow trout using inland saline water in Western Australia. Australas J Environ Manage 2003; 10(1): 15-24. [http://dx.doi.org/10.1080/14486563.2003.10648569]

[6] Doupé RG, Alder J, Lymbery AJ. Environmental and product quality in finfish aquaculture development: an example from inland Western Australia. Aquacult Res 1999; 30(8): 595-602. [http://dx.doi.org/10.1046/j.1365-2109.1999.00371.x]

[7] Altinok I, Grizzle JM. Effects of low salinities on Flavobacterium columnare infection of euryhaline and freshwater stenohaline fish. J Fish Dis 2001; 24(6): 361-7. [http://dx.doi.org/10.1046/j.1365-2761.2001.00306.x]

[8] Partridge GJ, Lymbery AJ, George RJ. Finfish mariculture in inland Australia: a review of potential water sources, species, and production systems. J World Aquacult Soc 2008; 39(3): 291-310. [http://dx.doi.org/10.1111/j.1749-7345.2008.00169.x]

[9] Pennell W, Barton BA, Eds. Principles of salmonid culture. Elsevier 1996.

[10] Finstad B, Staurnes M, Reite OB. Effect of low temperature on sea-water tolerance in rainbow trout, *Salmo gairdneri.* Aquaculture 1988; 72(3): 319-28. [http://dx.doi.org/10.1016/0044-8486(88)90220-7]

[11] Johnsson JI, Clarke WC, Blackburn J. Hybridization with domesticated rainbow trout reduces seasonal variation in seawater adaptability of steelhead trout (*Oncorhynchus mykiss*). Aquaculture 1994; 121(1): 73-7. [http://dx.doi.org/10.1016/0044-8486(94)90009-4]

[12] Tsintsadze ZA. Adaptational capabilities of various size-age groups of rainbow trout in relation to gradual changes of salinity. J Ichthyol 1991; 31(3): 31-8.

[13] Sottovia JL. Trout culture in marine waters 1983.

[14] Ford R. Norwegian salmon and trout farming. Mar Fish Rev 1984; 46(3): 44-7.

[15] Cowey CB, Walton MJ. Intermediary metabolism 1989.

[16] Saunders RL. Potential interaction between cultured and wild Atlantic salmon. Aquaculture 1991; 98(1-3): 51-60. [http://dx.doi.org/10.1016/0044-8486(91)90370-M]

[17] Ellis T, Scott S, Bromage N, North B, Porter M. What is stocking density? Trout News 2001; (Jul): 35-7.

[18] Welfare FF. Briefing Paper 2, Fisheries Society of the British Isles.

[19] Conte FS. Stress and the welfare of cultured fish. Appl Anim Behav Sci 2004; 86(3): 205-23. [http://dx.doi.org/10.1016/j.applanim.2004.02.003]

[20] Council FA, Britain G, Spedding CR. Report on the welfare of farmed fish. Farm Animal Welfare

Council 1996.

[21] Nafisi M, Sharifiyan M, Dehmobad D. Culture of rain bow trout (oncorhynchus mykiss) brackish water earth ponds system. Research project final report, IFRO 1998; 45.

[22] Cowey CB. Use of synthetic diets and biochemical criteria in the assessment of nutrient requirements of fish. Journal of the Fisheries Board of Canada 1976; 33(4): 1040-5. [http://dx.doi.org/10.1139/f76-132]

[23] Charles PM, Sebastian SM, Raj MC, Marian MP. Effect of feeding frequency on growth and food conversion of *Cyprinus carpio* fry. Aquaculture 1984; 40(4): 293-300. [http://dx.doi.org/10.1016/0044-8486(84)90170-4]

[24] Sampath K. Preliminary report on the effects of feeding frequency in *Channa striatus*. Aquaculture 1984; 40(4): 301-6. [http://dx.doi.org/10.1016/0044-8486(84)90171-6]

[25] Chiu YN, Sumagaysay NS, Sastrillo MA. Effect of feeding frequency and feeding rate on the growth and feed efficiency of milkfish, *Chanos chanos*, Forsskal, juveniles. Asian Fish Sci 1987.

[26] Merchie G, Lavens P, Verreth J, *et al.* The effect of supplemental ascorbic acid in enriched live food for *Clarias gariepinus* larvae at start feeding. Aquaculture 1997; 151(1): 245-58. [http://dx.doi.org/10.1016/S0044-8486(96)01505-0]

[27] Ahmadi MR, Alizadeh M. Effect of dietary protein and energy levels on rainbow trout (*oncorhynchus mykiss*) reared in brackish water. Iran J Fish Sci 2004; 4(1): 77-88.

[28] Murai T. Protein nutrition of rainbow trout. Aquaculture 1992; 100(1): 191-207. [http://dx.doi.org/10.1016/0044-8486(92)90371-Q]

[29] Wilson RP. Utilization of dietary carbohydrate by fish. Aquaculture 1994; 124(1): 67-80. [http://dx.doi.org/10.1016/0044-8486(94)90363-8]

[30] Kaushik SJ. Influence de la salinité sur le metabolisme azote et le besoin en arginine chez la truite arc-en-ciel (Doctoral dissertation, Ph. D. Thesis, L'Universite Bretagne Occidentale, Brest 230 pp).

[31] Lever C, Lymbery AJ, Doupé RG. Preliminary comparisons of yield and profit achieved from different rainbow trout, *Oncorhynchus mykiss*, production systems in inland Western Australia. J Appl Aquacult 2004; 16(1-2): 63-73. [http://dx.doi.org/10.1300/J028v16n01_05]

[32] Okumuş İ, Mazlum MD. Evaluation of commercial trout feeds: feed consumption, growth, feed conversion, carcass composition and bio-economic analysis. Turk J Fish Aquat Sci 2002; 2(2)

[33] Nafisi M, Sharifiyan M, Akhondi A, Alizadeh M. Increase in production of rain bow trout (oncorhynchus mykiss) brackish water earth ponds through aeration Research project final report. IFRO 2002.

[34] Boyd CE. Pond water aeration systems. Aquacult Eng 1998; 18(1): 9-40. [http://dx.doi.org/10.1016/S0144-8609(98)00019-3]

[35] Hollerman WD, Boyd CE. Nightly aeration to increase production of channel catfish. Trans Am Fish Soc 1980; 109(4): 446-52. [http://dx.doi.org/10.1577/1548-8659(1980)109<446:NATIPO>2.0.CO;2]

[36] Shireman JV, Aldridge FJ, Rottmann RW. Growth and food habits of hybrid carp, Ctenopha*ryngodon idella*× *Aristichthys nobilis*, from aerated and non aerated ponds. J Fish Biol 1983; 23(5): 595-604. [http://dx.doi.org/10.1111/j.1095-8649.1983.tb02939.x]

[37] Martinez-Cordova LR, Porchas-Cornejo MA, Villarreal-Colmenares H, Calderon-Perez JA. Winter culture of yellowleg shrimp *Penaeus californiensis* in aerated ponds with low water exchange. J World Aquacult Soc 1998; 29(1): 120-4. [http://dx.doi.org/10.1111/j.1749-7345.1998.tb00309.x]

[38] Abdalla A, Romaire RP. Effects of timing and duration of aeration on water quality and production of channel catfish. J Appl Aquacult 1996; 6(1): 1-9. [http://dx.doi.org/10.1300/J028v06n01_01]

[39] Yi Y, Lin CK. Effects of biomass of caged Nile tilapia (*Oreochromis niloticus*) and aeration on the growth and yields in an integrated cage-cum-pond system. Aquaculture 2001; 195(3): 253-67. [http://dx.doi.org/10.1016/S0044-8486(00)00558-5]

[40] Das A, Uchimiya H. Oxygen stress and adaptation of a semi-aquatic plant: rice (Oryza sativa). J Plant Res 2002; 115(5): 315-20. [http://dx.doi.org/10.1007/s10265-002-0043-9] [PMID: 12579355]

[41] Kolkovski S, Kolkovski J. Herbal medicine in aquaculture. Int Aqua Feed 2011; 14(2): 28-31.

[42] Castine SA, McKinnon AD, Paul NA, Trott LA, de Nys R. Wastewater treatment for land-based aquaculture: improvements and value-adding alternatives in model systems from Australia. Aquacult Environ Interact 2013; 4: 285-300. [http://dx.doi.org/10.3354/aei00088]

[43] Alizadeh M, Rajabipour F, Sarsangi H. Wastewater reuse of rainbow trout earth ponds to increase production in earth farms Research project final report. IFRO 2004.

[44] Yang Y, Zhou YB, Song XX, *et al.* Integrated Control Strategy of Schistosomiasis in The People's Republic of China: Projects Involving Agriculture, Water Conservancy, Forestry, Sanitation and Environmental Modification. Adv Parasitol 2016; 92: 237-68. [http://dx.doi.org/10.1016/bs.apar.2016.02.004] [PMID: 27137449]

[45] Berglund BE. Early agriculture in Scandinavia: Research problems related to pollen□analytical studies. Norw Archaeol Rev 1985; 18(1-2): 77-90. [http://dx.doi.org/10.1080/00293652.1985.9965412]

[46] Alizadeh M, Bitaraf M, Sarsangi H, Mohamadi M, Mashaiee N, Rajabipour F. Utility improvement of brackish water cultural earth ponds of rainbow trout (oncorhynchus mykiss) through establishing enclosure medium (pen) Research project final report. IFRO 2009.

[47] Glebe BD, Saunders RL. Genetic factors in sexual maturity of cultured Atlantic salmon (*Salmo salar*) parr and adults reared in sea cages. Can Spec Publ Fish Aquat Sci 1986; 89: 24-9.

[48] Ritter JA, Farmer GJ, Misra PK, Goff TR, Bailery JK, Baum ET. Parental influences and smolt size and sex ratio effects on seasonal variation of plasma lipids correlated with vitellogenesis. Comp Biochem Physiol 1988; 118: 183-90.

[49] Chadwick EM, Randall RG, Leger C. Ovarian development of Atlantic salmon (*Salmo salar*) smolts and age at first maturity. Can Spec Publ Fish Aquat Sci 1986; 89: 15-23.

[50] Thorpe JE. Developmental variation in Salmonid populations 1989. [http://dx.doi.org/10.1111/j.1095-8649.1989.tb03073.x]

[51] Falahati-Marvast A, Mojazi-Amiri B, Abtahi B, Alizadeh M. Comparative study on ovarian development of rainbow trout reared in brackish water 2001.

[52] Albrektsen S, Torrissen O. Physiological changes in blood and seminal plasma during the spawning period of maturing rainbow trout held under different temperature and salinity regimes, and the effect on survival of the broodstock and the eyed eggs

[53] Bon E, Corraze G, Kaushik S, Le Menn F. Effects of accelerated photoperiod regimes on the reproductive cycle of the female rainbow trout: I—seasonal variations of plasma lipids correlated with vitellogenesis. Comp Biochem Physiol Part A Physiol 1997; 118(1): 183-90. [http://dx.doi.org/10.1016/S0300-9629(96)00415-X]

[54] Alizadeh M. Cultured breeders of rainbow trout in brackish water, Bafgh,Yazd. IFRO newsletter 2008; (54 & 55): Spring & Summer 2008

CHAPTER 10

Fishing Landings of Crustacean Decapods Their Culture and Problems Associated with Diseases

Kostas Kapiris*

Hellenic Centre of Marine Research, Institute of Marine Biological Resources and Inland Waters, 46,7 km Athens-Sounio, P.O. Box 712, 19013, Anavissos, Athens, Hellas

Abstract: The total amount of crustaceans, mainly the decapods, from fishery and culture consist a serious part of the global fisheries and aquaculture production. Richest and poorest countries are included between the top ten shrimp-producing countries in the world. The management of some shrimp fisheries is effectually carried out and provides the potential profits of fisheries' management. Several problems are connected with shrimps culture, like the increasing demand for this seafood, growing number of unfamiliar diseases, a narrow capability to front the several diseases, inadequate post-harvest treatment of "wild" organisms and an increasing legislative charge regarding the marketing of live animals, indicating a clear warning for the future sustainability of this cultivation.

Keywords: Aquaculture, Crustaceans, Decapods, Deseases, Fishery.

INTRODUCTION

General Remarks on Crusracea

Crustacea present a great range of morphological diversity, more than any other plants or animals. This morphological diversification is the cause that makes the study of crustaceans so enthusiastic. In addition to this this great variation "provokes" some problems regarding their taxonomy and phylogeny. The current estimation of the already depicted species is about 52,000 [1]. This estimation is surely uncertain, as a recent estimation only of ostracods number is 10,000-15,000 whilst the total number of peracarids associated to the reefs has been estimated more than 54,000 [2]. This great number of Crustacea places them in fourth place, after insects, molluscs, and chelicerates, in terms of species diversity. But, in general terms, the morphological inequality is weighty in the Crustacea than in any other taxon.

* **Corresponding author Kostas Kapiris:** Hellenic Centre of Marine Research, Institute of Marine Biological Resources and Inland Waters, 46,7 km Athens-Sounio, P.O. Box 712, 19013, Anavissos, Athens, Hellas; Tel: ++210-9856712; E-mail: kkapir@hcmr.gr

Panagiotis Berillis (Ed)

Decapods, like crabs, lobsters and shrimps are, in the middle of, the most known crustaceans, but this taxonomic order also embraces a great diversification of other species. Due to the great variety of Crustacean's morphology is extraordinarily hard to form an acceptable comprehensive definition [3].

The most important crustaceans for man are primarily decapods. Decapods' fishery has been developed in many areas of the planet and consists valuable source of food. The most economic important decapod is probably the true lobster (Homarus species). Its catch has been decreased in early 20th century due to its overfishing in both the North American and the European waters. Freshwater crustaceans include crayfishes, some river prawns and crabs. Many species have only local market value. Other crustaceans, like the large barnacle (*Balanus psittacus*) (Subclass: Cirripedia) measuring up to 27 centimetres (11 inches), is a very delicious food in South America, while the stalked barnacle (Mitella pollicipes) is eaten in France and Spain also.

Crustaceans play very important roles in the aquatic ecosystems and food webs. For instance, some zooplanktonic organisms—like the copepod *Calanus* and the krill *Euphausia*—graze on the phytoplancton floating in the sea and in turn are eaten by fishes, seabirds, and whales. On the other point, benthic (bottom-dwelling) crustaceans are considered as a source of food for fishes and some whales. Crabs are important predators and the continuing fights between them and their preys induce the evolution of newer adaptations. For example, the massive and decorated shells of marine molluscs are thought to be a protective response to the predatory activities of these decapods; in addition to this the crabs start to present larger and stronger pincers.

Crustaceans also can be parasites, for instance some copepods which parasitize other aquatic animals, like whales and sea anemones. Larger crustaceans are often parasitized by smaller crustaceans, like isopods. Freshwater crustaceans can serve as intermediate hosts, like in the lung fluke, Paragonimus (a flatworm, Phylum Platyhelminthes) [3].

Marine Decapod Crustaceans' Fishery

Marine decapod crustaceans consist the 6.9% of worldwide fish and shellfish landings biomass [4] presenting an upward trend in recent decades [5]. The increased decapod crustaceans catches are mostly referable to the increased fishery effort as a result of decreased fish stocks in a global level [4, 5].

Forty-three percent of all shrimps were cultivated while the majority, or 57%, was caught in the fishery, mainly in the shallower waters (Fig. **1**). Crustaceans make up an incommensurate share of the world's seafood value, due to their high market value. Shrimps consist the most valuable wild fishery crustacean products, followed by the crabs, lobsters and krill.

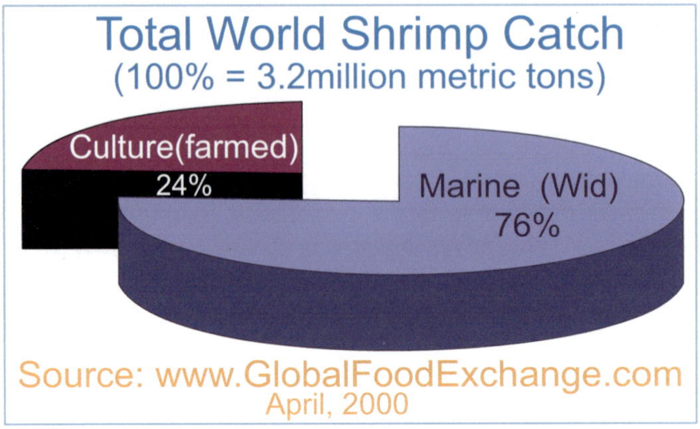

Fig. (1). Total world shrimp catch from fishery and from aquaculture activities.

Crustacean Fishery

A recent FAO study [6] showed that tropical **shrimps caught by commercial trawlers** generally have high discard rates and counting for over 27% of total estimated discards in all the marine fisheries of the world, reaching to 1.8 million tonnes/ year. Garcia (1989) points out that shrimp fishery composes the major source of fishermen conflict in the tropical zone. In spite of this, it is difficult to deny that shrimp fisheries are vital and produce substantial benefits. About 3.4 million tonnes of wild shrimps are yearly caught and this quantity is the most important internationally traded fishery commodity, consisting about 18% of the total value of all this commercial activity. In many tropical developing countries, shrimps occur the most valuable fishery export. The employment aspect is also a very important sector. For example, many years ago when Indonesia, stopped the shrimps' fishery trawling in its waters, a total of 25,000 people lost their jobs.

Shrimp trawling could consist a serious threat to food security, since the bycatch and discards of this fishery is a waste of valuable aquatic food protein source (Fig. **2**). Is is worth saying that the high levels of bycatch and discards in many world's fisheries, including the capture of ecologically important species and juveniles of economically valuable species, present important concerns.

The commercial trawler is a major characteristic gear of most large scale shrimp fishery, although a considerable interest exists to develop an alternative catching method. Recently, most shrimp technology efforts have been towards to improve the above mentioned gear rather than developing new industrial fishery technologies [7].

Fig. (2). A shrimps trawler of Gulf of Mexico-style working in the United States.

Fishing of shrimps generally takes place by means of fixed nets in estuary mouths, during the offshore migration of juveniles, or by otter trawlers in offshore waters where the estuarine nursery areas exist (Fig. **3**). Generally speaking, penaeid shrimps are not catchable by traps. Powered otter trawlers for shrimps have now been adopted worldwide as the main method for industrial scale catching shrimps. These gears can only operate on smooth sand and mud bottoms, adjacent to nursery areas, and typically harvest approximately 20-50% of the shrimps stock. Other gears are generally buried in the sediment, particularly during the day.

Fig. (3). Small wooden trawlers in Southeast Asia.

The bycatch caught in shrimp fisheries [8], by trawlers consists of over several hundred teleosts and outweigh of the shrimp catch by 20 to 1 or more. Thus the ecological damage of the wasted, populations – mainly of endangered and rare species - are threatened, since bycatch is considered as one of the most pressing and controversial aspect of shrimp fishery and much of stock management attention is focused on its reduction [9].

In Fig. (**4**) is given the relative shrimps' production of the ten most productive countries. China and four other Asian countries (India, Indonesia, Malaysia and Thailand) fish the 55% of the world shrimp catch. The most common species of cold-water shrimps, the northern prawn, consists only 12% of the world's shrimp catch. There are large differences among countries concerning the characteristics of shrimp fisheries (Fig. **4**). Major differentiations occur between warm and cold-water shrimps fisheries concerning species, scale of fishing operations, fishing gear, fishery strategies and stocks'management measures. The scope of the most shrimp fisheries worldwide is to produce food for man but in many areas shrimp catches can be used for bait. Recreational shrimping is a significant activity in some developed countries and the capture of broodstock and postlarval shrimp for their farming is important in several countries.

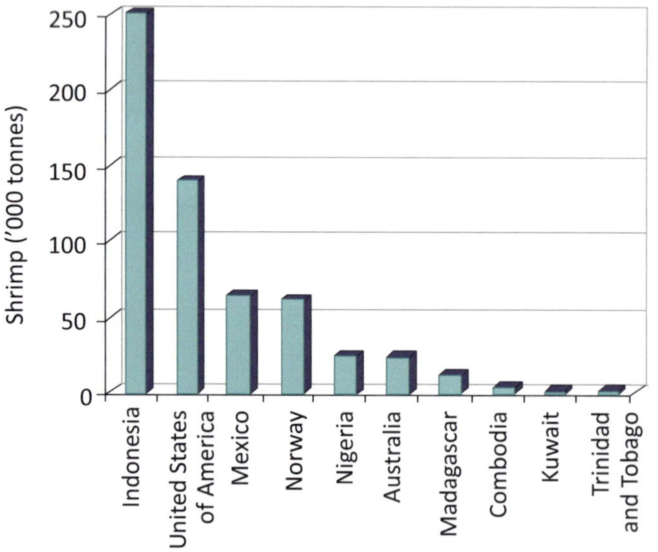

Fig. (4). Mean annual shrimp catches in ten countries in the period 2000-2005.

The major features of various geographic regions and the variety of important shrimp fishing conditions are given in Table **1**.

Table 1. Main characteristics of the shrimp fisheries in a global level.

Country's annual catch[1] and Shrimp fishery characteristics
Australia 24,000 tonnes
Australia has a significant involvement with shrimp fishing and its associated activities. Shrimp fishery produces about 24,000 tonnes of shrimp, mainly in the tropical, subtropical and temperate waters of the land, and ranges in scale from recreational fisheries to large-scale fishery activities using vessels up to 40 m in length. This particular country also produces shrimps from aquaculture and is involved in both the export and import of shrimp in various forms.
Cambodia 3,500 tonnes
Marine fisheries in Cambodia are of minor importance compared with freshwater fisheries, but shrimp fishery is important along its short coast. Trawlers and secondarily other gears, catch from 3 000 to 4 000 tonnes annually. Shrimp is important for domestic consumption and is the most important fishery export of the country.
Indonesia 250,000 tonnes
Indonesia's shrimp catch holds the third place in the world. Shrimp farming is a significant sector in the country, with over 65 000 participating households. Shrimp production, coming from both fishery and aquaculture, has reached over 400 000 tonnes per year, and this fishery activity plays most valuable fishery export in the country's fishery export.
Kuwait 1,500 tonnes
The shrimp fleet of Kuwait consists of two components: 35 steel-hulled double-rigged Gulf of Mexico-type trawlers and 34 dhow trawlers. Three species are the most economically important: green tiger prawn (*Penaeus semisulcatus*) (60 percent of catches), Jinga shrimp (*Metapenaeus affinis*) (30 percent of catches) and Kiddi shrimp (*Parapenaeopsis stylifera*) (10 percent of catches)
Madagascar 12,000 tonnes
Recently industrial, artisanal and traditional fishers in Madagascar have caught from 10 000 to 13 000 shrimps' tonnes. Employment associated with shrimp fishing is essential to the country, and total shrimp production (both captured and farmed) is the most valuable fishery export. In past years, about 5 000 tonnes of shrimp were produced by the aquaculture units.-Shrimps caught in Madagascar have a special identity in Europe and demands a higher price than similar products from Asia and Latin America. About two-thirds of the shrimp landings come from the export-oriented industrial trawl fleet (70 trawlers). A substantial amount of biological, economic and social research associated with shrimp fishery is carried out in the country.
Mexico 65,000 tonnes
Shrimp fishery in Mexico takes place in the Pacific, Gulf of Mexico and Caribbean, by ~~both~~ artisanal and industrial fleets. A huge number of small fishing vessels use many fishery gears to catch shrimp. The larger offshore shrimp vessels, numbering about 2,212, trawlers are using two nets (Pacific side) and four nets (Atlantic side). In 2003, shrimp production in Mexico of 123,905 tonnes came from three sources: 21.26% from artisanal fisheries, 28.41% from industrial fisheries and 50.33% from aquaculture. Shrimp is the basis of the most important fishery products in Mexico regarding the value, exports and employment. Catches of Mexican Pacific shrimps appear to have reached their maximum. It is generally recognized that the overcapacity of the fleet level is a problem in the various shrimp fleets.
Nigeria 25,000 tonnes

(Table 1) contd.....

Country's annual catch[1] and Shrimp fishery characteristics
Nigeria shrimp catch ranges of between 15,000 and 30,000 tonnes/year. There are about 225 industrial shrimp trawlers and a very large number of fishermen inshore using small trawls, beach seines and stow nets. Nigeria, shrimp export is responsible for a substantial amount of employment and is an important significant source of food in coastal areas.
Norway 62,000 tonnes
The total shrimps' catch in Norway ranges between 60,000 and 70,000 tonnes annually, and occupies the 14th largest producer of shrimp in a global basis. Shrimps fishery, however, is not nearly as important as for other species, like herring, blue whiting, cod and saithe. In 2003, shrimp represented about 4% of the economical value of all Norwegian fishery product exports. The main shrimp stocks exploited by fishers from Norway are in the Barents Sea, Skagerrak and the North Sea. The poor profitability of many types of shrimp vessels in Norway makes up a major problem. The management policy of Norwegian shrimp fishery, in national and international waters, is operated in order to avoid both overfishery and bycatch of cod and/or other important species.
Trinidad and Tobago 800 tonnes
Shrimp fishing is currently carried out in Trinidad and Tobago by about 102 artisanal trawlers, ten semi-industrial trawlers and 20-25 industrial trawlers. In the period 1999- 2004, the total annual shrimp catches counted about 825 tonnes. Like in other countries bycatch consists a large amount of shrimps fishery and this effect is one of the most important sources of conflict between the trawl fishery (professional fishermen) and other fisheries' types in the country.
United States of America 140,000 tonnes
Two main types of shrimp fisheries operate in the USA: the fishermen that target warm-water shrimps off the southeast Atlantic coast and the Gulf of Mexico, and those for cold-water shrimps mainly in the NE and NW parts of the country. In terms of value, shrimp is the second most important fishery in the United States, after crabs. The total landings for United States shrimp fisheries have been about 140,000 tonnes per year in the last years, with the warm-water fisheries responsible for over 90% in 2004. United States' shrimps market is now the largest in the world, followed by the European Union. Despite this, real and nominal prices for shrimps have declined, primarily due the lowest values of the imported ones.

[1] These are the average annual catches from 2000 to 2005, rounded to the nearest thousand tonnes.

Shrimp fishery is connected with several economical advantages but also with various high costs and risks. These benefits and costs of the shrimps' culture have been broadly discussed and many comparisons have been already done between them. There is not any final result till now but the complete perspective seems to be positive regarding an integrated and sustainable shrimps culture.

Most **crab** fishery worldwide takes place using traps, because are simple to use (mainly in deeper layers) and labor-efficient, offer more safety to the crabs caught and permits the product to be sold live actually in a higher market price. Many other fishery gears are used to catch crabs, such as trawlers, tangle netting, dredges, trotlines, and drop nets. The majority of the large crab fisheries occurs in tropical and temperate waters of the Northern Hemisphere and in Arctic waters also.

Lobsters and **crayfishes** support significant and commercial fisheries. The most marketable lobsters are marine species and fished from tropical to cold temperate waters, while the freshwater ones are primarily taken from tropical and subtropical regions. This type of fishery dominated by three groups, Homarus species (large-clawed lobsters), Nephrops species (small-clawed lobsters) and the palinurid group (spiny or rock lobsters, without claws). Catches of each of these groups are in the order of 60,000-80,000 tonnes/year. Fishing is generally by baited traps (for Homarus, most palinurid species and freshwater crayfishes), although some are taken by trawlers (Nephrops), and diving (tropical species of the genus Panulirus). The major clawed lobster fishery is for *Homarus americanus* and takes part off eastern Canada and the USA. A similar size fishery for *Nephrops norvegicus* occurs off the European Atlantic coasts and in Mediterranean also. Spiny lobster fisheries for Panulirus species occur through the tropics, with major fisheries in the Caribbean (*P. argus*) and Western Australia (*P. cygnus*). Smaller but not so intense fisheries for Jasus species occur in the temperate waters off southern Australia, New Zealand, and South Africa. The major freshwater crayfish fishery takes place in the southern states of USA. Stocks of these species have generally been resilient to fishing, with the exception of some Jasus species off Africa which have been significantly reduced over time.

Fishing pressure can cause modifications in population structure of both fishery resources and ecosystem function [10], presenting thus trophodynamic consequences [11]. Distinct abiotic and biotic factors may further work upon the distribution and community structure of marine benthic communities, like crustaceans. Some of these abiotic factors are depth profile [12], salinity [13], type of substratum [14] and rainfall patterns [15]. On the other side, important biotic factors encompass environmental productivity [16] and biological proceedings [17].

The management measures applied to the valuable tropical shrimps fisheries concentrate primarily on minimum trawl mesh sizes, local and temporal closures optimizing, thus, the quantity and size of decapods caught, in the frame of the sustainable fishery. Another measure for the management control of the stocks is the effort to decrease of the bycatch and discards catches. The most active measures are the fishing effort control and the transferable effort quotas. For the longer-lived decapods, for example more consistent cold-water pandalid shrimps and krill fisheries, catch quotas are the most appropriate measures and often utilized. Due to their larger body size (and value) and the dominant method of capture (trapping), the common stocks' management focus in crab fisheries is the regulation on legal minimum size and the protection of the spawning females. Gear design directives specifying 'escape gaps' reducing the capture of small size crabs have become a common management tool for this type of fisheries.

Similarly the management techniques for high-value lobster stocks regulates on legal minimum size, gear controls and mature female protection in the trap fisheries. With reference to the past years, limited management measures for sustaining lobster fisheries have been applied. The first measurements were the, 'transferable trap quotas' in the 1960s to the Australian spiny lobster fisheries. Total catch quotas, applied more currently through 'individually transferable quotas', have also been used to control fishing, mainly for the more persictent, longer-lived cold-water lobster fisheries.

Crustacean Aquaculture

Generally speaking the products from aquaculture currently contribute about 48% of aquatic animal food destined for human consumption. The additional supply, including crustaceans, required to meet the request gap should appear from aquaculture. Demand for fish is going up in all regions, due to growing population and to health considerations of fish consumption. There is a clear need for increased supply of fish protein, so that to meet the health needs and general social aspirations for global food security. In order to fulfil these goals and ambitions, significant challenges need to be addressed. A number of important sustainability issues will affect the potential contribution of crustaceans to global food security. The following three sectors are deemed worthy of attention and consideration, namely: sustainable shrimps' fishery, trade and aquaculture.

In the period 2000–2008, the State of World Fisheries and Aquaculture (SOFIA) [18] reported that aquaculture production of all major species continued to be increased. Crustacean production grew at an average annual rate of almost 15% in this period, in comparison to the finfishes and molluscs whose production grew at a lower speed than the period 1990–2000 (Fig. **5**). The rapid increase in crustacean production largely reflects the dramatic increase in white leg shrimp (*Penaeus vannamei*) culture in China, Thailand and Indonesia.

According to SOFIA [10], freshwater fishes continued to dominate in 2008 presenting a production of 28.8 million tons (54.7%) having a total value of US$ 40.5 billion (41.2%), followed by molluscs (13.1 million tons), crustaceans (5 million tons, 23.1% valued at US$ 13.1 billion), diadromous fishes (3.3 million tons), marine fishes (1.8 million tons) and other aquatic animals (0.6 million tons).

Table **2** shows the significant values of crustacean aquaculture production, in terms of quantity and commercial value in 2008 and 2009. Marine shrimps displayed the highest values in terms of quantity and value, followed by marine crabs, lobsters and freshwater crustaceans. Farmed crustaceans construct a significant component of current growth in the aquaculture sector. In the close future their contribution to overall aquatic protein production will be undoubtedly increased. Crustaceans are also important to the approximate annual US$ 300 million world global trade in ornamental aquatic species. The industry connected to the production and consumption of aquatic crustaceans is accurately globalised, with disconnection of net producers and net consumers requiring significant transboundary movements of live animals and their products.

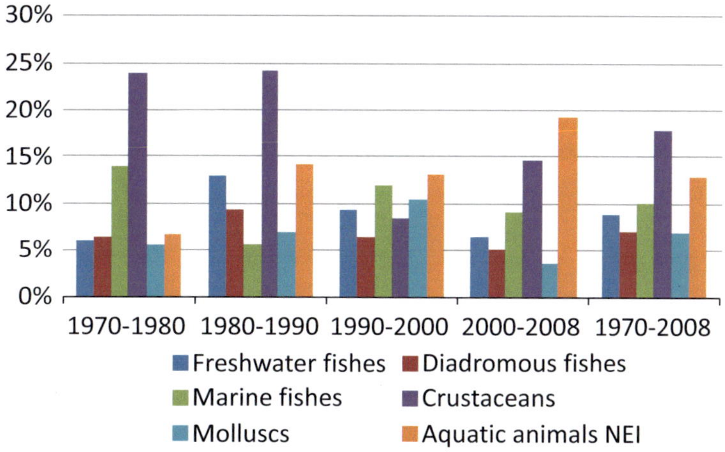

Fig. (5). Trend in world aquaculture production growth: average annual growth rate for major species groups in the period 1970–2008.

Table 2. Marine and freshwater crustacean aquaculture production in 2008-2009 [19].

	2008		2009	
	Quantity (Ton)	Value (USD'000)	Quantity (Ton)	Value (USD'000)
Marine crustaceans				
Marine shrimp/prawns	3,403,195	14,374,250	3,495,972	14,647,939
Marine cabs	240,789	747,938	246,523	767,408
Other marine species	1206	12,007	1511	14,328
Sub-total	3,645,190	15,134,195	3,744,006	15,429,675
Freshwater crustaceans				
Freshwater crab	518,365	3,608,126	574,247	3,997,060
Crayfishes and other species	418,242	1,868,073	526,670	2,410,691
Freshwater shrimp/ prawns	437,257	2,197,683	459,669	2,297,172
Sub-total	1,373,864	7,673,882	1,560,586	8,704,923

Crayfish (Crustacea, Decapoda, Astacidea) farming in Europe has been focusing on producing stocklings of native crayfish and recently also market size crayfish [20] using both native and translocated species. The crayfish farming in Southern USA has long been focusing on rearing red swamp crawfish (*P. clarkii*) to market size on large farms or rice fields [21].

Most of the present commercial market size freshwater crayfish production is derived from semiintensive pond systems [21]. Largest producers in late 1980's were USA (60,000 tonnes), Spain (8,000 tonnes) and Sweden (1,000 tonnes), and the total production in Europe was 11,000 tonnes annually [22, 23].

The crayfish production in USA is obtained from polyculture type semi-intensive farms, where former rice fields have been converted into red swamp crayfish (*P. clarkii*) production units [21], with sophisticated synchronized rearing of crayfish and the plants to supply food and nutrients for the crayfish [24]. Ponds are drain or trap harvested, and most of the catch is consumed locally. USA also exports large quantities of crayfish to European markets [25]. In Finland, Sweden and also elsewhere in Europe, the commercial freshwater crayfish production is based on semi-intensive monoculture, with the crayfish being normally the only species in the farm ponds or raceways [20].

The shrimp aquaculture introduces some serious indexes for some countries: (i) contribution to gross domestic product (GDP) (*e.g.* 1% of GDP for Madagascar, 0.25% of GDP for Norway, 0.2% of GDP for Trinidad and Tobago); (ii) consumption (*e.g.* 2.2 kg/person/yr for Australia, 1.9 kg for the USA, 1.7 kg for Norway, 0.66 kg for Mexico); (iii) employment (*e.g.* about 5% of all fishing employment in Australia, 1.2 million people in Nigeria involved in occupations associated with shrimp fishery and aquaculture); (iv) catch value (*e.g.* US$ 500 million in Indonesia, US$ 425 million in the USA, US$ 300 million in Mexico); and (v) exports. Shrimp represents the most valuable fishery export for some countries such as Indonesia (US$ 887 million), Mexico (US$ 346 million), Madagascar (US$ 68.2 million) and Nigeria (US$ 49 million).

Sustainable Shrimp Aquaculture

Many important ecological and practical subjects and problems have been resolved in order to motivate the shrimps aquaculture, like deforestation of mangroves to create land for the building of shrimp ponds, effluent discharge of waste water coming from shrimp farms use of antibiotics, use of fishmeal for nutrition, general system intensification, food safety, transboundary aquatic animal diseases, certification schemes and trade have to take into account [26, 27]. The human safety and health protection remain the most important limitation for an efficient production of shrimp under intensive aquaculture systems. Several serious diseases in shrimps' aquaculture and substantial environmental challenges had affected their production rate. Is worth saying that these two above mentioned parameters are closely interlinked as environmentally not safe practices increasing the exposure to diseases.

The aquaculture of three shrimps in the world presents some particular points (see Fig. **6**). The culture of *P. vannamei* (previously *Litopenaeus vannamei*) has attracted producers due to a number of competitive advantages in terms of the following parameters: faster growth rate, safe high stocking density, low salinity tolerance, water-cool tolerance, protein requirements and increased potentiality of breeding and domestication, lower number of diseases in comparison to *P. monodon*.

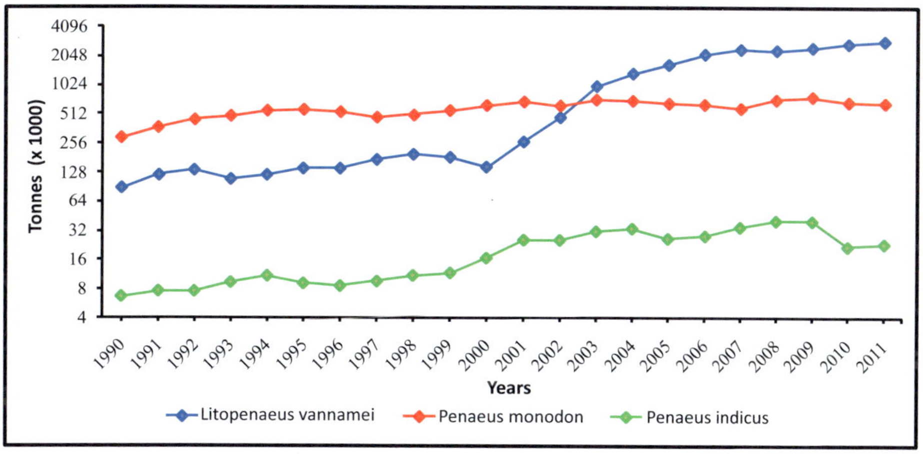

Fig. (6). Trends in three economically important shrimps' production from 1990 to 2011.

In the western hemisphere the culture of white shrimp (*P. vannamei*) occupies percentage above 90% of the total shrimp culture [28]. The most efficient and practical method of its culture is the semi-intensive systems in coastal areas. Successful culture of white shrimp depends on the quality of used sea water and the use of a wastewater treatment manufactory to prevent pollution of adjacent areas. A number of factors have limited expansion of white shrimp culture, as the high cost of coastal real estate and the habitual appearance of viral diseases, such as white spot (WSSV), which has brought the industry near collapse [29].

The most productive countries regarding *P. vannamei* production in the period 2005-2009 are shown in Table **3**, including producers from Asia (China, Thailand, Indonesia and Vietnam) and Latin America (Ecuador, Mexico, Brazil, Colombia, Nicaragua and Peru). China and Thailand remained the top two producers from Asia and Ecuador and Mexico from Latin America since 2005. Vietnam is a special case. It used to be among the top five since 2005 and showed a great drop in production in 2008 and 2009. Indonesia and Brazil showed a decrease in production in 2009.

Table 3. Aquaculture production (tons) of *Penaeus vannamei* in the period 2005–2009 [30].

Producers	2005	2006	2007	2008	2009
China	702,484	887,838	1,065,644	1,062,765	1,118,142
Thailand	374,487	480,061	508,446	501,394	535,000
Ecuador	118,500	149,200	150,000	150,000	179,100
Indonesia	103,874	141,649	164,466	208,648	170,969
Mexico	90,008	112,495	111,787	130,201	125,778
Brazil	63,134	65,000	65,000	70,251	65,188
VietNam	100,000	150,000	153,000	38,600	36,000
Colombia	19,000	21,600	20,300	18,400	18,100
Nicaragua	9633	10,860	11,097	14,690	17,362
Peru	8324	9257	11,657	13,314	13,425
Others	60,811	62,155	55,737	57,083	48,470
World	1,650,255	2,090,115	2,317,134	2,265,346	2,327,534

In the case of **China** (Fig. **7**) a clear decline in production from about 210,000 tons to 87,000 tons during the period between 1992 and 1993 has been shown; severe shrimp diseases occurred in 1993 [31].

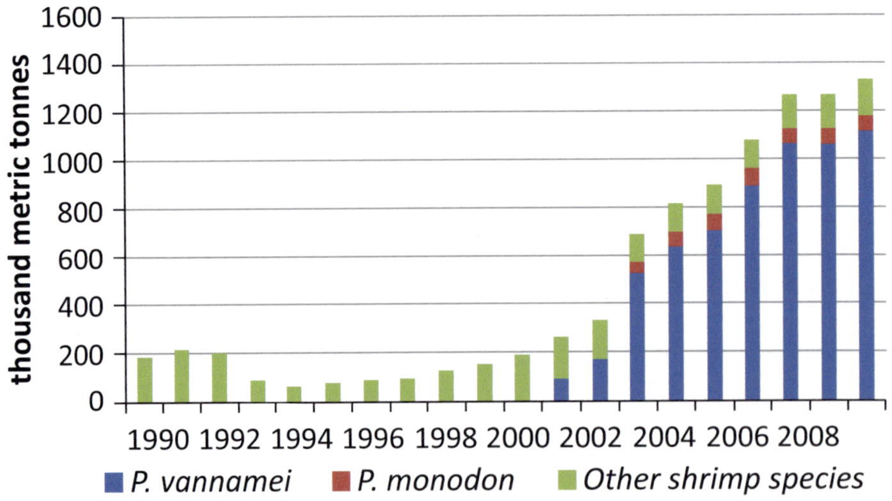

Fig. (7). Shrimp culture production in China from 1990 to 2008.

In **Ecuador**, there was a drop in production from 1999 to 2001 (Fig. **8**); this decrease could be attributed to the significant outbreak of white spot disease on *P. vannamei* during this period [31].

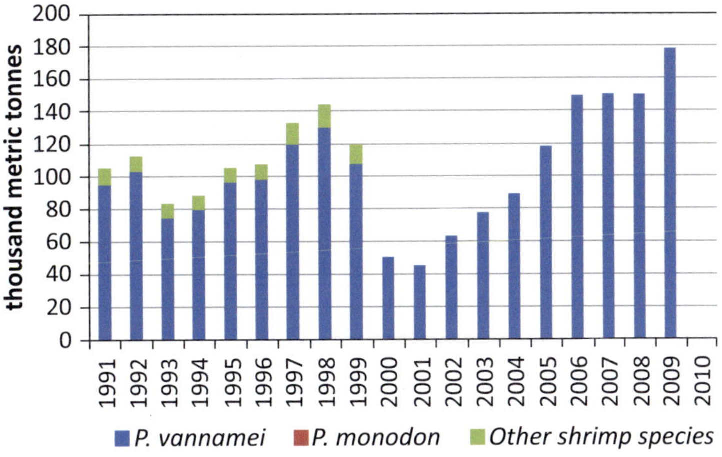

Fig. (8). Shrimp culture production in Ecuador from 1991 to 2009.

In **India**, the occurrence of yellowhead disease and white spot disease during 1994–1995 had as consequence to decrease of production [31]. The next observed decrease in production (2006-2008) was due to several factors like: decrease in farm gate price forced farmers to shift to Pangasius culture (Fig. **9**); detection of nitrofurans from shrimp exports resulted to return of consignments and forced farmers to shift to finfish farming.

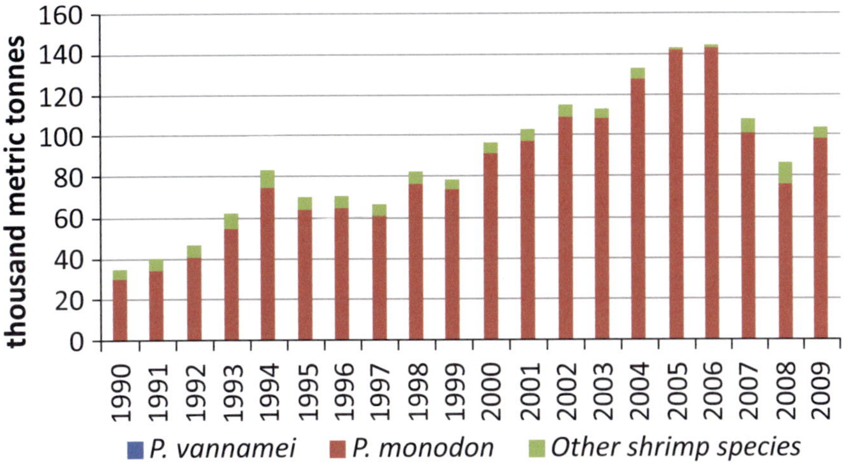

Fig. (9). Shrimp culture production in India from 1990 to 2009.

Viet Nam is currently the highest producer of tiger shrimp *P. monodon*, however the culture of *P. vannamei* is now growing. As mentioned earlier, there is no information why there was a significant drop in production in 2008–2009 (Fig. **10**). In the recent years (since 2010) there is an unknown disease impacting shrimps in the Mekong Delta (since 2010), the main source (80%) of white leg shrimp production in the country. This unknown disease affects both *P. monodon* and *P. vannamei* and the results in the next productions are still unknown.

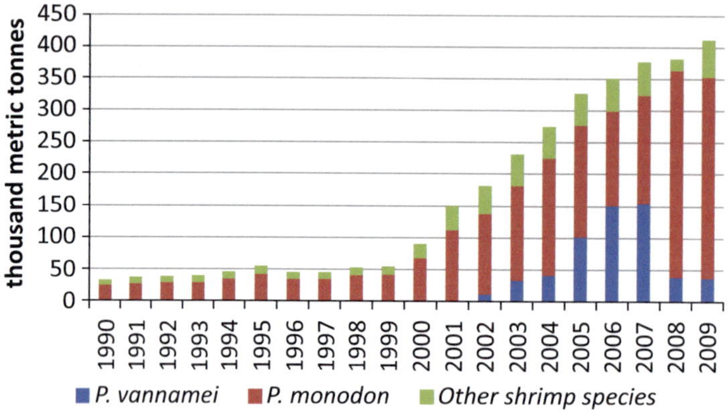

Fig. (10). Shrimp culture production in Viet Nam from 1990 to 2009.

There was a decrease in production in **Brazil** from 2003 to 2005 (Fig. **11**). During this period, the country was suffering from infectious myonecrosis virus (IMNV) outbreaks in *P. vannamei* in 2002 [32]. Reported losses during the first outbreak in Brazil was US$ 20 M. IMNV causes significant losses (100% prevalence; 40–70% mortalities) in farmed populations in marine and low salinity brackish water.

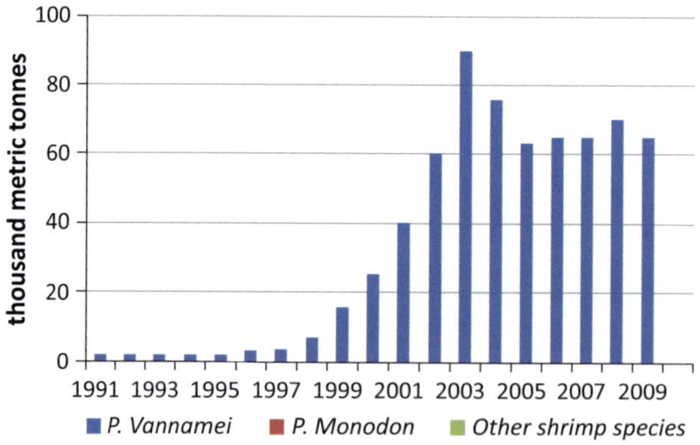

Fig. (11). Shrimp culture production in Brazil from 1991 to 2009.

In **Indonesia**, the disease Infectious Myonecrosis Virus (IMNV) has been reported during 2006–2008 [33] but there was not any important result in the production (Fig. **12**). The disease was reported to be spreading to several areas in 2009, and here we can see a drop in production figures reported by the country.

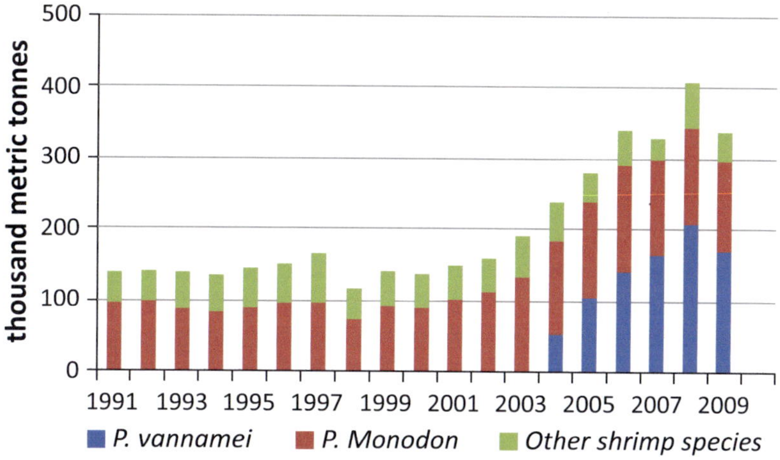

Fig. (12). Shrimp culture production in Indonesia from 1991 to 2009.

Various Diseases in Decapods' Culture

Viral infections have been coped in both wild and cultured crustaceans populations very early (in the 1960's) [34]. Till then more than 50 viruses have been described from a miscellaneous range of crustaceans. The results of these diseases had a crucial negative impact on the shrimp farming. Since 1981, some unfamiliar viruses were known in both Asia and America, causing mass mortalities affecting the economical sustainability of this type of industry [35]. There were number of important common characteristics to the cultured shrimp diseases that distinguish them from those imposing the terrestrial or aquatic vertebrates.

Finally, decapod crustaceans are infected by a wide variety of metazoan parasites. These include worm infections by members of the Trematoda, Cestoda, Acanthocephala, Nemertea and Turbellaria, and the parasitic crustaceans belonged to Copepoda, Rhizocephala and Isopoda.

Due to their larger size and predictable pathology, the DNA viruses (encapsulated by the bacilliform viruses of the hepatopancreas of many crustaceans, the white spot syndrome virus of penaeids and the virus infection of spiny lobsters, PAV1) have been described in details in most host groups studied (for examples see [36, 37]).

In general, disease control is most effective in fish and crustaceans farms. Apart from the nursery phase, disease prolepsis rather than control is eventually the unique way of action, and medication rarely appears as an object in published costs. However, significant financial losses appear during the growing phase of the organisms as a result of disease and can often be associated with many parameters, like poor diet, water quality or pond bottom's oceanographic parameters.

In the previous years, the epidemic diseases caused serious ecological and economical problems to European crayfishes and other cultured species. The recent years the current huge scale of shrimps farming and their great importance in the international trade in live, fresh and frozen species has re-pointed out the international need for effective disease diagnosis, prolepsis and control. Recently the relevant researches focus on the comprehension of the crustacean defence mechanisms, the effect of stressors on their physiology and to improve their strategies to promote disease resistance; including, thus, genetic selection and engineering. Causes of non-infectious diseases could be various, like the dietary adequacies or exposure to metabolic wastes and toxic chemicals.

The natural tolerance of the hosts can be affected from various parameters, like its life cycle stage, its moult stage, its genetic structure and the degree of stress during culture. Examples of stressors known to crush the crustacean immune responses include environmental factors, such as temperature, salinity and oxygen levels and industrial pollutants (*e.g.* heavy metals and pesticides) [38].

Some known diseases, such as white spot syndrome virus (WSSV) and Taura syndrome virus (TSV), often have minor impact on mud crabs (Scylla serrata, S. tranquebarica, S. paramamosain and S. olivacea), they are known to be effective microbial carriers and potentially viral reservoirs of these diseases. While the presence of WSSR in mud crabs can be assessed using polymerase chain reaction (PCR) testing, it has only rarely been reported that the virus can cause mortalities in these decapods. In fact, the farming of mud crab in areas where shrimp farming has been demolished by WSSV has been recommended as an alternative farming opportunity [39].

The sector of the shrimp's aquaculture has been rapidly increased the last years and this important sector of economy has been excessively developed. Unfortunately, the outbreak of viral diseases has also increased and, thus, the economic risks are dramatically related with this industry development [39]. Among the infectious diseases of cultured shrimp, certain virus-caused diseases are considered as the most significant. The pandemics due to the penaeid viruses WSSV (White spot) and TSV (Taura Syndrome), and secondarily to IHHNV

(Infectious Hypodermal and Hematopoietic Necrosis virus) and YHV (Yellow Head), caused damages in the penaeid shrimp industry billions of dollars in lost cultures, jobs and export revenue. The socioeconomic impacts of these diseases caused by these pathogens have been penetrated in countries in which shrimp farming constitutes a significant industry (Figs. **13, 14**).

Fig. (13). White muscle disease in bigger size of *L. vannamei.*

Fig. (14). Black gill disease observed during sampling.

One of the most common problems in the marine and freshwater crustacean aquaculture units are the shell diseases and external abnormalities or malformations. These external diseases have been mentioned in many natural crustacean populations, mainly in many economically important species, such as the Alaskan king crab, portunid crabs, shrimps and lobsters [40]. The shell diseases are characterized by various types of carapace erosive abrasions [41] and among them the most common kind of this type disease is known as 'brown spot' or 'black spot' [40]. These diseases in crustaceans have been associated with many contaminating factors.

In general, the sustainability of marine and freshwater crustaceans culture depends from various and different parameters. These factors affect the sustainability and the successful culture of decapods and among them are the novel diseases; the relatively limited ability to mitigate the disease process in farmed crustaceans, the increasing demand for crustacean seafood and an increasing legislative bother on the trading of live animals and their products. The linking between the global experts in the culture, capture and trading of crustaceans with pathologists, epidemiologists, ecologists, therapeutics specialists and policy makers in the field of food security will allow these problems to be better identified and addressed.

The development of a health management programme is a recommended strategy to minimize the risks of disease affecting any particular farm, including decapods culture. Such a programme should be farm-specific and customized to address basic processes and practices. A health management programme should include all key functional issues such as:

- pond or pen management;
- health monitoring;
- water quality management;
- feed management;
- rearing practices;
- record-keeping;
- source of health stock.

The source of stock has become a significant argument for international aquaculture. Stock of many aquaculture species has been moved between countries and several areas, with some of these moves resulting in the transfer of diseases with that stock [39]. The benefits of translocating stock need to be balanced against the risks involved. Importing stock should only be undertaken with appropriate securities, such as adequate health checking and quarantine procedures, in place to minimize the risk of introducing disease, as recommended in Article 9 of the Aquaculture Development section of the FAO Code of Conduct for Responsible Fisheries (the Code).

CONFLICT OF INTEREST

The author (editor) declares no conflict of interest, financial or otherwise.

ACKNOWLEDGEMENT

Declared none.

REFERENCES

[1] Martin JW, Davis GE. An Updated Classification of the Recent Crustacea. Natural History Museum of Los Angeles County 2001; p. 39.

[2] Kensley B. Estimates of species diversity of free living isopod crustaceans on coral reefs. Coral Reefs 1998; 17: 83-8. [http://dx.doi.org/10.1007/s003380050100]

[3] Green J, Conlan KE, Gordon I, Anderson W, Newman WA, Bousfield EL. https://www.britannica.com/animal/crustacean

[4] Yearbook of fishery statistics summary tables. FAO 2011; pp. 1-120.

[5] The state of world fisheries and aquaculture. FAO Fisheries and Aquaculture Department 2012; pp. 1-209.

[6] Kellcher K. Discards in thc world's marine fisheries – an update. FAO Fisheries Technical Paper 470. Rome 2005, FAO.

[7] Gillett R. Global Study of Shrimp Fisheries FAO Fish Tech Pap No 475, 331. Rome: FAO 2008.

[8] Eayrs S. A Guide to Bycatch Reduction in Tropical Shrimp-Trawl Fisheries. 1st ed. Rome: FAO 2007; p. 108.

[9] Harrington J, Myers R, Rosenberg A. Wasted fishery resources: discarded bycatch in the USA. Fish Fish 2005; 6: 350-61. [http://dx.doi.org/10.1111/j.1467-2979.2005.00201.x]

[10] FAO The state of world fisheries and aquaculture. FAO Fisheries and Aquaculture Department 2010; 1-197.

[11] Jennings S, Kaiser MJ. The effects of fishing on marine ecosystems. Adv Mar Biol 1998; 34: 201-351. [http://dx.doi.org/10.1016/S0065-2881(08)60212-6]

[12] Leibold MA. A graphical model of keystone predators in food webs: Trophic regulation of abundance, incidence and diversity patterns in communities. Am Nat 1996; 147: 784-812. [http://dx.doi.org/10.1086/285879]

[13] Munoz JE, Manjón-Cabeza ME, Raso JE. Decapod crustacean assemblages from littoral bottoms of the Alborán Sea (Spain, west Mediterranean Sea): Spatial and temporal variability. Sci Mar 2008; 72: 437-49.

[14] Gillett R. Global study of shrimp fisheries, FAO fisheries technical paper 475. Food and Agriculture Organization of the United Nation, Rome 2008; 1-331.

[15] Lavrado HP, Falcao AP, Carvalho-Cunha P, Silva SH. Composition and distribution of Decapoda in Guanamara Bay, RJ. Nauplius 2008; 8: 15-23.

[16] Teikwa ED, Mgaya YD. Abundance and reproductive biology of the penaeid prawns of Bagamoyo coastal waters, Tanzania. Western Indian Ocean Journal of Marine Science 2003; 2: 117-26.

[17] Follesa MC, Porcu C, Gastoni A, Mulas A, Sabatini A, Cau A. Community structure of bathyal decapod crustaceans off South-Eastern Sardinian deep-waters (Central-Western Mediterranean). Mar Ecol (Berl) 2003; 30: 188-99. [http://dx.doi.org/10.1111/j.1439-0485.2009.00323.x]

[18] FAO. The State of World Fisheries and Aquaculture. Rome: FAO 2010; p. 197.

[19] Josupeit H. Shrimp production and trade.The Shrimp Book. United Kingdom: Nottingham University

Press 2010; pp. 835-56.

[20] Ackefors H, Lindqvist OV. Cultivation of freshwater crayfishes in Europe.Freshwater crayfish aquaculture in North America, Europe, and Australia Families Astacidae, Cambaridae, and Parastacidae. New York, USA: Food Products Press 1994; pp. 157-216.

[21] Huner JV. Cultivation of freshwater crayfishes in North America. 1994.

[22] Huner JV, Gydemo R, Haug J, Järvenpää T, Taugbøl T. Trade, marketing and economics. 1987.

[23] Holdich DM. A review of astaciculture: freshwater crayfish farming. Aquat Living Resour 1993; 6: 307-17.
[http://dx.doi.org/10.1051/alr:1993032]

[24] Huner JV. Ecological observations of red swamp crayfish, *Procambarus clarkii* (Girard, 1852), and white river crayfish, *Procambarus zonangulus* (Hobbs & Hobbs, 1990), as regards their cultivation in earthen ponds. Freshwater Crayfish 1995; 10: 456-68.

[25] Holdich DM. 'Crawfish state' Louisiana plays host for astacologists. Fish Farmer 1990; 13(4): 32-3.

[26] Chamberlain GW. History of Shrimp Farming.The Shrimp Book. United Kingdom: Nottingham University Press 2010; pp. 1-34.

[27] Wurmann C, Madrid RM, Brugger AM. Shrimp farming in Latin America: currents status, opportunities, challenges and strategies for sustainable development. Aquac Econ Manag 2004; 8: 117-41.
[http://dx.doi.org/10.1080/13657300409380358]

[28] Jory DE, Dixon HM. White shrimp spot virus in the western hemisphere. Aquacult Mag 1999; 25: 83-91.

[29] Global Capture Production Statistics Dataset FA. 1950–2009 and Global Aquaculture Production Statistics dataset (Quantity and Value) 1950–2009. Rome: FAO 2011.

[30] Bondad-Reantaso MG, Subasinghe RP, Arthur JR, *et al.* Disease and health management in Asian aquaculture. Vet Parasitol 2005; 132(3-4): 249-72.
[http://dx.doi.org/10.1016/j.vetpar.2005.07.005] [PMID: 16099592]

[31] Lightner DV, Pantoja CR, Poulos BT, *et al.*

[32] Senapin S, Phewsaiya K, Briggs M, Flegel TW. Outbreaks of infectious myonecrosis virus (IMNV) in Indonesia confirmed by genome sequencing and use of an alternative RT-PCR detection method. Aquaculture 2007; 266: 32-8.
[http://dx.doi.org/10.1016/j.aquaculture.2007.02.026]

[33] Vago C. Perspectives sur la classification des virus d'invertébrés. Entomophaga 1966; 11: 347.
[http://dx.doi.org/10.1007/BF02373323]

[34] Walker PJ, Mohan CV. Viral disease emergence in shrimp aquaculture: origins, impacts and the effectiveness of health management strategies. Rev Aquacult 2009; 1: 125-54.
[http://dx.doi.org/10.1111/j.1753-5131.2009.01007.x]

[35] Takahashi Y, Itami T, Kondo M, *et al.* Electron microscope evidence of bacilliform virus infection in kuruma shrimp (Penaeus japonicus). Fish Pathol 1994; 29: 121-5.
[http://dx.doi.org/10.3147/jsfp.29.121]

[36] Behringer DC, Butler MJ IV, Shields JD, Moss J. Review of Panulirus argus virus 1-a decade after its discovery. Dis Aquat Organ 2011; 94(2): 153-60.
[http://dx.doi.org/10.3354/dao02326] [PMID: 21648244]

[37] Le Moullac G, Haffner P. Environmental factors affecting immune response in Crustacea
[http://dx.doi.org/10.1016/S0044-8486(00)00422-1]

[38] Flegel TW. 2006. Detection of major penaeid shrimp viruses in Asia, a historical perspective with emphasis on Thailand. Aquaculture 2000; 258: 1-33.

[http://dx.doi.org/10.1016/j.aquaculture.2006.05.013]

[39] Noga EJ, Smolowitz R, Khoo LH. Pathology of shell disease in the blue crab, *Callinectes sapidus* Rathbun, (Decapoda: Portunidae). J Fish Dis 2000; 23: 389-99. [http://dx.doi.org/10.1046/j.1365-2761.2000.00249.x]

[40] Johnson PT. Diseases caused by viruses, bacteria, rickettsia, and fungi.The biology of Crustacea. New York: Academic Press 1983; pp. 1-78. [http://dx.doi.org/10.1016/B978-0-12-106406-8.50012-1]

[41] Shelley C, Lovatelli A. Mud crab aquaculture – A practical manual. FAO Fisheries and Aquaculture Technical Paper No 567 Rome, FAO 2011; 1-78.

CHAPTER 11

Histological Methods to Assess the Effect of Diet and a Single Meal on the Liver and Intestine of Rainbow Trout: Fishmeal and Fishoil Replacement With Plant Protein and Oil

Panagiotis Berillis[1,*], **Sam Martin**[2] and **Eleni Mente**[1]

[1] *Department of Ichthyology and Aquatic Environment, School of Agricultural Sciences, University of Thessaly, Volos, Greece*

[2] *School of Biological Sciences, University of Aberdeen, Tillydrone Avenue, AB24 2TZ Aberdenn, UK*

Abstract: Rainbow trout (*Oncorhynchus mykiss*) is a highly commercial sport and market fish. Rainbow trout feed formulations are using fish oil, fish meal, grains and other in-gredients. In recent years the amount of fish meal has been reduced to less than 50 percent. This was achieved by using alternative protein sources such as soybean meal. The aim of this study was to assess the changes of intestine histology of rainbow trout fed three different iso-energetic and isoproteic diets with a partial substitution of fish meal and fish oil with vegetable protein and vegetable oil. The goblet cell number per μm of intestinal fold appeared to not be affected by diet, while goblet cell diameter appeared smaller in rainbow trout fed a diet of 31% organic vegetable protein. Rainbow trout fed the above diet had a smaller submucosa layer. Rainbow trout fed a diet of 36% vegetable protein and 6% vegetable oil had smaller intestinal folds, but the brush border height and muscular layer width were unaffected by the diet. Changes in the rainbow trout intestinal histology following a single meal were studied. Intestinal samples were taken before feeding and at 6 h and 12 h after feeding. At 6 h after feeding, the protein synthesis rate in trout is in its peak and the digestion process is in progress. Goblet cells then secrete their mucous and their size is decreased. The similarity in the size of the goblet cells before feeding and at 12 h after feeding sugest that rainbow trout should be fed twice per day and that the second feeding should take place between 6 and 12 hours after the first feeding. In fish intestine the goblet cells' post-prandial changes may occur in response to a single meal because they have a potential role in the digestion process. Research on how dietary composition and, in particular, plant ingredients modify intestine and liver histology over time and whether fish can adapt to plant protein and oil feeding are very important for a sustainable fish supply.

* **Corresponding author Panagiotis Berillis:** Department of Ichthyology and Aquatic Environment, School of Agricultural Sciences, University of Thessaly, Volos, Greece; Tel: (+30)2421093248; E-mail: pveril@uth.gr

Panagiotis Berillis (Ed)

Keywords: Fish meal substitution, Intestinal histology, Rainbow trout, Vegetable proteins.

INTRODUCTION

In recent years the aquaculture production has dramatically increased, and it is predicted to grow from 67 M tons in 2012 to 140 M tons by 2050 [1]. These figures reflect the growing global human population and an increased demand for fish protein [2]. The aquaculture expansion is limited by the reliance of the aquafeed production on wild fish in order to fulfil the demands of aquaculture. To attain maximum growth rates for salmonids and other carnivorous fish, a high protein diet is required. For carnivorous fish, fishmeal is used in high-protein feed. For various reasons, fish meal and fish oil have historically been the dominant raw materials in the production of fish feeds. The overexploitation of fishery resources and climate change resulted in a reduced supply and, consequently, alternative protein sources such as plant protein meal are used in aquafeeds [2]. The most common plant protein ingredients are soybean, soybean meal, and increasingly, soybean protein concentrate. However, their use in some species is restricted because of intestinal inflammation and the high content of non-starch polysaccharides and other carbohydrates that are poorly utilized by aquatic animals. The long-term metabolic consequences of feeding fish with a diet of high-plant-protein, low-fishmeal, is still under research. Further information must be collected in order to understand the way in which dietary composition influences protein metabolism in fish. A better understanding of the potential interactions between dietary protein sources and the way that they may regulate the processes involved in protein metabolism must be obtained.

Rainbow trout (*Oncorhynchus mykiss*) is a highly commercial sport and market fish. Many countries have reported rainbow trout farming production. The relative output of some of them is insignificant compared to the production from the larger systems located in the primary producing areas in Europe, Chile, Australia, North America and Japan [3]. Rainbow trout is the dominant freshwater salmonid farmed in Europe and North America [4]. Feeds for rainbow trout have been modified over the years, and the cooking-extrusion processing of foods now provides compact nutritious pelleted diets for all life stages. In the past, trout were fed with trash fish, slaughterhouse by-products, offal and wastes. Feeding trout in that way is rather inconvenient and very polluting both to the rearing tanks/ponds and to the surrounding environment. In the modern trout farming industry, efficient pelleted dry feeds (0.6–1.1 FCR) have replaced the traditional feeds [5]. Pellets made in this way absorb high amounts of added fish oil and permit the production of high-energy feeds, with over 16% fat. In high-energy feeds the dietary fat levels now exceed 22 percent as the dietary protein levels in that feeds have increased from 35% to 45%. High-quality expensive feeds are chosen by many farmers for the first stages. In that stages the fish are most vulnerable and sensitive and little feed is used [5]. Rainbow trout feed formulations are using fish oil, fish meal, grains and other ingredients. In recent years the amount of fish meal has been reduced to less than 50 percent. This was achieved by using alternative protein sources such as soybean meal. These high-energy diets are efficiently converted by the rainbow trout in high ratios, sometimes close to 1:1 [3].

The gastrointestinal tract main functions are to store food and water, to absorb water, osmolytes and nutrients from the external medium, to process the ingested food and water, and to excrete waste [6]. The food is mixed and processed in the stomach and then the bolus is pushed into the intestine where absorption occurs. There are large differences in the morphology and physiology of the intestinal regions between the vertebrate groups and between different feeding strategies within the same vertebrate group. The intestine length of fish varies from 0.4 to >38 times the body length [6]. The percentage of plant material in the diet is the major determining factor for intestinal length, where the intestines of herbivorous fish are generally longer than those of carnivorous fish [7, 8]. The intestine is divided to three regions: the proximal intestine, which contains pyloric caeca, and the mid and distal intestines. Enterocytes perform nutrient absorption in all regions of the intestine. Absorption is achieved by passive and facilitated diffusion and by active transport [9]. Functionally, the anterior region and the pyloric caeca are the primary sites for nutrient uptake [10]. The epithelial layer consists mainly of absorptive columnar cells, which include enterocytes, goblet cells (mucus-secreting cells) and endocrine cells. Numerous extensions (microvilli) can be found on the apical surface of the enterocytes. The whole apical surface of the epithelium is referred to as the brush border membrane. The microvilli increases the area for absorption and membrane-bound digestive and absorptive enzymes by extended the surface area of the apical membrane [8]. Adjacent enterocytes are joined together at the apical end of the lateral surface by junctional complexes. In most fishes, the muscular coat (smooth muscle) of the intestine has become well developed to ensure peristaltic activity and contains two layers: the circular muscle layer and the longitudinal muscle layer [11].

Goblet cells are common components of the post-gastric mucosa in fish, and they are the dominant mucous cells in the intestine [12, 13]. Mucus, which fills the upper part of the cells, is discharged through an apical pore. The nucleus are located at the bottom part. Goblet cells can synthesize neutral and sulphate mucins and sialomucins containing sialic acid [14]. The mucus secreted by goblet cells is responsible for the lubrication of the undigested materials for onward progression into the rectum. It also protects the mucosa of the digestive tract. Intestinal mucins may have a role in osmoregulation [14]. Post-prandial changes in the goblet cells in fish intestines may respond to a single meal because they have a potential role in the digestion process.

The aim of this study was to assess any possible changes in the intestine histology of rainbow trout fed three different iso-energetic and isoproteic diets with a partial substitution of fish meal and fish oil with plant protein and plant oil. Changes in intestinal histology following a single meal before feeding and at 6 h and 12 h after feeding were also studied.

EXPERIMENTAL METHODOLOGY

The rearing conditions were as described in Mente *et al.* [15, 16]. Briefly, rainbow trout *Oncorhynchus mykiss*, mixed sex, weighing approximately 44.98±1.08 g, were stocked in triplicates (120 juveniles per dietary treatment) in 250 l freshwater tanks. The temperature was maintained at 12 °C, pH at 7.60 and oxygen saturation at 90% under a natural photoperiod [15, 16]. Three different iso-energetic and isonitrogenic diets were used for fish feeding (three tanks per diet). The dietary treatments were as follows: Diet A was a low marine fishmeal /fish oil diet (22% fishmeal, 8% fish oil) with soya cake, peas, wheat, corn gluten and 2% vegetable oil, Diet B was a high marine fishmeal/fish oil diet (56% organic fishmeal, 13% organic fish oil and organic peas, organic soya cake) and Diet C contained 40% fishmeal, 10% fish oil, 6% vegetable oil, soya cake, peas, wheat and corn gluten. All diets had a 10% of a vitamin & mineral premix and diet B was free of synthetic amino acids (Table **1**). Fish were fed *ad libitum* by hand twice daily at 09:00 and 15:00 for 5 weeks. At the end of the experiment, fish were fasted for 24h and 18 fish, two from each tank, six per treatment, were used for the present histological study.

In addition, the last day of the experiment (after a 24h fastening, the prefeeding group), intestine samples were collected following a single meal with diet B before feeding, and at 6h and 12h after feeding. One fish from each tank, three fish in total, used as the prefeeding group for measurements of the intestine histology. Fish were sacrificed by anesthesia (diluted 1:1 phenoxyethanol in ethanol) and a blow to the head. The remaining trouts were fed normally and a group of 6 trouts was selected randomly, sacrificed and removed at 6h and 12h after feeding, respectively. Ultra structure of liver and intestine was also examined under transmission electron microscope.

For light microscopy, midgut samples were fixed for 24 h in 10% buffered formalin at 4°C. Dehydration process was performed by samples immersion in graded series of ethanol. Immersion in xylol and paraffin wax embedding followed. 5-7μm sections were mounted, deparaffinized and the rehydrated. Two slides were prepared per sample. One slide stained with Hematoxylin and Eosin and the other one stained with Alcian blue. Slides mounted with Cristal/Mount. Slides were examined under light microscope and random cross sections of the midgut photographed by a digital camera. These images were used in order to measure the mean number of the goblet cells per μm of intestinal fold, the size of the goblet cells, the intestinal folds height, the epithelium brush border length, the statum compactum length, the submucosa length and the muscular layer length. All measurements were done according Baeza-Arino *et al.* (2016). For the post feeding study the goblet cells area per intestinal fold and the goblets cells diameter were measured before feeding, 6h and 12h after feeding. All measurements were made by the ZEN microscope software of ZEISS.

Table 1. Ingredient and chemical composition of the three diets [15, 16].

Diets	**A**	**B**	**C**
Ingredients (g kg-1)			
Fishmeal	220	-	400
Organic certified fishmeal	-	560	-
Fish oil	80	-	100
Organic certified fish oil	-	130	-
Vegetable oil	20	-	60
Organic vegetable protein (pea and soya)	-	310	-
Vegetable protein (pea, soya, corn gluten meal)	560	-	360
Wheat	110	-	80
Chemical composition (%)			
Crude protein	45	43	45
Total lipid	22	23	21
Ash	7.1	10	8
GE (kJ.kg^{-1}DM)[1]	21.9	21.5	21.7

[1] GE, Gross energy calculated from nutrients assuming gross energy content of 23.6 kJ g^{-1} for protein, 39.5 kJ g^{-1} for lipid, 17.2 kJ g^{-1} for carbohydrate [17].

For electron microscopy, tissue samples of liver and intestine were first fixed for 24 h at 4°C in 2.5% glutaraldehyde diluted in a 0.1-M sodium cacodylate solution. The samples were washed in 0.1-M sodium cacodylate solution before and after fixation. Post-fixation for 2 h at 4°C was performed by immersing the tissue samples in 1% osmium tetroxide, buffered in 0.1-M sodium cacodylate. The samples were washed in the same buffer before and after post-fixation and then rinsed in distilled water. The dehydration process took place by immersing the samples in a graded ethanol series. An overnight immersion in a 1:1 mixture of propylene oxide and the embedding resin followed. The final step of the embedding took place in capsules containing agar resin. Resin polymerisation was completed after 48 h at 60°C. Ultrathin sections (60 to 80 nm) were cut with a Reichert Supernova ultramicrotome (DeKalb, IL, USA). Sections were mounted on copper grids, stained with uranyl acetate and lead citrate and examined in a Philips CM10 electron microscope (Amsterdam, The Netherlands).

THE EFFECT OF THE THREE DIETS ON FISH GROWTH AND IN THE INTESTINAL HISTOLOGY OF RAINBOW TROUT

At the beginning of the trial there were no significant differences among tanks for the initial weight of the fish (44.98±1.08, p>0.05). Fish final weights were 65.7 ± 2.9g, for Diet A, 54 ± 2.9g for Diet B and 32.60 ± 1.91g for diet C (Mente *et al.* 2017a,b). Survival rate was lower, 75% for Diet C in comparison to Diets A and B which was 90 and 92%. The effects of different diet on midgut parameters are shown at Fig. **1**) and Table **2**. Goblet cells number per um of intestinal fold appeared not to be affected by the diet, while goblet cells diameter appeared smaller (18.29 ± 4.33 μm) at rainbow trout fed with diet B. Rainbow trouts fed with diet C had the smaller intestinal folds (264.63 ± 66.17 μm) but brush border height and muscular layer width were unaffected by the diet. Submucosa layer appeared smaller (32.46 ± 4.66 μm) at rainbow trouts fed diet B. Fish fed diet A and B had normal supranuclear absorptive vacuoles in the enterocytes, while in fish fed with diet C absorptive vacuoles cannot be observed.

The mean number of the goblet cells per μm of intestinal fold after a single meal of diet B before feeding and at 6h and 12h after feeding was similar (0.07 ± 0.004, 0.06 ± 0.004 and 0.06 ± 0.003, respectively) ($p > 0.05$, Table **3**, Fig. **2**). Regarding the size of the goblet cells the results were 18.25 ± 5.57 μm before feeding, and 14.46 ± 3.35 μm and 17.53 ± 4.00 μm at 6 h and 12 h after feeding respectively ($p < 0.05$, Table **3**). Smaller goblet cells appeared at the intestine at 6 h after feeding while before feeding and at 12 h after feeding the goblet cells had similar size ($p > 0.05$).

Liver ultrastructure (Fig. **3**) after a single meal with diet B revealed accumulation of glycogen in the hepatocytes and small number of lipid droplets with no signs of steatosis. Microvilli in the brush border of the intestine appeared normal and supranuclear absorptive vacuoles in the enterocytes were found. No pathological signs or signs of enteritis were detected. The most effective diets were diets A and B. Fish fed with diet C lost weight in comparisons to the other diets. The small amount of lipid droplets that were detected in the hepatocytes of fish fed diet B indicate that hepatocytes managed to oxidize fatty acids and there was not excessive dietary intake of lipids.

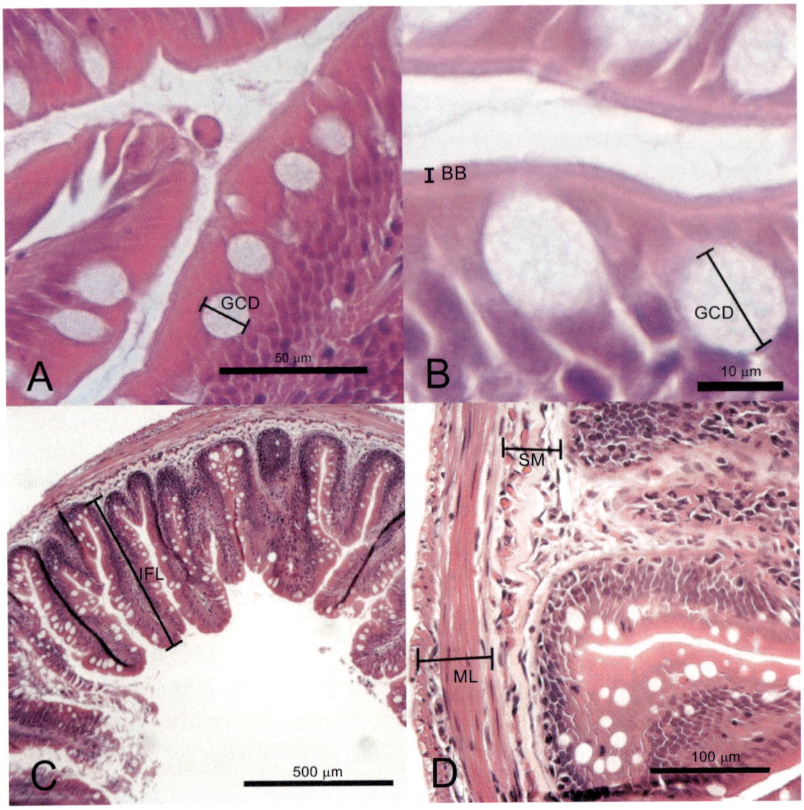

Fig. (1). Detail of thickness on intestine layers in midgut of rainbow trout fed different diets. (**A**) Detail of Goblet cell diameter (GCD) of rainbow trout fed diet A. (**B**) High magnification detail of brush border length (BBL) of rainbow trout fed diet B. (**C**) Detail of intestinal fold length (IFL) of rainbow trout fed diet B. (**D**) Details of submucosa (SM) and muscular layer (ML) of rainbow trout fed diet C. Hematoxylin and Eosin stain.

Table 2. Effect of different diet on midgut parameters. Data are presented as means ± S.E. Means in a column followed by the same superscript are not significantly different ($p > 0.05$).

Diet	Goblet cells diameter (μm)	Goblet cells per 100 um of folds	Intestinal fold length (μm)	Intestinal fold width (μm)	Brush border height (μm)	Muscular layer width (μm)	Submucosa layer width (μm)
A	22.00^{a} ± 0.63 (144)	4.78^{a} ± 0.55 (18)	423.97^{a} ± 33.53 (18)	208.62^{a} ± 11.74 (18)	2.46^{a} ± 0.12 (18)	70.11^{a} ± 2.23 (18)	39.14^{a} ± 1.58 (18)
B	18.29^{b} ± 0.39 (124)	6.67^{b} ± 0.33 (18)	348.95^{a} ± 27.82 (18)	212.73^{a} ± 10.44 (18)	2.43^{a} ± 0.08 (18)	63.15^{a} ± 3.31 (18)	32.46^{b} ± 1.10 (18)

(Table 2) contd.....

Diet	Goblet cells diameter (μm)	Goblet cells per 100 um of folds	Intestinal fold length (μm)	Intestinal fold width (μm)	Brush border height (μm)	Muscular layer width (μm)	Submucosa layer width (μm)
C	21.92[c, a] ± 0.43 (112)	5.28[a, b] ± 0.51 (18)	264.63[b] ± 15.60 (18)	127.31[b] ± 8.15 (18)	2.48[a] ± 0.07 (18)	65.87[a] ± 2.58 (18)	48.67[c] ± 2.04 (18)

Table 3. Changes in goblet cells diameter and number of goblet cells per 100 μm of intesti-nal fold following a single meal of diet B before feeding, 6 hours and 12 hours after feeding. Data are presented as means ± S.E. Means in a column followed by the same superscript are not significantly different ($p > 0.05$).

Time	Mean diameter (μm) of goblet cells. Goblet cells total numbers are given within parenthesis	Mean number of goblet cells. The number of intestinal folds is given within parenthesis
Before feeding	17.97a ± 0.79 (50)	7.1a ± 0.43 (12)
6 hours after feeding	14.46b ± 0.47 (50)	5.88a ± 0.44 (17)
12 hours after feeding	17.54a ± 0.57 (50)	5.9a ± 0.31 (10)

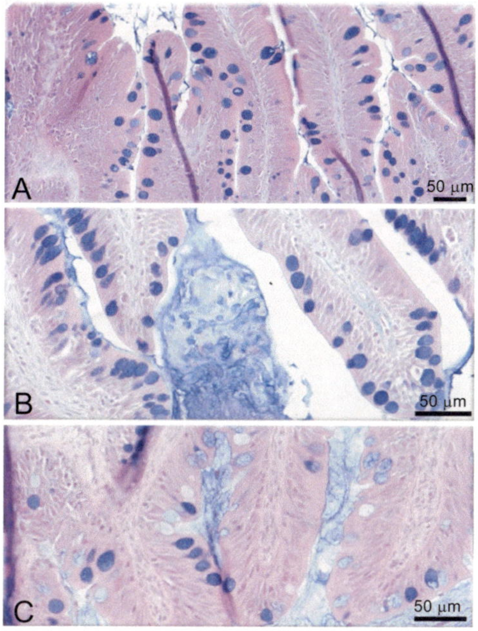

Fig. (2). Rainbow trout intestine **A**: Before feeding. Goblet cells are painted blue. **B**: 6 h after feeding. The goblet cells are smaller in size than the ones before feeding. **C**: 12 h after feeding. The goblet cells have similar size with the ones before feeding. Alcian blue stain.

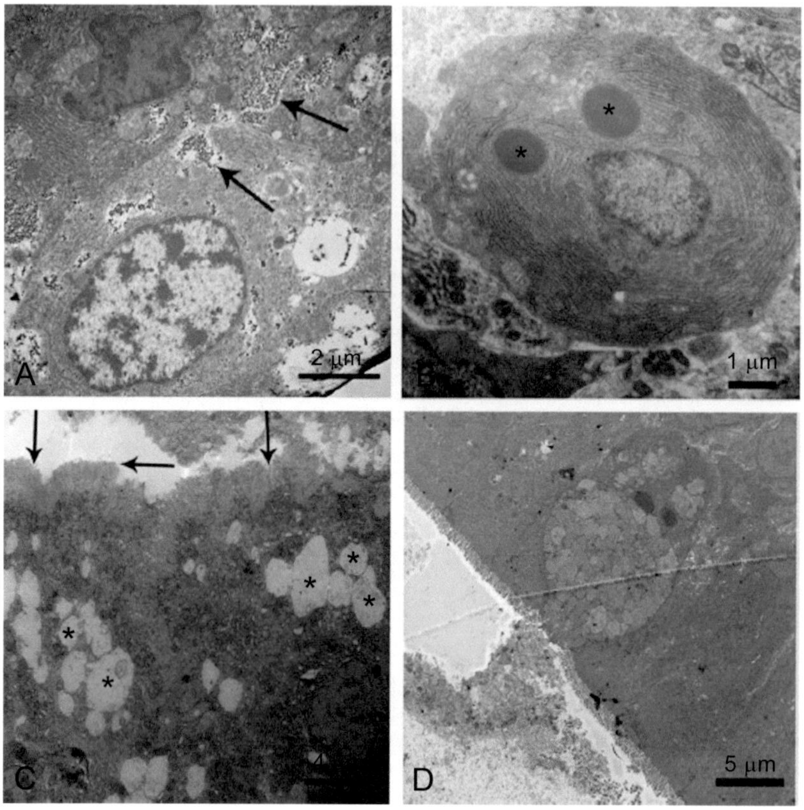

Fig. (3). Rainbow trout liver and intestine ultrastracture fed diet B. **A**: 6 h after feeding. Glycogen (arrows) was acumulated in the hepatocytes. **B**: 6 h after feeding. Some lipid droplets (stars) could also found in hepatocytes. **C**: 6h after feeding. Supranuclear absorptive vacuoles (stars) in the enterocytes were found. Microvilli (arrows) in the brush border appeared normal. **D**: 12 hours after feeding. Goblet cells were similar in size and numbers with the ones before feeding.

DISCUSSION

It is generally accepted that there are variations in the sensitivity of different fish species to dietary soybean meals. Plant based ingredients are not were absorbed by salmonids due to their carnivorous feeding habits. Atlantic salmon and rainbow trout (although to a lesser extent than Atlantic salmon) are highly sensitive to diets with soybean meals. Histological analysis of the digestive system is considered a good indicator of the nutritional status of fish [18]. When discussing the soybean products inclusion in commercial diets, it is very important to evaluate their effects on intestinal histology by microscopy. Structural and functional intestinal changes may explain the deleterious effects of soybean meal on nutrient utilization and disease resistance [19 - 21]. A 100% replacement of dietary fish oil with vegetable oils has a negative effect on the growth of most marine finfish species [22, 23], but a large fraction (60–75%) of fish oil can be substituted without significantly affecting growth and feed efficiency, if the essential fatty acid requirements are met.

The most effective diet was diet A. Rainbow trout that were fed diet A gained more weight than did rainbow trout fed the other two diets. Diet B contained only organic certified ingredients. Fish fed this diet gained half as much weight as did those fed diet A. The small amount of lipid droplets that were detected in the hepatocytes indicate that hepatocytes managed to oxidize fatty acids, and there was not an excessive dietary intake of lipids. Diet C was insufficient because there was reduced survival, and the fish fed that diet lost weight by the end of the experiment. The intestine in salmonids is not only a reacylation site or a site for packaging dietary lipids, but it also has important roles in fatty acid metabolism [24, 25]. Enteritis in salmonids has been described as a shortening of the simple and complex intestinal folds with a widening of the lamina propria and submucosa, shortened microvilli of the brush border membrane, an elevated number of goblet cells and a dramatic decrease or even absence of the normal supranuclear absorptive vacuoles in the enterocytes [26 - 28]. Studies have shown that the Atlantic salmon (*Salmo salar* L.) is more susceptible to enteritis than the rainbow trout (*Oncorhynchus mykiss* L.) [29]. In our results, fish that were fed diet C had shorter and thinner intestinal folds, enlarged submucosa and an almost complete absence of the normal supranuclear absorptive vacuoles in the enterocytes. These are signs of abnormal nutrient absorption but could also be signs of enteritis. Diet C contained the highest amount of vegetable oil. Dietary vegetable oils can induce serious histological changes in the enterocytes of fish. These changes are originated mostly from supranuclear lipid droplet formation, possibly due to altered reacylation mechanisms and decreased phospholipid synthesis [18, 30, 31].

The muscular layer and brush border did not seem to be altered with the different diets, but the submucosa layer thickness showed significant differences with the different diets. Rainbow trout fed diet C showed the thickest submucosa layer. This layer, which is composed of connective tissue, controls the expansion of the intestine when voluminous food is consumed in carnivorous fish [32] and is responsible for most of the absorption. Diet C contained 6% vegetable oil and 36% vegetable proteins. Diet A contained 2% vegetable oil 56% vegetable proteins. Diet B contained 31% organic vegetable protein and no vegetable oil. Rainbow trout fed diet C had lost weight by the end of the experiment (the initial weight was 44.98±1.08, and the final weight was 32.60±1.91 g). The thicker submucosa layer that was observed in the rainbow trout fed diet C could be associated with insufficient nutrition. Sitja-Bobadilla *et al.* [33] conducted a study on fish meal substitution of 20%, 75% and 100% with a vegetable mixture in gilthead sea bream and found that the submucosa layer was dilated with connective tissue forming loose layers. Baeza-Arino *et al.* [34], in a study on *Sparus aurata* fed diets in which fish meal was substituted by a mixture of vegetable proteins by 30%, 60% and 90%, determined that the submucosa layer thickness was not affected by the different diets.

Although not statistically significant, rainbow trout fed diet A had longer intestinal folds than rainbow trout fed diet B. The intestinal fold widths for diets A and B were similar. Rainbow trout fed diet C had significantly smaller and thinner intestinal folds, which is logical because these fish lost weight in the 5-week experimental feeding. For diets A and B, it seems that with more vegetable proteins, the intestinal folds became longer. Baeza-Arino *et al.* [34] also found that the intestinal folds in the distal intestine of Sparus aurata individuals were longer when fish were fed a diet with a substitution of fish meal for vegetable proteins. In fact, the intestinal folds tended to be longer as the vegetable protein amount rose. Boonyaratpalin *et al.* [35], Santigosa *et al.* [36] and Øverland *et al.* [37] found that when barramundi, Atlantic salmon and gilthead sea bream were fed diets with a substitution of fish meal for a vegetable source, the distal intestine showed shorter intestinal folds. According to Baeza-Arino *et al.* [34], the intestinal fold width could be related to the number of enterocytes, thus making these villi appear more or less bulky. Enterocytes [38], which are present in the epithelium, make a bulge and serve to strengthen the epithelial layer. Our results showed that fish fed diet A or B had similar intestinal fold widths and that the absorption of the nutrients by the enterocytes was likely similar.

Goblet cells are very important for the fish health and nutrition. According to van den Ingh *et al.* [26], a sign of enteritis can be the increased amount of goblet cells in the epithelium of the intestine. Enteritis is usually related to distal parts of the gastrointestinal tract and is also characterized by goblet cell hyperplasia and hypertrophy. Bozic *et al.* [39] observed that there was an increment in the number of intestinal goblet cells in carp after starvation. The mucus secreted by goblet cells lubricates undigested materials for onward progression into the rectum. For the fish fed diet B, the goblet cells (18.29 ± 0.39 μm) were statistically smaller ($p=1.19x10-6$) than were those of the fish fed diets A or C (22.00 ± 0.63 μm and 21.92 ± 0.43 μm, respectively). This result can be explained as follows: diets A and C produced larger amounts of undigested materials because of the higher quantities of the plant ingredients (they are not digestible and contain anti-nutritional factors). Thus, more intestinal lubrication was needed, and bigger goblet cells were formed. Fish fed diet B had more goblet cells per 100 μm of intestinal fold (6.67 ± 0.33) than did fish fed diet A (4.78 ± 0.55). Skalli *et al.* [40] reported between 1 and 1.8 goblet cells per 100 μm of intestinal fold length for rainbow trout. Urán *et al.* [41] reported 2 to 4.5 goblet cells per 100 μm of intestinal fold length for Atlantic salmon. Martínez-Llorens *et al.* [42] substituted fishmeal with carob seed germ meal in the sea bream diet and measured 5.8 to7.5 goblet cells per 100 μm of intestinal fold length. There was a reduction in the number of goblet cells fed a carob seed germ meal diet. Our goblet cell numbers are higher than Skalli's [40] but are similar to Martínez-Llorens' [42] because we had a reduction of goblet cell number for diet A, which had a larger amount of vegetable meal than diet B did.

The study of changes of intestine histology following diet containing a single meal showed that 6 h after feeding the size of the goblet cells decreased. Goblet cells returned to their original size 12 h after feeding. It seems that at 6 h after feeding, which is the peak of the protein synthesis rates in trout [15], the digestion process is in progress, and goblet cells secrete their mucous by decreasing their size. 12 h after feeding the digestion process is towards the end and the protein synthesis rates are decreasing [15]. Thus, the production of mucous by the goblet cells is not needed. The similar size of the goblet cells before feeding and at 12 h after feeding confirms that rainbow trout should be fed twice per day and that the second feeding should take place between 6 and 12 hours after the first feeding.

Intestinal histology could be a helpful tool for evaluating different diets on fish growth and welfare, particularly if the substitution of fish meal and fish oil by plant ingredients occurs. The goblet cell numbers, intestinal fold length, intestinal fold width, brush border height, muscular layer width and submucosa layer width are factors that can be evaluated for intestinal health. The study of changes of intestine histology following a single meal can be very helpful for estimating the feeding frequency. This study gives more insight into the feeding management of trout aquaculture. The long-term metabolic consequences of feeding fish with low fishmeal and fish oil and a high plant protein and plant oil diet is still under consideration, and further knowledge is required regarding the way in which the dietary composition influences gastrointestinal health in fish.

CONFLICT OF INTEREST

The author (editor) declares no conflict of interest, financial or otherwise.

ACKNOWLEDGEMENT

Eleni Mente was awarded a visiting fellowship by Marine Alliance for Science and Technology Scotland (MASTS).

REFERENCES

[1] Waite R, Beveridge M, Brummett R, *et al.* Improving Productivity and Environmental Performance of Aquaculture 2014.

[2] FAO. The State of World Fisheries and Aquaculture: Opportunities and Challenges, Food and

Agriculture Organization of the United Nations, Rome. 2014.

[3] FAO. Cultured Aquatic Species Information Programme Oncorhynchus mykiss. Rome: FAO Fisheries and Aquaculture Department 2005.

[4] Ellis T, North B, Scott AP, Bromage NR, Porter M, Gadd D. The relationships between stocking density and welfare in farmed rainbow trout. J Fish Biol 2002; 61: 493-531. [http://dx.doi.org/10.1111/j.1095-8649.2002.tb00893.x]

[5] FAO. Small-scale rainbow trout farming. Rome: Food and Agriculture Organization of the United Nations 2011.

[6] Jutfelt F. The intestinal epithelium of salmonids: transepithelial transport, barrier function and bacterial interactions Department of Zoology, Zoophysiology. Sweden: Göteborg University 2006.

[7] Buddington RK, Krogdahl A, Bakke-Mckellep AM. The intestines of carnivorous fish: structure and functions and the relations with diet. Acta Physiol Scand Suppl 1997; 638: 67-80. [PMID: 9421581]

[8] Clements KD, Raubenheimer D. Feeding and Nutrition.The Physiology of Fishes. Boca Raton, FL, USA: Taylor and Francis Group 2006; pp. 47-82.

[9] Collie NL, Ferraris RP. Nutrient fluxes and regulation in fish intestine. Biochemistry and Molecular Biology of Fishes 1995; 4: 221-39. [http://dx.doi.org/10.1016/S1873-0140(06)80012-3]

[10] Nordrum S, Bakke-McKellep AM, Krogdahl A, Buddington RK. Effects of soybean meal and salinity on intestinal transport of nutrients in Atlantic salmon (*Salmo salar* L.) and rainbow trout (*Oncorhynchus mykiss*). Comp Biochem Physiol B Biochem Mol Biol 2000; 125(3): 317-35. [http://dx.doi.org/10.1016/S0305-0491(99)00190-X] [PMID: 10818266]

[11] Genten F, Terwinghe E, Danguy A. Atlas of fish histology. Science Publishers 2009. [http://dx.doi.org/10.1201/b10183]

[12] Buke D. The anatomy and histology of the carnivorous fish the pike Esox lucius L. J Fish Biol 1971; 31: 421-31. [http://dx.doi.org/10.1111/j.1095-8649.1971.tb05914.x]

[13] Groman GB. Histology of the striped bass. 1982.

[14] Khojasteh SM. The morphology of the post-gastric alimentary canal in teleost fishes: a brief review. Int J Aquat Sci 2012; 3: 71-88.

[15] Mente E, Pierce GJ, Antonopoulou E, Stead D, Martin SA. Postprandial hepatic protein expression in trout *Oncorhynchus mykiss* a proteomics examination. Biochemistry and Biophysics Reports 2016; 9: 79-85. [http://dx.doi.org/10.1016/j.bbrep.2016.10.012]

[16] Mente E, Nikouli E, Antonopoulou E, Martin SA, Kormas KA. Core *vs.* diet -associated and postprandial bacterial communities of the rainbow trout (*Oncorhynchus mykiss*) midgut and feaces. Unpublished mimeo.

[17] Brafield AE. Laboratory studies of energy budgets.Fish energetics: New perspectives. London, Sydney: Croom Helm Ltd. 1985; pp. 257-81. [http://dx.doi.org/10.1007/978-94-011-7918-8_10]

[18] Caballero MJ, Izquierdo MS, Kjorsvik E, *et al.* Morphological aspects of intestinal cells from gilthead seabream (*Sparus aurata*) fed diets containing different lipid sources. Aquaculture 2003; 225: 325-40. [http://dx.doi.org/10.1016/S0044-8486(03)00299-0]

[19] Ringø E, Lødemel JB, Myklebust R, Kaino T, Mayhew TM, Olsen RE. Epithelium-associated bacteria in the gastrointestinal tract of Arctic charr (Salvelinus alpinus L.). An electron microscopical study. J Appl Microbiol 2001; 90(2): 294-300. [http://dx.doi.org/10.1046/j.1365-2672.2001.01246.x] [PMID: 11168733]

[20] Ringø E, Myklebust R, Mayhew TM, Olsen RE. Bacterial translocation and pathogenesis in the digestive tract of larvae and fry. Aquaculture 2007; 268: 251-64. [http://dx.doi.org/10.1016/j.aquaculture.2007.04.047]

[21] Harper GM, Dimitroglou A, Ringø E, Merrifield DL. Electron microscopy: an important tool to assess the effects of dietary components on the gastrointestinal tract of fish. Asian Aquaculture Network. Practical 2011; 2: 22-4.

[22] Sales J, Glencross B. A meta-analysis of the effects of dietary marine oil replacement with vegetable oils on growth, feed conversion and muscle fatty acid composition of fish species. Aquacult Nutr 2011; 17: 271-87. [http://dx.doi.org/10.1111/j.1365-2095.2010.00761.x]

[23] Nasopoulou C, Zabetakis I. Benefits of fish oil replacement by plant originated oils in compounded fish feeds. A review. Food Sci Technol 2012; 47: 217-44.

[24] Bell MV, Dick JR, Porter AE. Pyloric ceca are a major site of 22:6n-3 synthesis in rainbow trout (*Oncorhynchus mykiss*). Lipids 2003; 39: 39-44. [http://dx.doi.org/10.1007/s11745-003-1029-5] [PMID: 12669818]

[25] Fonseca-Madrigal J, Bell JG, Tocher DR. Nutritional and environmental regulation of the synthesis of highly unsaturated fatty acids and of fatty-acid oxidation in Atlantic salmon (*Salmo salar* L.) enterocytes and hepatocytes. Fish Physiol Biochem 2006; 32: 317-28. [http://dx.doi.org/10.1007/s10695-006-9109-2]

[26] van den Ingh TS, Krogdahl Å, Olli JJ, Hendriks HG, Koninkx JG. Effects of soybean-containing diets on the proximal and distal intestine in Atlantic salmon (*Salmo salar*): a morphological study. Aquaculture 1991; 94: 297-305. [http://dx.doi.org/10.1016/0044-8486(91)90174-6]

[27] van den Ingh TS, Olli JJ, Krogdahl Å. Alcohol-soluble components in soybeans cause morphological changes in the distal intestine of Atlantic salmon, *Salmo salar* L. J Fish Dis 1996; 19: 47-53. [http://dx.doi.org/10.1111/j.1365-2761.1996.tb00119.x]

[28] Baeverfjord G, Krogdahl A. Development and regression of soybean meal induced enteritis in Atlantic salmon, *Salmo salar* L., distal intestine: a comparison with the intestines of fasted fish. J Fish Dis 1996; 19: 375-87. [http://dx.doi.org/10.1111/j.1365-2761.1996.tb00376.x]

[29] Refstie S, Korsøen ØJ, Storebakken T, Baeverfjord G, Lein I, Roem AJ. Differing nutritional responses to dietary soybean meal in rainbow trout (*Oncorhynchus mykiss*) and Atlantic salmon (*Salmo salar*). Aquaculture 2000; 190: 49-63. [http://dx.doi.org/10.1016/S0044-8486(00)00382-3]

[30] Caballero MJ, Obach A, Rosenlund G, Montero D, Gisvold M, Izquierdo MS. Impact of different dietary lipid sources on growth, lipid digestibility, tissue fatty acid composition and histology of rainbow trout, *Oncorhynchus mykiss*. Aquaculture 2002; 214: 253-71. [http://dx.doi.org/10.1016/S0044-8486(01)00852-3]

[31] Olsen RE, Dragsnes BT, Myklebust R, Ringø R. Effect of soybean oil and soybean lecithin on intestinal lipid composition and lipid droplet accumulation of rainbow trout, *Oncorhynchus mykiss* Walbaum. Fish Physiol Biochem 2003; 29: 181-92. [http://dx.doi.org/10.1023/B:FISH.0000045708.67760.43]

[32] Baecker R. Über die Nebennieren der Teleostier. Z Mikrosk Anat Forsch 1928; 15: 204-73.

[33] Sitjà-Bobadilla A, Peña-Llopis S, Gómez-Requeni P, Médale F, Kaushik S, Pérez-Sánchez J. Effect of fish meal replacement by plant protein sources on non-specific defence mechanisms and oxidative stress in gilthead sea bream (*Sparus aurata*). Aquaculture 2005; 249: 387-400. [http://dx.doi.org/10.1016/j.aquaculture.2005.03.031]

[34] Baeza-Ariño R, Martínez-Llorens S, Nogales-Mérida S, Jover-Cerda M, Tomás-Vidal A. Study of

liver and gut alterations in sea bream, *Sparus aurata* L., fed a mixture of vegetable protein concentrates. Aquacult Res 2014; 47: 460-71.
[http://dx.doi.org/10.1111/are.12507]

[35] Boonyaratpalin M, Suraneiranat P, Tunpibal T. Replacement of fish meal with various types of soybean products in diets for the Asian seabass, Lates calcarifer. Aquaculture 1998; 161: 67-78.
[http://dx.doi.org/10.1016/S0044-8486(97)00257-3]

[36] Santigosa E, Sánchez J, Médale F, Kaushik S, Pérez-Sánchez J, Gallardo MA. Modifications of digestive enzymes in trout (Oncorhynchus mykiss) and sea bream (*Sparus aurata*) in response to dietary fish meal replacement by plant protein sources. Aquaculture 2008; 282: 68-74.
[http://dx.doi.org/10.1016/j.aquaculture.2008.06.007]

[37] Øverland M, Sørensen M, Storebakken T, Penn M, Krogdahl Å, Skrede A. Pea protein concentrate substituting fish meal or soybean meal in diets for Atlantic salmon (*Salmo salar*)—effect on growth performance, nutrient digestibility, carcass composition, gut health, and physical feed quality. Aquaculture 2009; 288: 305-11.
[http://dx.doi.org/10.1016/j.aquaculture.2008.12.012]

[38] Baker JR. The structure and chemical composition of the Golgi element 1944.

[39] Bozić F, Srebocan E, Kozarić Z. Starvation-induced pathobiology in the gut of carp (*Cyprinus carpio* L.). Berl Munch Tierarztl Wochenschr 2001; 114(3-4): 134-8.
[PMID: 11314585]

[40] Skalli A, Castillo M, Andree KB, Tort L, Furones D, Gisbert E. The LPS derived from the cell walls of the Gram-negative bacteria Pantoea agglomerans stimulates growth and immune status of rainbow trout (*Oncorhynchus mykiss*) juveniles. Aquaculture 2013; 416: 272-9.
[http://dx.doi.org/10.1016/j.aquaculture.2013.09.037]

[41] Urán Carmona PA, van Houcke J, Rombout JH, *et al.* Comparison of methodologies for measuring the degree of soybean induced enteritis in Atlantic salmon (Salmo salar L). Bali, Indonesia: World Aquaculture 2005.

[42] Martínez-Llorens S, Baeza-Ariño R, Nogales-Mérida S, Jover-Cerdá M, Tomás-Vidal A. Carob seed germ meal as a partial substitute in gilthead sea bream (*Sparus aurata*) diets: amino acid retention, digestibility, gut and liver histology. Aquaculture 2012; 338: 124-33.
[http://dx.doi.org/10.1016/j.aquaculture.2012.01.029]

CHAPTER 12

Developing a Methodology for the Mariculture of Bath Sponges in Larymna Gulf, North Evoic, Greece

Jane Castritsi - Catharios[1,*], Constantin Vamvakas[2], Ilias Baras[3] and **George Ekonomou[1]**

[1] *Department of Ichthyology and Aquatic Environment, School of Agricultural Sciences, University of Thessaly, Volos, Hellas, Greece*

[2] *Laboratory of Aquaculture & Artemia Reference Center, Faculty of Bioscience Engineering, University of Ghent, Ghent, Belgium*

[3] *Selonda Aquaculture SA, Navarchou Nikodimou 26-30, 10556 Athens, Hellas, Greece*

Abstract: Climate change, pollution, overfishing and the periodic appearance of the sponge disease, dramatically reduced the stock of sponges, the number of species and the diversity of their forms. Demand continues to grow despite their high prices (150 euros/kg). Price depends on type, quality, necessary processing and geographical origin. The possibility of producing secondary metabolites & collagen from sponges makes them a valuable source for the production of raw materials of high added value. An experimental culture was performed in an integrated aquaculture fish production unit in Larymna (Greece) on ropes between fish cages (arranged vertically), on iron frames placed 0.5 m above the bottom (arranged horizontally) and in onshore tanks. Thus far results from the sea experiments (vertical and horizontal arrangements) have been encouraging, while those from the onshore facilities have been less so. Three experiments were performed with two species of sponges in order to develop a suitable method of sponges' mariculture in parallel with fish farming with the following objectives: (a) to study the robustness of the sponges, (b) to investigate the method of harvesting in order to ensure sustainability of their biotopes, (c) to investigate the modes of transport from the harvesting areas to the experimental unit, (d) to examine a biological decontamination technology in closed bays and protected areas thanks to the high filtration capacity of the sponges.

Keywords: Fish cages, *Hippospongia communis*, Sponge mariculture, *Spongia officinalis officinalis*.

* **Corresponding author Jane Castritsi – Catharios:** Department of Ichthyology and Aquatic Environment, School of Agricultural Sciences, University of Thessaly, Volos, Hellas, Greece; Tel: (+30)6972234770; E-mail: cathario@uth.gr

Panagiotis Berillis (Ed)

INTRODUCTION

Sponges are a zoic group with a special interest both for basic science and for applied research. In fact, it may be the most important phylum in the Animal Kingdom, because it is the key that allows us to better understand the organization of other metazoans and to study the structure and function of their symbiotic organisms.

The *Porifera* Phylum differs from other Phyla of invertebrates, even from those deriving from similar ecological niches, since they maintain a cellular interdependence, like *Protozoa*.

Sponges are sedentary organisms that react strongly to changes observed in the abiotic environment and have a relatively low spread potential due to the short life of their planktonic larvae [1]. The entire life cycle of sponges is centered on the filtration of large quantities of water, that can reach three times their volume [2], consequently becoming a unique seawater filter. Their uniqueness is due to their hydrophoric system, which allows them, together with thousands of their incurrent pores and their choanocells, to absorb and to infiltrate the surrounding water. Their filtering capacity is counter proportional to their age. Aristotle was the first to discover the existence of sensation in sponges and, therefore, he classified them as animals. This view was shared by Plinios and Aelianos, who both found that sponges are able to perform contractile and dilatory motions when they receive a stimulus.

Sponges have always been a natural source of income for Greece, but the occasionally occurring "sponge disease" caused nearly complete desertification of the fishing grounds [3 - 10].

According to the sponge fishermen the fields are gradually recovering, which was expected since the disease is not transferred to a depth greater than 50 meters [3]. Since ancient times the sponges, were mainly used for cosmetic and domestic purposes, in biotechnical applications, in painting, *etc*. There is no doubt that no synthetic product can be commercially compared with natural sponges.

According to the literature, for several years efforts have been made on a global scale for the exploitation not only of the commercially known species but also for the wild, noncommercial ones. Recently, Ehrlich *et al*. [11] showed that collagen and chitin are the most investigated materials of biological origin with a wide field of applications in biomedicine (some demosponges and glass sponges possess chitin as a component of their skeleton). Also, collagen A and the short-chain collagen B [12 - 14] as well as the bioactivity of the secondary metabolites of sponges are raising the interest not only of the pharmaceutical industry

[15 - 17] but also of other industries, including those producing cosmetics [*e.g.* 18, 19].

The skeleton of commercial natural sponges belonging to the genera *Spongia, Hippospongia* and *Conscioderma* consists of a dense network of horny sponge fibers, a specific form of collagen, which presents properties that do not exist in synthetic sponges obtained by polymerizing vinylic and allylic derivatives [20].

Natural commercial sponges, depending on their quality, should be sold as two products: "normal" and "luxury" [21]. Their commercial value is determined by their elasticity, durability and softness. The determination of quality relies on the experience of a few sponge dealers. Various methods have been developed to quantify the quality of the sponge, such as the measurement of the tensile strength [22, 23]. This method has been applied to a number of other natural products, *e.g.* collagen fibers [24]. The quality of commercial sponges depends mainly on the abiotic parameters of the habitat which affect, in various ways, the form [25] distribution, metabolism and reproductive ability.

Specifically

Each ecosystem depends on several factors and, according to the general rule in nature, each ecosystem tends to a stability, to an equilibrium. This stability may be overruled either by natural or by anthropogenic factors. Each ecosystem is self - protected and appears the ability of self - adjustment and uniformity. In the case of sponges' culture the biotic and abiotic characteristics along with those of their substrate are among the main affecting parameters. By analyzing single ecological factors and mainly the abiotic ones, it appears that the most relevant factor to their biodiversity is temperature, since it acts on the oxygenation of the water and the primary production. Reiswig [26] reported, that the quality of the water filtered by a sponge increases with temperature. Simpson [27] found that the species *Microciona prolifera, in vivo* cultured, appeared the same growth rate at 15° C and 20° C. Under natural conditions the growth was discontinued at 10° C. The turbidity is another significant factor for the growth of sponges and depends on the hydrodynamic characteristics of the area. The inorganic salts and the organic material constitute the stocks of nitrogen, phosphorus and silicon, which are necessary for the growth of phytoplankton.

Lighting consists an inhibiting or stimulating factor for the growth of Demosponges and affects the dimension and, mainly, their shape [28]. The primary production is a parameter that is directly influencing the existence of a phytal system. The level of primary production plays an important role in the development rate and quality of sponges. Their survival and growth are exclusively depending on the three dimensional environment to which they are indissolubly bound. The chemical composition of the substratum, in nature, according to bibliographical data, does not affect the growth of sponges. From observations of Kalymnian divers made on wreckages it was found that the sponges are better binding on bottom plating, on wood then and less on the stones.

During the period 1991-2014, the global production of natural sea sponges was 13400 metric tons, while the demand was much higher [29]. It is obvious that the harvesting of sponges from natural populations does not meet the demand, without taking into account the continuous growing interest from the pharmaceutical and cosmetics industries. Thus, the only alternative to the growing problem of the market is the mariculture of sponges.

The mariculture of sponges may offer a solution to the environmental problem of organic pollution, as well as helping satisfy the growing demand for the production of large numbers of new bioactive substances [15]. In 1987 Venderal and Venderal [21] mentioned that a maximum of 200 metric tons of natural sponges (*Hippospongia* genus) is made available to the market annually. They also reported that "starting with pieces of sponge attached to vertical ropes, cultivation is easy."

The experimental culture of sponges on ropes started a century ago [30]. It evolved in the Mediterranean using new materials such as plastics [31 - 33]. The usage of larvae collector as a method of further cultivation was not successful. On the contrary the experiments which were based on the regenerative activity of the sponges were much promising [34, 35]. In these experiments, parts of sponges were cut and placed on a solid substrate until these were grown to the commercial size.

Targeting to a successful sponge mariculture, emphasis should be placed on: the planting area, the distance from other sponge fishing grounds (so that the time between harvesting and implantation will be minimized) and their diversity.

This study aimed to experimentally mariculture commercial sponges with various methods in order to gain significant experience in the evaluation of installation sites, the time and mode of transport, the way of cutting, the size of the implants, the way of binding, the distances between implants, the appropriate depth, the shading and the diet (for the case of land cultivation), the robustness of cultivated species and their simultaneous use as biological filters to clean sea areas with high organic loads.

MATERIALS AND METHODS

Place of Installation

In order to select the appropriate fish farm to host the experiment, an initial investigation was made of different areas in Greece, among the ones belonging to Selonda SA, which had floating cages and a hatchery. A number of samples was taken for testing the sea water from the units in the following prefectures: Corinth - farm "Selonda", Argolida, Magnesia - farm "Trikeri", Fthiotida - farms "Larymna" and "Gatza", and the abiotic and biotic parameters were measured. The following microbiological parameters were also measured: total bacteriological counts (TBC), coliforms, *Escherichia coli* and enterococci. A number of physicochemical parameters were also measured: pH, conductivity, nitrites, nitrates, ammonium, chlorides and phosphates and total suspended solids. Also microbiological and physicochemical parameters of the water coming from wells which supply the hatchery of Larymna (Tables **3** and **4**) were measured while Tables **1** and **2** exhibit correspondingly the physicochemical and microbiological parameters tested as well the methods used.

Field Sampling

Two series of samples were taken in the same site, the first on 18 June 2015 and the second on 22 July 2015 with a sponge-fishing boat, suitably equipped to collect the sponges and to transport them in the installation-unit for culture. In the first experimental trip in June, samples of sponges were collected by diving from the area of Zagora, in the east coasts of Pelion and at a depth 9-12 m. The characteristics of the fishing area were: water temperature at the surface 23.6° C, in the bottom 23.4° C, dissolved oxygen equal to 6.89 mg/lt and complete absence of currents. The substrate was rocks and coarse material, mainly coming from the mechanical disaggregation of rocks, *i.e.* gravels and sand, of spheroidal, rounded and angular forms (particle analysis with a proper sieve). In the tanks, where the sponges were placed for transport, a pump and a hose with filter were installed so that during the trip (six hours duration) the seawater in the tanks was renewed with a controlled flow and thus the temperature and the oxygen levels remained stable, minimizing the stress of the sponges during transfer.

Once the reception and the transport tanks were prepared on board, the first dive for the collection took place, which lasted 55 minutes, and a total of 35 sponges of the species *Spongia officinalis officinalis* (common name matapàs or mantapàs) was collected.

Table 1. Physicochemical Water tests.

Type	Method
pH	4500-H,B St. Met.
conductivity	2510 B St. Met.
Total hardness	2340 B St. Met.
chloride	4500 Cl, B St. Met.
nitrate	Fasmaphotometry 4500 NO_3^- - B St. Met.
nitrite	Fasmaphotometry 4500 NO_2,B St. Met.
ammonium	Fasmaphotometry 4500 NH_3 - F St. Met.
phosphates	4500 P, C St. Met.
Total suspended solids	2540-D St. Met.
calcium	ICP 320 A,B Modified St. Met.
magnesium	ICP 320 A,B Modified St. Met.
potassium	ICP 320 A,B Modified St. Met.
sodium	ICP 320 A,B Modified St. Met.
sulphates	4500 SO_4, E St. Met.
phosphorus	4500 P, E St. Met.
silicon	ICP
color	2120C, St. Met.
cloudiness	St. Met. 2130 B
BOD_5	5210-D, 4-94 St. Met.
COD	USEPA 1980, 45 (78) 26811-26812
Total disolved solids	St. Met. 2540 C
Total Nitrogen	Kjeldahl

Table 2. Microbiological Water tests.

Type	Method
TBC	ISO 6222:1999
Coliforms	ISO 9308-1:2000
Escherichia coli	ISO 9308-1:2000
Enterococci	ISO 7899-2:2000

The sponges were removed from their substrate with a very sharp knife. During this procedure the sponge-fishermen made sure that a small quantity of sponge remained on the initial substrate so that a new sponge would develop asexually. These samples did not come into contact with air; during the diving they were placed in plastic containers full of sea water (Fig. **1A**) and transported to the deck of the fishing boat and placed in the transport tanks.

The tanks were covered in order to avoid direct exposure to the sun, and samples were taken from the harvested sponges for further laboratory analysis.

The second dive took place at the same spot, the same day and lasted 45 minutes, during which 30 sponges were collected, 20 belonging to the species *Spongia officinalis officinalis* (matapas) and 10 to the species *Hippospongia communis* (common name kapadiko) (Fig. **1B**).

Fig. (1). A) Harvested sponges during their transport on board the fishing vessel, **B**) Healthy sponges (kapadiko and matapas) covered with exopinacoderm.

Sponge Cutting Process

The process of dissecting the sponges into segments suitable for implantation started during the trip. The cutting was done in the transport tank (Fig. **2**) using a piece of wood as a cutting table. At the same time, pieces were threaded onto fishing line in groups of 20. The process of dissecting and threading onto fishing line lasted 2.5 hours and the pieces were ready for implantation.

In total, 65 sponges were collected in the first experimental trip of which 55 were of matapas species and 10 of kapadiko species.

Throughout the transport from the collection point and during the implant preparation process, temperature and oxygen measurements were taken in the transport tanks (Table **5**).

Calculation of the Required Number of Implants

Some assumptions were made in order to calculate the necessary quantity of sponges from which the number of implants would come for the experimental cultures. One kilogram of sponges that have just come out of the sea corresponds to 5-6 individuals (orange insize) of an estimated size of 15x15x5 cm.

Fig. (2). Transport & cutting tank of sponges.

In total, for each long-line it was calculated that 80-85 implants were required with a volume of 150 -200 cm^3 each (for 4 ropes 40-60 sponges were needed).

For the culture in frames, it was calculated that 300 implants were required with a volume of 150-200 cm^3 each (for the 3 frames 50-60 sponges were needed).

For onshore tanks it was estimated that 120-150 implants were required for each tank, which derived from 15-20 sponges.

The total number of sponges for the three experiments was about 140-150 adult individuals of the kapadiko and matapas species. The exact number depended on the population in the nearest sponge fields and was collected during the two experimental trips. During the second trip on 22 July 2015, a significant number of small sponges was collected, which were added in the three types of the experiments following the same method as for the implants.

The entire process described above took about six months (sampling with the boat, analysis of the abiotic and biotic parameters, investigation of healthy sponge areas with divers, test cultures of phytoplankton and training of the company's scientific staff in relation to the culture of a new species for the unit).

Checking the sponge robustness

In order to check the robustness of the sponges before their implanting small specimen were soaked, without having been exposed to air, with dilute Rose Bengal solution and remained for several seconds. The same checks were performed during the cultivation. The pink color was a key indication that the tissues are alive.

Cultivation Methods

Marine Installations in Ropes (Long-lines)

The distance between the fish-cages was approximately 8 m. Between two cages a horizontal rope was fixed on which hung vertically four ropes of 18-20 m in length. The lower ends were connected to a rope similar to the one at the top and this was tied to the anchors of the cages. On each vertical rope implants were tied below 10 m, with 0.50 m between them (Fig. **3**) (20 implants/rope). In total, each long-line set took 80-85 implants. Each implant had a volume 150-200 cm^3.

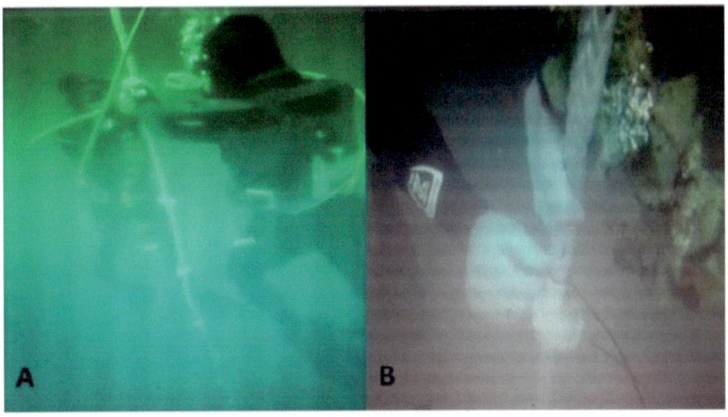

Fig. (3). A) Sponges - implants in a rope 8 months after installation, **B**) Binding the implants on the knobs of the vertical ropes.

Marine Installations in Frames on the Bottom

At a distance of about 200-400 m from the fish-cages three metal frames of 2x2 m were placed at the height of 0.5 m from the sea bed. These frames were fixed on the substrate.

Above each frame was fixed a metal mesh grid and on this mesh grid were attached 100 implants (Fig. **4**). Each implant had a volume of 150-200 cm^3, and in the three frames there were placed 30-40 individuals of (15x15x5 cm).

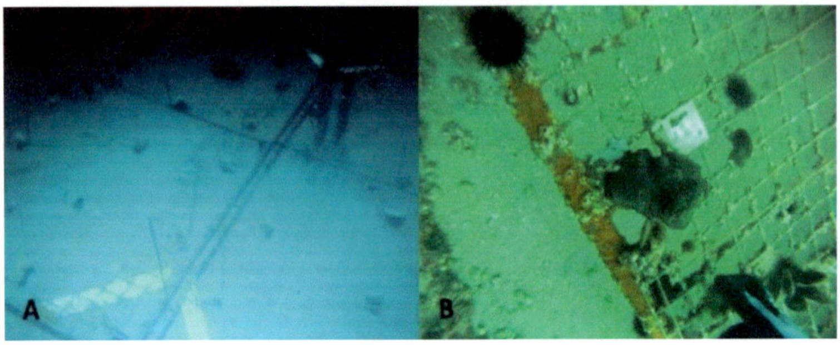

Fig. (4). A) Frames in the bottom with implants, **B**) Implant after 12 months.

Onshore Facilities

Three relatively shallow cylindrical tanks (depth about 1.20 m) with a volume of 4.5 m^3 were used. The water used was of good quality (according to the findings of the analysis) and it was derived from a bore. The food provided (*Nannocloropsis galbana*, *Nanocloris atomus* and *Isochrisis sp.*) was generous; it was derived from internal cultures in the hatchery. The tanks were covered with single & double shading (Fig. **5A**) (lighting plays an important role in the color and possibly in the elasticity of sponge fibers). At the bottom of each tank a plastic mesh was placed and were tied so that it could be lifted close to the surface for observations and measurements of the sponges (Fig. **5B**).

Fig. (5). A) Surface experimental tanks for cultivating sponge. The first bearing double shading, the second a single one and the third is uncovered. **B**) Plastic mesh close to the surface for observation of implants.

RESULTS AND DISCUSSION

Voultsiadou and Lagoutari [36] in 2008, reported that "the area for the installation of implants for the experiment should be done very carefully. The survival of the implanted sponges and their growth speed are dependent on the environmental factors" [37].

According to the analytical findings (Tables **3** and **4**) there are no significant differences in abiotic and microbiological parameters in the regions examined. The same resulted from the analysis of water bores that supply the onshore tanks. They were all within the expected limits and found suitable for rearing sponges.

After testing, the fattening unit in cages in the fish hatchery of "Larymna" was selected to be used for the experimental cultivation of the sponges. The "Larymna" unit is located in the bay of the same name in the Fthiotida prefecture. Apart from the fish farm cages, this farm has onshore facilities (hatchery) so access, storage and daily monitoring of the cultivation of sponges is possible.

Larimna bay is leeward, with good renewal of the water, although the abundance of suspended inert substances and organic material coming from the cages puzzled for its suitability in relation to the survival and development of the implants. Another advantage of this area was that before the disease which occurred in 1986 and according to a personal communication coming from former sponge fishermen, there were sponge fishing grounds inside the bay and in the wider area. An additional advantage of "Larymna" farm is that the fish hatchery is in a production process and contains an active live feed production department, fully controlled.

The easy transport of fresh samples of sponges in the laboratories in Volos (University of Thessaly) and Athens (University of Athens) for analysis, and the presence of sponge fishing grounds with healthy populations in the Pagasitikos Gulf, near the unit, contributed also to the choice of the location for the experiment.

Throughout the transport from the collection point and during the implant preparation process, temperature and oxygen measurements were taken in the transport tanks. Both water temperature and dissolved oxygen levels, as is clear from Table **5**, remained stable for six hours of travel.

In the onshore tanks, originally there was a continuous streaming system which supplied the tanks with 150 l of water per day. The abundant supply of food, either with phytoplankton (Fig. **6**) or with the water runoff from nurseries (dissolved organic matter from the farming of fish), created problems of water turbidity resulting in a dramatic increase in the mortality of implants. Thus it was necessary to immediately reduce both the amount of food and the frequency of feeding.

Table 3. Results of chemical and microbiological analyses.

FISH FARM	TYPE OF ANALYSIS	PARAMETER	RESULTS
TRIKERI	MICROBIOLOGICAL	*Escherichia coli*	<1
		Enterococci	<1
	PHYSICOCHEMICAL	pH	7.7
		Conductivity (μS/cm)	51461
		Total Hardness	388
		Cl^- (mg/lt)	22791
		NO_3^- (mg/lt)	2
		NO_2^- (mg/lt)	-
		NH_4^+ (mg/lt)	-
SELONDA	MICROBIOLOGICAL	TBC	41
		Coliforms	<1
		Escherichia coli	<1
		Enterococci	<1
	PHYSICOCHEMICAL	pH	8
		Conductivity (μS/cm)	49200
		NO_3^- (mg/lt)	2.1
		NO_2^- (mg/lt)	-
		NH_4^+(mg/lt)	-
		Total suspended solids	-
LARYMNA	MICROBIOLOGICAL	TBC	18
		Coliforms	<1
		Escherichia coli	<1
		Enterococci	<1
	PHYSICOCHEMICAL	pH	8.2
		Conductivity (μS/cm)	48967
		NO_3^- (mg/lt)	2
		(NO_2^-‘mg/lt)	-
		NH_4^+ (mg/lt)	-
		PO_4^{-3}(mg/lt)	-

(Table 3) contd.....

FISH FARM	TYPE OF ANALYSIS	PARAMETER	RESULTS
GATZA	MICROBIOLOGICAL	TBC	14
		Coliforms	<1
		Escherichia coli	<1
		Enterococci	<4
	PHYSICOCHEMICAL	pH	8.1
		Conductivity (μS /cm)	48632
		NO_3^-(mg/lt)	2.1
		NO_2^- (mg/lt)	0.04
		NH_4^+ (mg/lt)	-
		PO_4^{-3} (mg/lt)	-

Table 4. Chemical and microbiological analyses of drilling water.

PARAMETER	Γ1	Γ2	Γ3	Γ4	Γ5	Γ6
pH	7.4	7.4	7.7	7.4	7.5	7.5
Conductivity(μS/cm)	46223	46125	46516	45734	45539	43975
Total Hardness	355	340	341	324	329	312
Ca (mg/lt)	723	660	663	655	668	640
Mg (mg/lt)	1108	1083	1085	1015	1030	973
K (mg/lt)	458	440	428	405	415	395
Na (mg/lt)	11325	10938	10913	10450	10475	10475
Cl^-(mg/lt)	20838	20803	20874	20129	20129	19312
NO_3^-(mg/lt)	2.1	2.3	2.3	2.3	2.3	2.6
NO_2^- (mg/lt)	-	-	-	-	-	-
NH_4^+ (mg/lt)	-	-	-	-	-	-
SO_4^{-2}(mg/lt)	3290	3280	3280	3240	3200	3220
P (mg/lt)	-	-	-	-	-	-
Si (mg/lt)	4	3,8	3,7	3,8	3,8	3,6
Color	-	-	-	-	-	-
Turbidity	-	-	-	-	-	-
BOD_5	-	-	-	-	-	-
COD	-	-	-	-	-	-
Total suspended solids	-	-	-	-	-	-
Total dissolved solids (mg/lt)	36724	34884	36165	34153	36156	32647

(Table 4) contd.....

PARAMETER	Γ1	Γ2	Γ3	Γ4	Γ5	Γ6
Total N (mg/lt)	2.8	-	7	2.8	-	1.4

Table 5. Measurements of temperature and oxygen during the transport of sponges.

Hour	Water temperature	Oxygen content of water (mg/lt)
10:00	23.4° C	6.33
10:30	23.5° C	6.34
11:00	23.6° C	6.34
11:30	23.4° C	6.33
12:00	23.5° C	6.34
12:30	23.3° C	6.34
13:00	23.6° C	6.35
13:30	23.4° C	6.33
14:00	23.5° C	6.34
14:30	23.6° C	6.34
15:00	23.4° C	6.33
15:30	23.6° C	6.34
16:00	23.3° C	6.33
16:30	23.6° C	6.34
17:00	23.4° C	6.33

Fig. (6). A) Monoculture *Nannnocloropsis galbana*, **B**) Monoculture *Isochrysis sp.*

In order to maintain the quality of the water at the desired levels, the supplied feed was reduced to 250 cc/day/tank, whilst 10% of the water in the tanks was replaced with fresh seawater according to the "Berlin Method" [38]. Oxygen, ammonia and pH were stable throughout the experiment. Significant differences were observed in levels of light: tank D1 had better light because it was shaded by a single cover. In general, under the shading cover and depending on weather conditions, tank D1 had 700-2200 lux, while D2 (with double cover) was consistently about 20% lower.

After the above modifications of the initial conditions and for a period of 12 months, the survival of sponges (*Spongia officinalis officinalis* & *Hippospongia communis*) raised to 100%, which indicates very good adaptability to tank conditions (Fig. **7**)

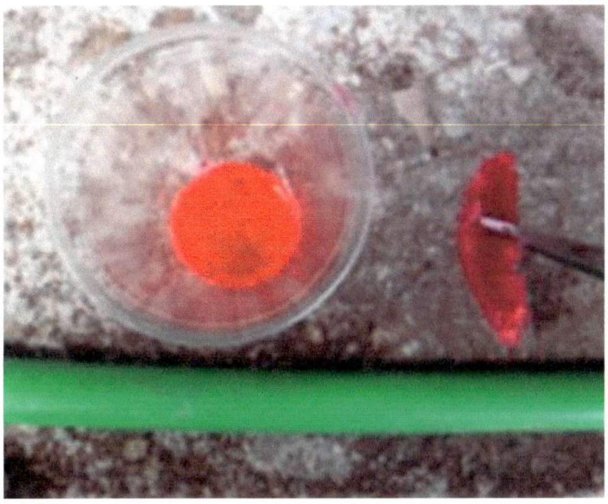

Fig. (7). Checking the robustness of the sponges Sponge specimen immersed in Rose Bengal solution.

However, the growth rate was minimal and in tanks with both single and double shading covers there was significant variation (decrease) in tensile strength [22, 23]. By contrast, the survival rate of the species *Tethya citrina* was 81.1%, with a significant reduction of the size of the implants [39]. Examination of the samples in the laboratory revealed morphological changes in the mesh of fibers (Fig. **8**).

Variations were observed in the thickness of the secondary fibers (Table **6**) compared to those collected directly from the natural environment. This may result in changes in the quality of the cultivated sponges, such as reduction of the aquiferous system [40] or changes in the elasticity and tensile strength, as a response to adverse environmental conditions.

In the vertical ropes between the cages the mortality, observed one month after the installation, was over 70% except for those implants that did not come from cutting sponges but were very small sponges which were removed from their natural substrate and placed using the same method directly in the long line system. These latter sponges remained alive after 12 months, showed no changes in shape, color of the outer membrane, mesh, or thickness of the secondary fibers (Table **6** stations: 14, 15, 18, 26, 27, 34) and they approximately doubled their volume (Fig. **9**). This result is particularly promising because if these sponges had remained in the bay of Larymna they would have had reduced metabolic activity with consequent slowing of growth, due to the large quantity of suspended inert substances which they would have continuously filtered from their environment.

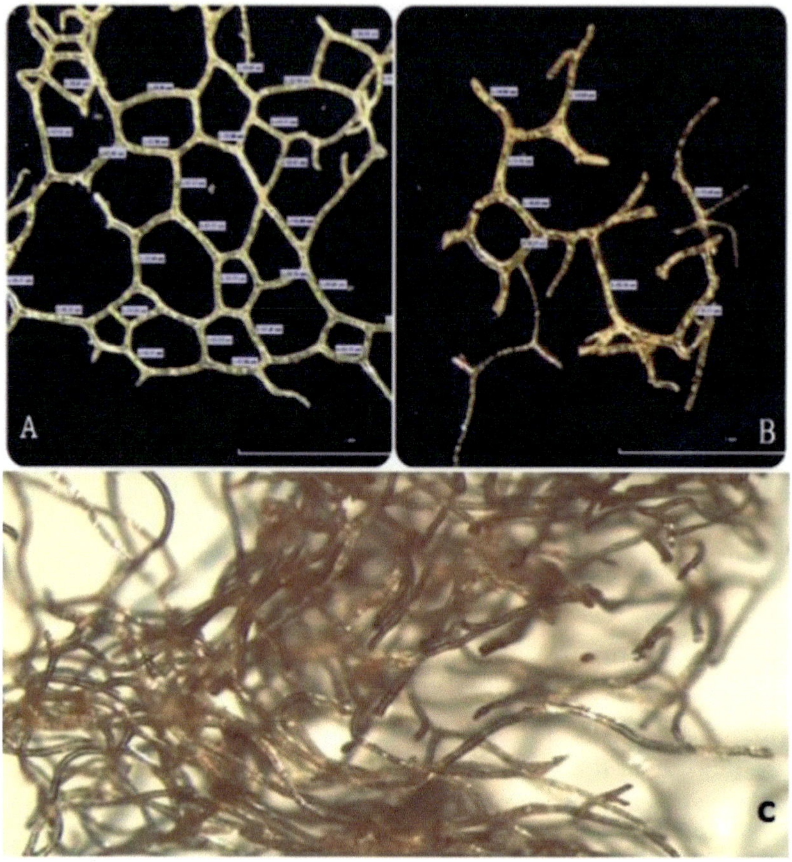

Fig. (8). Fibers from land tanks implants. **A**) Tanks with single shading cover, **B**) Tanks with double shading cover, **C**) Differentiation of the grid with a low tension resistance.

Table 6. Thickness of fibers (in microns) from implants of both types of sponges in land and marine areas.

Nr of stations	Single cover	Double cover	14	15	18	26	27	34
1	25.38	34.06	24.19	30.80	22.98	25.71	22.98	23.44
2	22.14	24.84	22.23	31.31	23.48	22.75	29.39	26.16
3	19.77	33.56	26.80	29.39	23.30	28.96	27.58	21.61
4	21.61	30.83	24.19	38.37	21.16	25.76	26.80	28.85
5	18.39	39.11	29.93	32.01	27.77	28.22	27.96	26.20
6	17.98	23.44	28.85	22.14	21.80	21.89	26.80	22.70

(Table 6) contd.....

Nr of stations	Single cover	Double cover	14	15	18	26	27	34
7	21.16	25.76	23.84	28.07	24.88	22.61	24.88	20.81
8	19.12	31.11	27.15	21.46	26.00	20.56	31.38	23.66
9	18.21	23.66	31.88	23.30	35.18	28.56	26.20	25.76
10	21.46	26.80	27.58	28.67	23.48	23.48	28.78	22.56
11	22.75	30.24	25.38	27.77	21.05	24.36	28.07	28.85
12	24.72	25.88	20.96	27.04	23.84	23.30	24.88	29.96
13	20.81	36.42	26.68	30.56	27.04	19.77	22.23	30.35
14	21.00	33.71	26.20	28.22	21.46	21.61	23.48	20.40
15	17.50	27.65	22.61	22.98	25.43	22.70	26.52	20.81
16	27.77	35.36	30.83	19.12	24.84	25.09	25.09	25.38
17	23.48	27.77	31.44	27.62	21.16	26.20	24.75	26.68
18	23.97	29.96	23.66	29.07	22.98	22.61	26.00	26.72
19	22.19	33.71	25.30	24.06	24.84	24.00	20.81	20.81
20	22.14	36.34	25.38	29.96	23.97	21.16	21.16	26.20
AV	21.58	30.51	26.25	27.60	24.33	23.97	25.79	24.90
SD	2.67	4.66	3.11	4.43	3.19	2.63	2.76	3.20

Fig. (9). Sponges – implants in rope 8 months after their installation.

After 12 months, the survival rate exceeded 90% in the culture on frames. Implants were absolutely normal (Fig. **10A**), covered entirely by the exopinacoderm (Fig. **10B**). No changes in shape (Fig. **4B**), mesh or thickness of fibers or in the aquiferous system were observed. The only problem was with predators, but this is manageable.

The international community accepts that the species of commercial sponges from the Mediterranean sea are of higher quality than those from other regions such as the Caribbean and some island groups of the Pacific ocean (Micronesia) *etc*.

In 2002 Greece established a quality standard certification for commercial species of Mediterranean sponges, AGROCERT, through the Certification & Inspection of Agricultural Products Agency (OPEGEP). This Standard has been prepared by a Technical Committee set up under the auspices of OPEGEP for training in Sectoral Voluntary Prototypes in the agricultural sector. The text for the standard was based on the original research and writings of Dr J. Castritsi - Catharios.

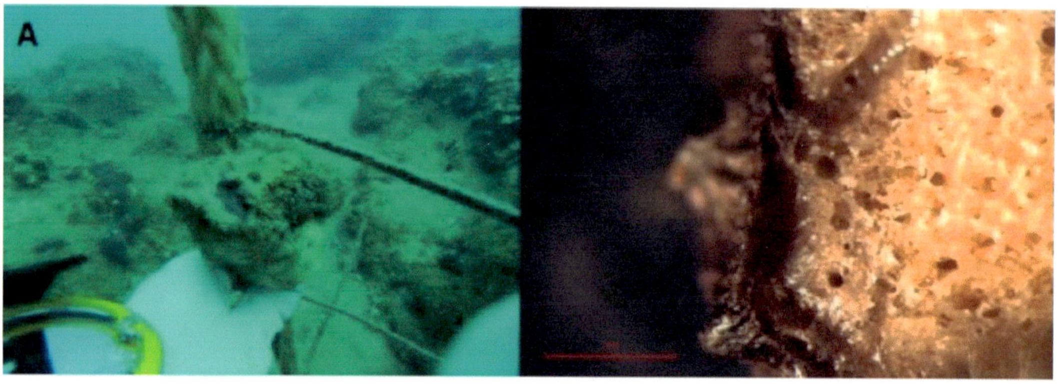

Fig. (10). A) Implants grown on frames, **B**) Implants growing on frame, outer membrane, conules and mesohyl with water circulation channels.

The increasing global demand for bath sponges cannot be satisfied only by harvesting from the wild. Aquaculture is the only solution to the problem. Indeed, if one considers that in accordance with the Treaty of Berne for the protection of wildlife, commercial sponges are among endangered species, this is another reason for the further development of their production through cultivation.

Particular attention should be paid to the siting of the installation for the culture of sponges, since the viability of the business depends on the growth rate of the sponges. The unit requires the presence of a diver, which raises production costs, but the cost of the installation is low. The initial cost for collecting the sponges is gradually reduced by using implants from the parent cultures. The implants should preferably be set in place in late winter so that with the coming of spring and the gradual increase in water temperature the healing and development of implants would be accelerated.

A further reason to promote the mariculture of sponges is that they comprise the main filters (biological detergents) in the marine environment since they take their food (heterotrophic bacteria, phytoplankton and detritus) from the suspended components. The ability of sponges to use bacteria as their staple food suggests another possibility of their use: as biofilters to clean enclosures and coastal zones where human activities have significantly contaminated the environment with large amounts of organic matter. Reiswig [2, 26] has demonstrated the ability of Demosponges to non selectively filter organic particles sized 0.1-50 mm, cleaning the water column within 24 h and retaining up to 80% of suspended particles. Also purification tests have been done with sponges, using live microorganisms [41]. Stabili *et al.* [42] demonstrated the high filtering action of *S. officinalis* by removing and substantially reducing the concentration of bacteria. They also reported that the retention yield was increased during the first hour, wherein the bacterial density is very high. Bacteria isolated from *S. officinalis* (sponge extract and sea water) collected from various regions of the Aegean Sea, with different levels and types of pollutants, were significantly higher in the sponge extract than in the corresponding column of seawater [43].

CONCLUSION

According to our experimental results and the literature [*e.g.* 32, 41, 44, 45] we conclude that sponges can be cultivated for the commercial exploitation in areas with a high organic load and the production of active substances such as secondary metabolites. In addition, they may contribute to the cleaning of the overlying water column, acting as natural biofilters. In particular, the findings of this study demonstrate the possibility of successful mariculture of sponges alongside with fish farming in cages, thus contributing to the sustainability of the marine environment with parallel production of spongin B (spongin short-chain collagen) and collagen A from the healthy and robust sponges *Spongia officinalis officinalis* (matapas) & *Hippospongia communis* (kapadiko).

CONFLICT OF INTEREST

The author (editor) declares no conflict of interest, financial or otherwise.

ACKNOWLEDGEMENT

The study was funded by the European Regional Development Fund (ERDF) of the European Union and by National Resources under the Operational Programme of Competitiveness and Entrepreneurship (EPAN II).

REFERENCES

[1] Voultsiadou E. Reevaluating sponge diversity and distribution in the Mediterranean sea. Hydrobiologia 2009; 628(1): 1-12. [http://dx.doi.org/10.1007/s10750-009-9725-9]

[2] Reiswig HM. Water transport, respiration and energetics of three tropical marine sponges. J Exp Mar Biol Ecol 1974; 14: 231-49. [http://dx.doi.org/10.1016/0022-0981(74)90005-7]

[3] Castritsi–Catharios J. Kalymnos and the secrets of the sea Contribution to the sponge fisheries. Athens, Greece: Commission of the European Communities 1998.

[4] Gaino E, Pronzato R, Corriero E, Buffa P. Mortality of commercial sponges: Incidence in two Mediterranean areas. Boll Zool 1992; 59: 79-85. [http://dx.doi.org/10.1080/11250009209386652]

[5] Pronzato R. Sponge–fishing, disease and farming in the Mediterranean sea. Aquat Conserv 1999; 9: 485-93. [http://dx.doi.org/10.1002/(SICI)1099-0755(199909/10)9:5<485::AID-AQC362>3.0.CO;2-N]

[6] Perez T, Garrabou J, Sartoretto S, Harmelin JG, Francour P, Vacelet J. Mortalité massive d'invertébrés marins: un événement sans précédent en Méditerranée nord-occidentale. C R Acad Sci III 2000; 323(10): 853-65. [http://dx.doi.org/10.1016/S0764-4469(00)01237-3] [PMID: 11098401]

[7] Cerrano C, Bavestrello G, Bianchi CN, *et al.* A catastrophic mass-monality episode of gorgonians and other organisms in the Ligurian Sea (Nonh-western Mediterranean), Summer 1999. Ecol Lett 2000; 3: 284-93. [http://dx.doi.org/10.1046/j.1461-0248.2000.00152.x]

[8] Webster NS. Sponge disease: a global threat? Environ Microbiol 2007; 9(6): 1363-75. [http://dx.doi.org/10.1111/j.1462-2920.2007.01303.x] [PMID: 17504474]

[9] Castritsi-Catharios J, Miliou H, Kapiris K, Kefalas E. Recovery of the commercial sponges in the central and southeastern Aegean Sea (NE Mediterranean) after an outbreak of sponge disease. Mediterr Mar Sci 2011; 12: 5-20. [http://dx.doi.org/10.12681/mms.50]

[10] Pérez T, Vacelet J. Effect of Climatic and Anthropogenic Disturbances on Sponge Fisheries.The Mediterranean Sea: Its history and present challenges. Springer Science & Business Media 2014; pp. 577-87. [http://dx.doi.org/10.1007/978-94-007-6704-1_35]

[11] Ehrlich H, Worch H. Sponges as natural composites: from biomimetic potential to development of new biomaterials.Porifera Research: Biodiversity, Innovation and Sustainability Rio de Janeiro: Museu National. Brazil 2007; pp. 303-12.

[12] Gross J, Sokal Z, Rougvie M. Structural and chemical studies on the connective tissue of marine sponges. J Histochem Cytochem 1956; 4(3): 227-46. [http://dx.doi.org/10.1177/4.3.227] [PMID: 13332235]

[13] Garrone R, Vacelet J, Pavans de Ceccatty M, Junqua S, Robert L, Huc A. Une formation collagene particuliere: les filaments des éponges cornées Irsinia. Etude ultrastructurale, phisco-chimique. J Microsc (Paris) 1973; 17: 241-60.

[14] Swatschek D, Schatton W, Kellermann J, Müller WE, Kreuter J. Marine sponge collagen: isolation, characterization and effects on the skin parameters surface-pH, moisture and sebum. Eur J Pharm Biopharm 2002; 53(1): 107-13. [http://dx.doi.org/10.1016/S0939-6411(01)00192-8] [PMID: 11777758]

[15] Sipkema D, Osinga R, Schatton W, Mendola D, Tramper J, Wijffels RH. Large-scale production of pharmaceuticals by marine sponges: sea, cell, or synthesis? Biotechnol Bioeng 2005; 90(2): 201-22.

[http://dx.doi.org/10.1002/bit.20404] [PMID: 15739169]

[16] Laport MS, Santos OC, Muricy G. Marine sponges: potential sources of new antimicrobial drugs. Curr Pharm Biotechnol 2009; 10(1): 86-105. [http://dx.doi.org/10.2174/138920109787048625] [PMID: 19149592]

[17] Duckworth A. Farming sponges to supply bioactive metabolites and bath sponges: a review. Mar Biotechnol (NY) 2009; 11(6): 669-79. [http://dx.doi.org/10.1007/s10126-009-9213-2] [PMID: 19585169]

[18] Takur NL, Müler WE. Biotechnological potential of marine sponges. Curr Sci 2004; 86: 1506-12.

[19] Wijffels RH. Potential of sponges and microalgae for marine biotechnology. Trends Biotechnol 2008; 26(1): 26-31. [http://dx.doi.org/10.1016/j.tibtech.2007.10.002] [PMID: 18037175]

[20] Lou X, Chirila TV, Clayton AB. Hydrophilic sponges based on 2-hydroxyethyl methacrylate. IV. Novel synthetic routes to hydroxyl – containing crosslinking agents and their effect on the mechanical strength of sponges. Int J Polym Mater 1997; 37(1-2): 1-14. [http://dx.doi.org/10.1080/00914039708031472]

[21] Venderal B, Venderal M. Evaluation de l' intérêt économique de la culture d' éponges commerciales sur les côtes méditerranéennes francaises. Aquaculture 1987; 69: 9-29.

[22] Castritsi-Catharios J, Magli M, Vacelet J. Evaluation of the quality of two commercial sponges by tensile strength measurement. J Mar Biol Assoc U K 2007; 87: 1763-71. [http://dx.doi.org/10.1017/S0025315407056986]

[23] Castritsi-Catharios J, Zaoutsos SP, Ekonomou G, Berillis P. Physical parameters and chemical composition of four sponge species. 2nd International Congress of Applied Ichthyology & Aquatic Environment, HydroMediT, Messolonghi, Greece 2016; pp. 156-9.

[24] Dombi GW, Haut RC, Sullivan WG. Correlation of high-speed tensile strength with collagen content in control and lathyritic rat skin. J Surg Res 1993; 54(1): 21-8. [http://dx.doi.org/10.1006/jsre.1993.1004] [PMID: 8429634]

[25] Meroz-Fine E, Shefer S, Ilan M. Changes in morphology and physiology of an East Mediterranean sponge in different habitats. Mar Biol 2005; 147: 243-50. [http://dx.doi.org/10.1007/s00227-004-1532-2]

[26] Reiswig HM. Particle feeding in natural populations of three marine demosponges. Biol Bull 1971; 141: 568-91. [http://dx.doi.org/10.2307/1540270]

[27] Simpson TL. The biology of the marine sponge *Microciona prolifera* (Ellis and Solander) II. Temperature-related, annual changes in functional and reproductive elements with a description of larval metamorphosis. J Exp Mar Biol Ecol 1968; 2: 252-77. [http://dx.doi.org/10.1016/0022-0981(68)90019-1]

[28] Bell JJ, Barnes DK. Sponge morphological diversity: A quantitative predictor of species diversity? Aquat Conservat Mar Freshwat Ecosyst 2001; 11(2): 109-21. [http://dx.doi.org/10.1002/aqc.436]

[29] FAO Fishery information. Data and statistics Unit. 2004.

[30] Moore HF. A practical method of sponge culture 1910.

[31] Verdenal B, Vacelet J. Sponge culture on vertical ropes in the Northwestern Mediterranean Sea.New Perspectives in Sponge Biology. Washington, DC: Smithsonian Institution Press 1990; pp. 416-24.

[32] Pronzato R, Bavestrello G, Cerrano C, *et al.* Sponge farming in the Mediterranean Sea: new perspectives. Mem Queensl Mus 1999; 44: 485-91.

[33] Corriero G, Longo C, Mercurio M, Mazzano CN, Lenbo G, Spedicato MT. Rearing performance of

Spongia officinalis on suspended ropes off the Southern Italian Coast (Central Maditerranean Sea). Aquaculture 2004; 238: 105-205.
[http://dx.doi.org/10.1016/j.aquaculture.2004.04.030]

[34] Storr JF. Ecology of the Gulf of Mexico 1964.

[35] Vacelet J. Bases historiques et biologiques d'une éventuelle spongiculture. Oceanis 1985; 11: 551-84.

[36] Voultsiadou E, Lagoutari E. Sponge mariculture Biotechnology- Sustainable usage of the marine wealth. University Studio Press 2008. (in Greek)

[37] Duckworth AR, Battershill CN, Bergquist PR. Influence of explant procedures and environmental factors on culture success of three sponges. Aquaculture 1997; 156: 251-67.
[http://dx.doi.org/10.1016/S0044-8486(97)00131-2]

[38] Fossa SA, Nilsen AJ. The modern coral reef aquarium 2000.

[39] Di Bari G, Cardone E, Liuzzi GM, Marzano N, Scoccia E, Corriero G. Biological variations in a long-term ex situ cultivation: a Mediterranean demosponge as model system. Mediterr Mar Sci 2015; 16: 73-81.
[http://dx.doi.org/10.12681/mms.872]

[40] Louden D, Inderbitzin S, Peny Z, de Nys R. Development of a new protocol for testing bath sponge quality. Aquaculture 2007; 271(1-4): 275-85.
[http://dx.doi.org/10.1016/j.aquaculture.2007.06.010]

[41] Milanese M, Chelossi E, Manconi R, Sarà A, Sidri M, Pronzato R. The marine sponge *Chondrilla nucula* Schmidt, 1862 as an elective candidate for bioremediation in integrated aquaculture. Biomol Eng 2003; 20(4-6): 363-8.
[http://dx.doi.org/10.1016/S1389-0344(03)00052-2] [PMID: 12919821]

[42] Stabili L, Licciano M, Giangrande A, *et al.* Filtering activity of *Spongia officinalis* var. adriatica (Schmidt) (Porifera, Demospongiae) on bacterioplankton: implications for bioremediation of polluted seawater. Water Res 2006; 40(16): 3083-90.
[http://dx.doi.org/10.1016/j.watres.2006.06.012] [PMID: 16884759]

[43] Kephalas E, Castritsi–Catharios J, Miliou E. Bacteria associated with the sponge *Spongia officinalis* as indicators of contamination. Ecol Indic 2003; 2: 339-43.
[http://dx.doi.org/10.1016/S1470-160X(03)00002-5]

[44] Pronzato R, Cerrano C, Cubeddu T, *et al.* Sustainable development in coastal areas: role of sponge farming in integrate aquaculture. 1998.

[45] Müller WE, Wimmer W, Schatton W, Böhm M, Batel R, Filic Z. Initiation of an aquaculture of sponges for the sustainable production of bioactive metabolites in open systems: Example, *Geodia cydonium.* Mar Biotechnol (NY) 1999; 1(6): 569-79.
[http://dx.doi.org/10.1007/PL00011812] [PMID: 10612682]

CHAPTER 13

The Establishment of Blue Crab *Callinectes sapidus* Rathbun, 1896 in the Lagoon Pogonitsa (Amvrakikos Gulf, Western Greece)

George N. Katselis[1,*] and **Constantin Koutsikopoulos**[2]

[1] *Department of Fisheries and aquaculture Technology, TEI of Western Greece, Messolonghi Greece*

[2] *Department of Biology, University of Patras, Greece, Patras*

Abstract: The presence of the blue crab *Callinectes sapidus* has been recorded late in the 2000s in the lagoons of Amvrakikos Gulf. Ninety four individuals from Pogonitsa lagoon (Amvrakikos Gulf, Western Greece) have been observed and measured during August 2016. Based on biometric and biological features, these crabs were mature while most of the females were ovigenous in various eggs age since extrusion. The estimated ages of the animals and the fecundity were in agreement with those that are reported in native and/or in well-established regions of species. The population of the blue crab seems to be increasing and can be considered as established in Pogonitsa lagoon as well as in other lagoons of Gulf.

Keywords: Amvrakikos Gulf, Blue crab, *Callinectes sapidus*, Invasive species, Ionian Sea.

INTRODUCTION

The blue crab *Callinectes sapidus* Rathbun, 1896 a species originating from the western Atlantic, has invaded, at least since the beginning of the 20th century, European waters in multiple independent introductions via ballast water [1] and it is ranked amongst the 100 worst invasive species in the Mediterranean [2]. In Mediterranean it distributed in different regions, especially in the Eastern part [3] while early records (the earliest dating back to 1948) from Greece refer exclusively to the Aegean Sea [4] particularly to the northern Aegean coasts where the species is currently abundant [5]. Until recently, limited data were available in scientific literature on the presence of the species on the Ionian coasts although its occurrence has been documented (at least from late 1990s) at several

* **Corresponding author George N. Katselis:** Department of Fisheries and aquaculture Technology, TEI of Western Greece, Hellas, Messolonghi Greece; Tel: ++302631058237; E-mail: gkatsel@teimes.gr

Panagiotis Berillis (Ed)

Ionian costal ecosystems from North Adriatic to South Greece [6, 7].

Amvrakikos Gulf (Western Greece, Ionian Sea) is a fjordlike hydrological regime (400 km^2), around which 14 brackish lagoons are located covering an area of about 86 km^2 (about the 20% of total Greek lagoon surface) In the north and northwestern coast of the gulf two rivers (*i.e.* Arachthos and Louros) are also discharged degrading the water quality of the gulf [8, 9] (Fig. **1**). The Gulf is characterised by a well-stratified two layer structure in the water column made up of a surface layer and a bottom layer that are separated by a strong pycnocline. At the entrance over the sill, there is a brackish water outflow in the surface water and a saline water inflow in the nearbed region [9]. The rivers and gulf and its surrounding lagoons have been also subjected to a high degree of human impacts [8, 9]. The system is protected under the Ramsar convention (www.ramsar.org) or part of the Natura 2000 network (http://ec.europa.eu/environment/nature/natura 2000/index_en.htm).

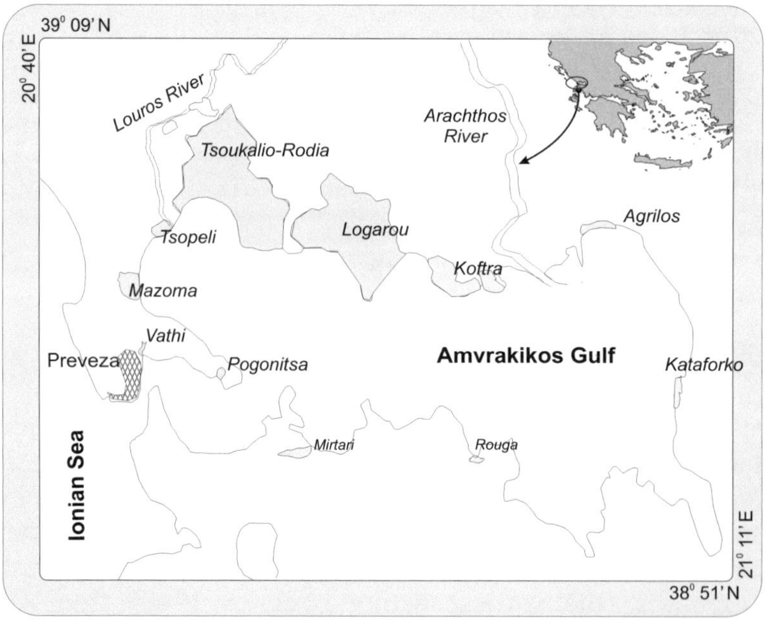

Fig. (1). Map of sampling area.

The blue crab appeared in the lagoons Tsoukalio-Rodia and Logarou (north coast Amvrakikos Gulf) late in the 2000s [7] and based on personal communications with the local fishermen, today it appears in all lagoon systems showing various abundances, while the greater abundance of species is reported to Pogonitsa lagoon. In this work, basic biometric and biological features of the blue crab, from samples taken from the Pogonitsa lagoon were studied.

MATERIALS AND METHODS

The Lagoon of Pogonitsa is situated in the northwestern coast of Amvrakikos Gulf, (38.964039°N, 20.815861°E) (Fig. **1**), covering a surface of about 0.45 Km^2, with a depth of 1.2m, sandy bottom, two communication channels by Amvrakikos, mean salinity 33 psu and water temperature annually ranging from 11 to 28°C [8]. Fishery exploitation in the lagoon, as in most Greek lagoons in general (and of Amvrakikos Gulf), is a traditional extensive culture based on seasonal ongoing migration movements of fry and adult fish specimens between sea and lagoons. A fence by cement shell installations and barrier traps separate the lagoon and the sea. Co-operatives of fishermen leasing the lagoons and exploiting the barrier traps also use trammels and fyke nets [8].

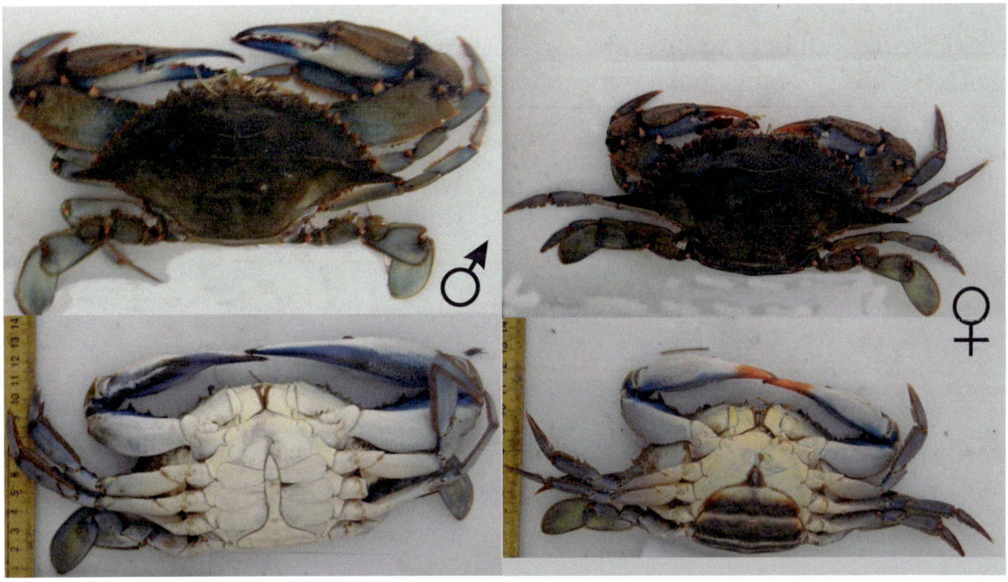

Fig. (2). Morphology of males and females of blue crab.

The data presented in this paper was collected in August 2016, based on the observation of 94 individuals of blue crabs caught in barrier traps on August 2nd and August 25th. During the sampling the salinity, water temperature by portable CTD and water current by electromagnetic flow meter (Valeport, model 802), were measured. Carapace measurements, including length (CL – distance between the center of the anterior interorbital margin and the center of the posterior margin) and width (CW – maximal distance between the posterior anterolateral spines) were measured (CL, CW; mm), while each specimen is weighed (W; g) sex identification (Fig. **2**), females stage of maturity and eggs age since extrusion by their colorization (eggs color yellow-orange: eggs age 1 to 7 days post-extrusion, eggs color brown-dark: 8 to 15 days post-extrusion) [10], were recorded. The egg mass was weighted and egg number per individual (absolute fecundity: FQ) was calculated using the gravimetric method. The significance of regressions CW-CL and CW-W were assessed by analysis of variance (ANOVA; R^2; df;F-ratio;P=0.05) [11].

RESULTS AND DISCUSSION

The salinity, the water temperature and the water flow during the sampling was around 29.3 psu, 29.8°C and 0.37 m/s, respectively. The sample consists of adults specimens, in majority, ovigerous females (Table **1**). The carapace width (CW) frequency distribution of the blue crab is shown in Fig. (**3**).

The carapace width of the 94 crabs ranged from 128.2 to 193.9 mm (mean value ± SD: 165.34±14.68 mm), the carapace length ranged from 58.2 to 81.2 mm (mean value ± SD: 69.6±5.25 mm) and the weight ranged from 129.2-442.4 g (mean value ± SD: 274.33±78.67g).

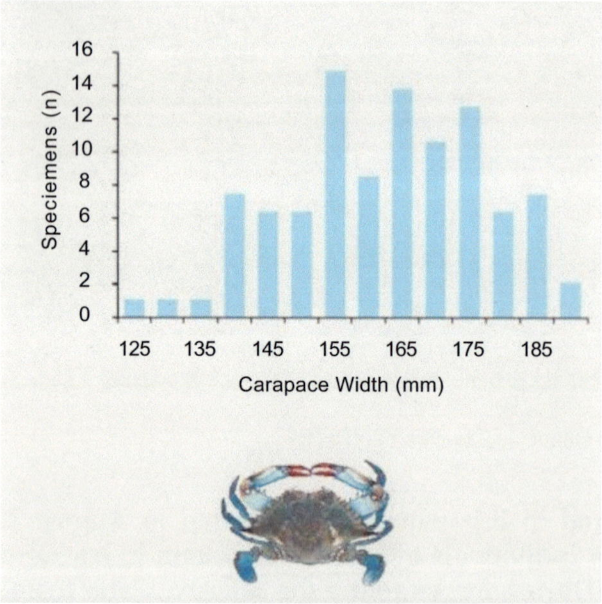

Fig. (3). Carapace width distribution of blue crab from Pogonitsa lagoon.

The carapace width-weight relationship of pooled sample was $W=0.0002CW^{2.805}$ ($R^2=0.66$; SE=0.18; df=1.93; F-ratio=186.1; $P<0.05$) and the carapace length-width relationship was $CL=3.38+0.40CW$ ($R^2=0.78$; SE=2.67; df=1.93 ;F-ratio=178.3;$P<0.05$) (Fig. **4**).

The sex ratio was 1.08 females: 1 male and between the samples ranged from 0.5 to 2.9 females:1 male. All females were mature while the 44 out of the 49 females (89.8%) were ovigerous (Table **1**).

Table 1. Biometric and biological features of blue crab samples. (CW: carapace width, CL: carapace length, W: weight, SD: standard deviation, F: female, M: Male, OF: ovigerous Females, n: number of specimens).

	Biometric and biological features						
Sampling date	**n**	**CW (SD) in mm**	**CL (SD) in mm**	**W (SD) in g**	**sex ratio (F:M)**	**OF (n)**	**OF/F (%)**
2/8/2016	43	161.26 (15.94)	68.03 (4.95)	253.99 (83.85)	32:11 (2.9:1)	32	100
25/8/2016	51	168.79 (12.69)	71.06 (5.74)	291.49 (70.36)	17:34 (0.5:1)	12	70.6
total	94	165.34 (14.68)	69.6 (5.25)	274.33 (78.67)	49:45 (1.08:1)	44	89.8
min-max		128.2-193.9	58.2-81.2	129.2-442.4			

The mean eggs mass was 49.06±10.38 gr, while the ovigerous females where shared between the age of eggs of 1-7 days post-extrusion (20 individuals) and 8-15 days post-extrusion (24 individuals). The absolute fecundity ranged from 1.30×10^6 to 3.20×10^6 eggs per ovigerous female (Table **2**).

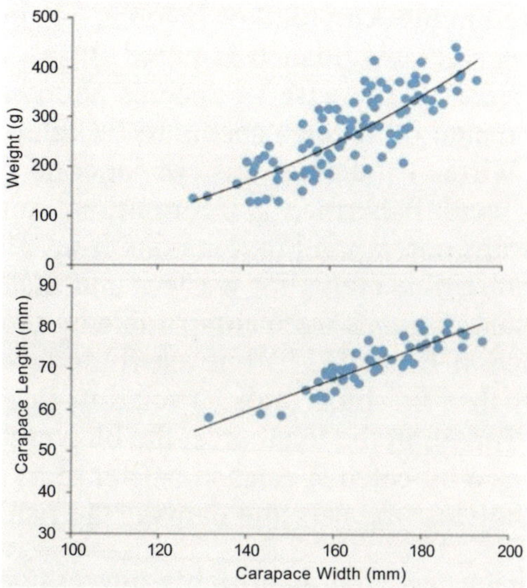

Fig. (4). Relationships of weight-carapace width and carapace length-carapace width of blue crab from Pogonitsa lagoon.

Table 2. Biological features of ovigerous blue crab females. W: total body weight, SD: standard deviation, We: eggs mass weight n: number of specimens, FQ: absolute fecundity).

		n with eggs age		
W (SD) in g	We (SD) in g	1-7 days	8-15 days	FQ (x10^6)
242.66 (48.96)	49.06 (10.38)	20	24	1.30-3.20

These findings are in agreement with those that are reported in native and/or in well-established regions of species. Indeed, the usual carapace width of mature stock in the native and/or in well-established regions of species was noted to be greater than 120mm [10, 12], the infinite size (CW_∞) between the 180 to 200 mm (close to the CW of the bigger individual in the present study) [12] while the species spawn in May or June, followed by a second spawning during August to September. Also, the mean egg mass and calculated absolute fecundity were comparable to that of various USA stocks [10]. Moreover, the sample characterized by the present of all egg-ages since extrusion supporting that this step of species reproductive process can be completed in the study area. Finally, local fishermen confirm the presence of ovigerous females and young specimens in the study lagoon. Taking into account the above, the population of the blue crab can be considered as established in Pogonitsa lagoon.

The essential ecological requirements of species (*e.g.* salinity, temperature) [10, 13, 14] are offered by Amvrakikos Gulf and the lagoons [8, 9]. Despite to fact that the blue crab appeared in the Amvrakikos lagoons late in the 2000s [7], recently a sensible change of species abundance is noted. In all lagoons, the fishermen's perception is a progressive increase of species abundance from early of 2010. Moreover, the abundance of species seems to be relatively high in the western (Vathi, Pogonitsa, Mirtari) than the eastern lagoons (Tsoukalio, Logarou and Koftra). Based on local fishermen's information, during August the species captures in barrier traps correspond to densities from 350 to 1200 ind.km^{-2} and 2 to 30 ind.km^{-2} of mature blue crabs, for western and eastern lagoons, respectively. The densities in western lagoons were comparable by them in some native regions of species (*e.g.* Hundon River [13]:) but ten folds lower than Chesapeake Bay [14]. These support that the blue crab population in Amvrakikos Gulf system could be exploited. Today, in the local area, the blue crab has low acceptance as a fishing product and as a by-catch species in barrier traps and nets, it's pestering for fishermen because they tear the nets and damage the fish.

The blue crab has been reported as a highly aggressive species and it has been selected amongst the 100 "worst invasive" species in the Mediterranean with impact on both biodiversity and socioeconomics [2]. Recently, as response to this, a "win to win" solution has been proposed which includes the control of blue crab population via the fishing exploitation [15, 16]. Further studies on the blue crab population in the Amvrakikos Gulf system would be of interest to provide richer information on population structure, dynamics of the species and their management.

CONFLICT OF INTEREST

The author (editor) declares no conflict of interest, financial or otherwise.

ACKNOWLEDGEMENTS

The study is part of project "*Identification, consequences and management of the anoxic zone of Amvrakikos Gulf (NW Greece)*" that co-funding by EEA Grants and Greek Ministry of Environment and Energy. We thank the fishermen of Amvrakikos lagoon's for the collaboration and the help on the sampling collection.

REFERENCES

[1] Nehring S. Invasion History and Success of the American Blue Crab Callinectes sapidus in European and Adjacent Waters.In the Wrong Place-Alien Marine Crustaceans: Distribution, Biology and Impacts. Invading Nature - Springer Series in Invasion Ecology 2011; pp. 607-24. [http://dx.doi.org/10.1007/978-94-007-0591-3_21]

[2] Streftaris N, Zenetos A. Alien marine species in the Mediterranean - the 100 "Worst Invasives" and their impact. Mediterr Mar Sci 2006; 7: 87-118. [http://dx.doi.org/10.12681/mms.180]

[3] Galil BS, Froglia C, Noël P. CIESM Atlas of Exotic Species in the Mediterranean 2008.http://www.ciesm.org/atlas/appendix2.html

[4] Serbetis C. Un nouveau crustacé commestible en mer Egeé *Callinectes sapidus* Rath. (Decapod brach.). Proc Gen Fish Counc Medit 1959; 5: 505-7.

[5] Kevrekidis K, Antoniadou C, Avramoglou K, Efstathiadis J, Chintiroglou C. Population structure of the blue crab *Callinectes sapidus* in Thermaikos Gulf (Methoni Bay). Proceedings of the 15th Pan-Hellenic Congress of Ichthyologists. Thessaloniki, Greece. 2013; pp. 113-6.

[6] Beqiraj S, Kashta L. The establishment of blue crab *Callinectes sapidus* Rathbun, 1896 in the Lagoon of Patok, Albania (south-east Adriatic Sea). Aquat Invasions 2010; 5(2): 219-21. [http://dx.doi.org/10.3391/ai.2010.5.2.16]

[7] Perdikaris C, Konstantinidis E, Gouva E, Klaoudatos D, Nathanailides C, Paschos I. Occurrence of the invasive crab species *Callinectes sapidus* Rathbun, 1896 in NW Greece. Walailak J Sci Technol 2016; 13: 1-12.

[8] Katselis G, Moutopoulos DK, Dimitriou EN, Koutsikopoulos C. Long-term changes of fisheries landings in enclosed gulf lagoons (Amvrakikos gulf, W Greece): Influences of fishing and other human impacts. Estuar Coast Shelf Sci 2013; 131: 31-40. [http://dx.doi.org/10.1016/j.ecss.2013.07.004]

[9] Ferentinos G, Papatheodorou G, Geraga M, *et al.* Fjord water circulation patterns and dysoxic/anoxic conditions in a Mediterranean semi-enclosed embayment in the Amvrakikos gulf, Greece. Estuar Coast Shelf Sci 2010; 88: 473-81. [http://dx.doi.org/10.1016/j.ecss.2010.05.006]

[10] Millikin RM, Williams BA. Synopsis of biological data on the blue crab, Callinectes sapidus Rathbun. NOAA/National Marine Fisheries Service, (NOAA Technical Report NMFS, 1). 1984. http://spo.nmfs.noaa.gov/tr1.pdf

[11] Zar JH. Biostatistical Analysis. 4th ed., New Jersey, USA: Prentice Hall, Inc. 1999.

[12] Sumer C, Teksam I, Karatas H, Beyhan T, Aydin CM. Growth and Reproduction Biology of the Blue Crab, *Callinectes sapidus* Rathbun, 1896, in the Beymelek Lagoon (Southwestern Coast of Turkey). Turk J Fish Aquat Sci 2013; 13: 675-84. [http://dx.doi.org/10.4194/1303-2712-v13_4_13]

[13] Chenery MA. Population dynamics of blue crab (Callinectes sapidus) in the Hudson River, New York. Msc Thesis, University of Maryland. 2002; http://hjort.cbl.umces.edu/crabs/docs/Chenery_02.pdf p. 178.

[14] Sharov AF, Vølstad JH, Davis GR, Davis BK, Lipcius RN, Montane MM. Abundance and exploitation rate of the blue crab (*Callinectes sapidus*) in Chesapeake Bay. Bull Mar Sci 2003; 72(2): 543-65.

[15] Mancinelli G, Chainho P, Cilenti L, *et al.* On the Atlantic blue crab (*Callinectes sapidus* Rathbun 1896) in Southern European coastal waters: time to turn a threat into a resource? Processing of Hydromedit 2016, 2nd Interna-tional Congress on Applied Ichthyology & Aquatic Environment 10 – 12 November 2016, Messolonghi, Greece, Editor: Chatziefstathiou M., published by: University of Thessaly - School of Agricultural Sciences, Department of Ichthyology and Aquatic Environment. 2016; http://hydromedit2016.apae.uth.gr pp. 180-184.

[16] Mancinelli G, Chainho P, Cilenti L, *et al.* On the Atlantic blue crab (Callinectes sapidus Rathbun 1896) in southern European coastal waters: Time to turn a threat into a resource? Fisheries Research 2017; 194: 1-8. [http://dx.doi.org/10.1016/j.fishres.2017.05.002]

CHAPTER 14

CyHV-2 Outbreak Associated with *Aeromonas* spp. in Crucian Carp (*Carassius carassius*) in Piedmont (Italy)

Caruso Claudio[1,*], **Pastorino P.**[2], **Prato R.**[1], **Burioli E.A.V.**[2], **Peletto S.**[3], **Righetti M.**[2], **Bona M.C.**[2], **Foglini C.**[2], **Acutis P.L.**[3], **Masoero L.**[1] and **Prearo M.**[2]

[1] *Department of Virology, Istituto Zooprofilattico Sperimentale del Piemonte, Liguria e Valle d'Aosta, via Bologna 148, 10154 Turin, Italy*

[2] *Fish Diseases Laboratory, Istituto Zooprofilattico Sperimentale del Piemonte, Liguria e Valle d'Aosta, via Bologna 148, 10154 Turin, Italy*

[3] *Department of Genetics and Immunobiochemical, Istituto Zooprofilattico Sperimentale del Piemonte, Liguria e Valle d'Aosta, via Bologna 148, 10154 Turin, Italy*

Abstract: Cyprinid Herpesvirus 2 (CyHV-2) is an emerging pathogen which can cause huge economic losses in fish farming worldwide. This study describes a disease outbreak in crucian carp (*Carassius carassius)*, reporting for the first time a CyHV-2 infection in Piedmont (Italy), since it has never been described in this area. Our findings also highlighted that both *Aeromonas hydrophila* and *Aeromonas sobria* have been isolated in association with CyHV-2; we have speculated that motile *Aeromonas* spp. and CyHV-2 could be regarded as a potential "infectious pathogenic-complex". However, this evidence-based hypothesis must be corroborated by further studies and investigations. In our opinion, it is very important, through experimental studies and case report, to determine the impact of CyHV-2 on *Carassius carassius*, in order to establish proper commercial regulation and biosecurity measures to prevent the introduction and widespread of diseases.

Keywords: *Aeromonas* spp., *CyHV-2*, *Crucian carp.*

INTRODUCTION

The main fish species farmed both for food production and for ornamental purposes can be affected by several viral diseases such as Infectious Hematopoietic Necrosis (NEI) and Viral Haemorrhagic Septicaemia (VHS) in

* **Corresponding author Caruso Claudio:** Department of Virology, Istituto Zooprofilattico Sperimentale del Piemonte, Liguria e Valle d'Aosta, *via* Bologna 148, 10154 Turin, Italy; Tel: 0039 - 0112686247; E-mail: claudio.caruso@izsto.it

trout farms, diseases caused by EVA, EVEX and *Anguillid Herpesvirus-1* in eels farms, diseases caused by Iridovirus in sturgeon and several infections by Herpesvirus in Cyprinid farms.

The emergence of viral diseases in fish farms, also highly connected to the aquatic environment, is a considerable problem to be assessed because of the high morbidity and economic impact related to the spread in breeding tanks. Viral fish diseases are usually stress-related and temperature-dependent and epidemiological pattern of the disease in a fish farm could be highly influenced by some variables including seasons and population density.

Cyprinid Herpesvirus 2 (CyHV-2) is an emerging pathogen responsible of huge economic losses in fish farming worldwide [1, 2]. It is considered one of the most pathogenic agents for goldfish *(Carassius auratus)* in which it causes Herpesviral Haematopoietic Necrosis Virus of Goldfish (HVHNV) or Goldfish Haematopoietic Necrosis Virus (GFHNV) [2, 3]; CyHV-2 could also infect crucian carp (*Carassius carassius*) and common carp (*Cyprinus carpio*) [5].

CyHV-2 includes a member of the genus *Cyprinivirus,* which also comprises the Carp pox virus (CyHV-1) and KHV (Cyprinid Herpesvirus 3) (CyHV-3), belonging to the family *Alloherpesviridae.* Clinical signs, host range and pathogenic role of *Cyprinivirus* (CyHV-1, CyHV-2 and CyHV-3) share both common and different features [6]. CyHV-2 can cause epizooties typically during the spring and autumn, affecting both young and adult goldfish. The disease usually rapidly develops, showing a mortality rates up to 90-100% [7] (Podok *et al.*, 2014). Typical signs and gross lesions of CyHV-2 include lethargy, anorexia, bleeding and pale gills, ascites and splenomegaly [8] (Goodwin *et al.*, 2009). External lesions could be relevant but also more specific lesions may be detected in liver, spleen and kidney; spleen often displays white granular nodules.

Histopathological findings typically include mild to severe multifocal diffuse coagulative necrosis in kidney and spleen tissue; viral latency is suggested by the presence of CyHV-2 DNA in healthy goldfish for long time [7, 8]. Detection of CyHV-2 has been reported worldwide from 1995 in goldfish [7 - 11] and gibel carp [4, 12]. It has been reported in Japan [3, 9], USA [13, 14], Taiwan [15], Australia [16], New Zealand [17], United Kingdom [18] and China [1, 19].

Since the lack of published data, the epidemiology of CyHV-2 infection in Italy is not still clear.

CyHV-2 has been described [20] in an ornamental fish farm in Veneto (Northeast Italy) and also reported in Central Italy, specifically in a tributary of Arno River [21] and in a fish pond located in Tuscany [22]. Despite this, also the role of bacteria, such as *Aeromonas* spp. in association with CyHV-2 infection needs to be clarified. During routine investigation, aimed to establish the cause of mortality in freshwater fish farms in Italy, we reported CyHV-2 infection in a private lake in Piedmont region; this is the first detection of CyHV-2 in Northwest Italy.

Following a disease outbreak of *C. carassius*, a sample of affected animals has been processed to determine the possible cause of illness. Clinical signs and anatomopathological lesions observed in several subjects gave rise to concerns over the possibility of infectious disease. This study describes symptoms, gross lesions, and molecular results of CyHV-2 infection in association with motile *Aeromonas* spp. detected in affected crucian carp.

MATERIAL AND METHODS

During the period between spring and autumn 2014, a disease outbreak was observed in a private lake located in the province of Turin, Piedmont (Italy). The 60 m^2 lake was for sport fishing purpose; crucian carp population consisted in both juveniles and adult subjects. A low density presence of ichthyophagous birds, such as cormorants and herons were usually observed in the area near the lake. During the period in which episodes of mortality have been reported, water temperature was around 17°C± 0.5°C. The amateur farmer referred fish illness, with a low but constant mortality over the time. Specifically, history reported the death of about one or two adult subjects for week, starting from March until November 2014; the mortality rate decreased with the decrease of water temperature and no mortality episodes were reported during colder months.

Four samples of affected fish were collected and examined just after death. Symptoms reported were lethargy, reduced motility and anorexia. A complete parasitological examination has been performed for all four subjects. Inoculation into Flexibacter Maritimus Medium (FMM) and Cytophaga Agar (CY) from skin lesions and into Tryptic Soy Agar (TSA) and Columbia Blood Agar Media (BA) from brain and kidney was performed. Plates were incubated at 22±2°C for 3 days. Isolates were identified by morphological, phenotypical and biochemical methods (API 20E; API 20NE, bioMèrieux, Italia). *Cyprinivirus* infection has been strongly considered, taking into account clinical signs and gross lesions,.

Detection of CyHV-2 in the samples was carried out by a Pan *Alloherpesvirinae*–PCR assay, based on a set of degenerated primers (Forward HV and ConsLower by Waltzek *et al.* [23]) that amplified a fragment from 441 to 474 base pairs length, targeting the DNA polymerase gene. PCR mix was carried out with Platinum® Taq PCR DNA Polymerase (Invitrogen) in a final volume of 25 μl containing: 1 x PCR Buffer, 0.8 mM dNTP, 1.6 mM $MgCl_2$, 1 μM of each primer, 1.25 U of Taq polymerase, 5 μl of DNA and 9.45μl of H_2O nuclease free.

Cycling conditions were the following: incubation at 95°C for 5 min., followed by 30 cycles at 95°C for 30 sec. and 51°C for 30 sec. and 72°C for 2 min, with a final elongation step of 5 min at 72°C (Thermocyclers 2720 APPLIED BIOSYSTEM). The PCR products were subjected to electrophoresis on 2% agarose gel (Sigma, USA) and visualized by transilluminator GEL-DOC (Bio-Rad, USA), after staining with GelGreen Nucleic Acid Stain (Invitrogen, USA). The PCR products were separated on 2% agarose gels and the bands were cut from the gel and purified using the Nucleo Spin Extract II kit (Macherey–Nagel, Düren, Germany). Purified DNA was used for sequencing both strands with an ABIPRISM Big Dye Terminator v.3.1 cycle sequencing kit.

RESULTS

Necropsy performed by the Fish Diseases Laboratory of the Istituto Zooprofilattico of Piemonte, Liguria and Aosta Valley, revealed pale gills and ascites in all subjects and a large ulcerated wound around the opercular area in a single subject. In two affected fish a pale skin area has been also observed. No macroscopic lesions in visceral organs were highlighted during necropsy.

Parasitological analysis showed, in two subject and always at a low intensity, the presence of *Dactylogyrus* spp. and *Gyrodactylus* spp. from gills lamellae and skin scraping.

Microbiological analysis allowed the isolation of *Aeromonas hydrophila* and *Aeromonas sobria* in the kidney of the subjects.

In three brain samples, PCR results showed a band of expected size (about 450 bp, Fig. **1**). Consequently, positive samples have been submitted to further genetic sequence analysis, in order to obtain a proper identification of the virus. Genetic sequencing and Blast analysis retrieved maximum similarity with CyHV-2 (100% similarity with NCBI Reference Sequence NC_019495.1).

DISCUSSION

This study describes a disease outbreak in *C. carassius*, reporting for the first time a CyHV-2 infection in Northwestern Italy, which has never been detected in this area. CyHV-2 has been identified by molecular assay in three out of four homogenated tissues and in three brain sample, respectively. Fish could have been infected by CyHV-2 in several ways. Taking into account the import of both juveniles and adult fish from farms located in the whole Italian territory, it could be speculated that introduction of infected subjects with CyHV-2 could be regarded as the most likely means of transmission. Alternatively, it may be possible, although very unlikely in our opinion, the introduction of virus by ichthyophagous birds, acting as passive mechanical vector [24]. Since the CyHV-2 reactivation is water temperature related, fish death recorded just during the spring and the summer season could be easily justified. A recent study conducted by Ito & Maeno [10], clarified the influence of water temperature on the development of HVHN in goldfish (*C. auratus*) after experimentally induced infection with CyHV-2; the authors demonstrated that cumulative mortalities of the 15°C, 20°C, 25°C and 30°C groups were 10%, 90%, 90% and 60%, respectively. Therefore, the temperature range of 20°C-25°C is considered highly permissive for HVHN. In the light of this finding, it could be supposed that CyHV-2 infection in *C. carassius* follows the same dynamic course as described in *C. auratus*. Indeed, in our case, mortality episodes started with a water temperature of about 17°C ± 0.5°C; this permissive but not optimal temperature explained the low pathogenic outbreak (about one or two subject for a week) with a constant mortality over the time. In this work, a complete parasitological and bacteriological investigations was also carried out on all the subjects, to obtain a complete diagnostic scenario. Regarding the detection, at low intensity, of both *Dactylogyrus* spp. and *Gyrodactylus* spp. in the gills and skin, it could be affirmed that they could not represent the primary cause of mortality but, eventually, a contributory one.

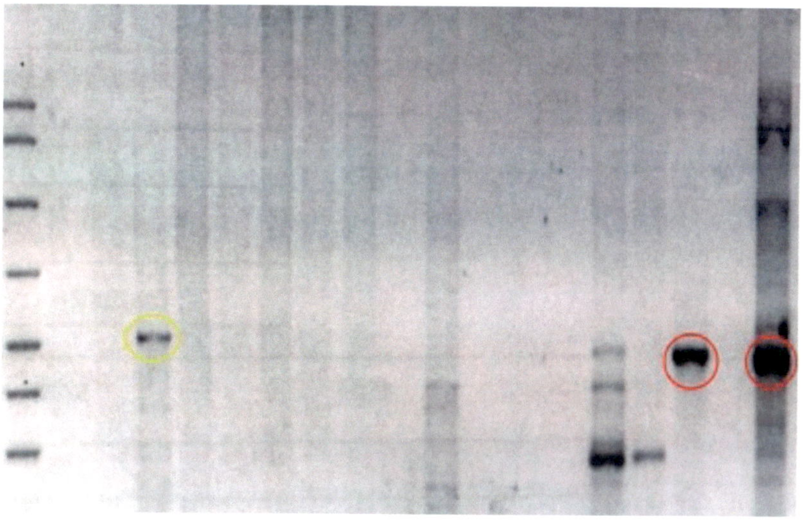

Fig. (1). PCR run in 2% agarose gel - positive bands corresponding to length of 450bp (in red circle positive controls diluition 1/50 and pure, in yellow circle a brain sample).

To the other hand, bacteriological analysis output revealed the presence of *A. hydrophila* and *A. sobria* in the kidney of all affected subjects. Similar findings were previously reported also by other authors [25]. Indeed, in a previous work, Fichi *et al.* [21] highlighted the presence of *A. sobria* isolated from the brain, kidney and spleen of affected crucian carp, while in the gills, transmission electron microscopy revealed herpesvirus-like particles, subsequently identified as CyHV-2; although the authors concluded that it was not possible to attribute a pathogenic role to CyHV-2 in that mortality event, they arise the question about its potential role in this species. Whereas *A. hydrophila* is the most often identified bacterium in motile *Aeromonas* septicaemia, several studies confirmed *A. sobria* as a causative agent of disease in fish. Our findings highlighted that both *A. hydrophila* and *A. sobria* have been isolated in association with CyHV-2; we supposed that motile *Aeromonas* spp. and CyHV-2 could be regarded as a potential "infectious pathogenic-complex", especially in *Carassius* spp. However, this evidence-based hypothesis must be corroborate by further study and investigations. Pretto *et al.* [20] isolated CyHV-2 from a batch of *C. auratus* imported from Singapore; the authors speculated that the disease was likely present in a latent phase and viral reactivation was caused by stress factors, such as temperature change occurred during the transport.

In conclusion CyHV-2 is likely to be an important, but rarely detected, pathogen worldwide [14]; in our opinion it is very important, through experimental studies and case report, to determine the impact of CyHV-2 on *Carassius* spp. population in order to establish proper commercial regulation and bio-security measures to prevent the introduction and widespread of diseases.

CONFLICT OF INTEREST

The author (editor) declares no conflict of interest, financial or otherwise.

ACKNOWLEDGEMENTS

Declared none.

REFERENCES

[1] Xu J, Zeng L, Zhang H, Zhou Y, Ma J, Fan Y. Cyprinid herpesvirus 2 infection emerged in cultured gibel carp, *Carassius auratus gibelio* in China. Vet Microbiol 2013; 166(1-2): 138-44. [http://dx.doi.org/10.1016/j.vetmic.2013.05.025] [PMID: 23850442]

[2] Ma J, Jiang N, LaPatra SE, *et al.* Establishment of a novel and highly permissive cell line for the efficient replication of cyprinid herpesvirus 2 (CyHV-2). Vet Microbiol 2015; 177(3-4): 315-25. [http://dx.doi.org/10.1016/j.vetmic.2015.04.006] [PMID: 25912023]

[3] Ito T, Kurita J, Ozaki A, Sano M, Fukuda H, Ototake M. Growth of cyprinid herpesvirus 2 (CyHV-2) in cell culture and experimental infection of goldfish *Carassius auratus*. Dis Aquat Organ 2013; 105(3): 193-202. [http://dx.doi.org/10.3354/dao02627] [PMID: 23999703]

[4] Li L, Luo Y, Gao Z, *et al.* Molecular characterisation and prevalence of a new genotype of Cyprinid herpesvirus 2 in mainland China. Can J Microbiol 2015; 61(6): 381-7. [http://dx.doi.org/10.1139/cjm-2014-0796] [PMID: 25900131]

[5] Davison AJ, Kurobe T, Gatherer D, *et al.* Comparative genomics of carp herpesviruses. J Virol 2013; 87(5): 2908-22.
[http://dx.doi.org/10.1128/JVI.03206-12] [PMID: 23269803]

[6] Podok P, Xu L, Xu D, Lu L. Different expression profiles of Interleukin 11 (IL-11), Intelectin (ITLN) and Purine nucleoside phosphorylase 5a (PNP 5a) in crucian carp (*Carassius auratus gibelio*) in response to Cyprinid herpesvirus 2 and *Aeromonas hydrophila.* Fish Shellfish Immunol 2014; 38(1): 65-73.
[http://dx.doi.org/10.1016/j.fsi.2014.03.001] [PMID: 24636855]

[7] Goodwin AE, Sadler J, Merry GE, Marecaux EN. Herpesviral haematopoietic necrosis virus (CyHV-2) infection: case studies from commercial goldfish farms. J Fish Dis 2009; 32(3): 271-8.
[http://dx.doi.org/10.1111/j.1365-2761.2008.00988.x] [PMID: 19309419]

[8] Goodwin AE, Merry GE, Sadler J. Detection of the herpesviral hematopoietic necrosis disease agent (Cyprinid herpesvirus 2) in moribund and healthy goldfish: validation of a quantitative PCR diagnostic method. Dis Aquat Organ 2006; 69(2-3): 137-43. b
[http://dx.doi.org/10.3354/dao069137] [PMID: 16724557]

[9] Jung SJ, Miyazaki T. Herpesviral haematopoietic necrosis of goldfish, *Carassius auratus* (L.). J Fish Dis 1995; 18: 211-20.
[http://dx.doi.org/10.1111/j.1365-2761.1995.tb00296.x]

[10] Ito T, Maeno Y. Effects of experimentally induced infections of goldfish *Carassius auratus* with cyprinid herpesvirus 2 (CyHV-2) at various water temperatures. Dis Aquat Organ 2014; 110(3): 193-200. a
[http://dx.doi.org/10.3354/dao02759] [PMID: 25114043]

[11] Ito T, Maeno Y. Susceptibility of Japanese Cyprininae fish species to cyprinid herpesvirus 2 (CyHV-2). Vet Microbiol 2014; 169(3-4): 128-34. b
[http://dx.doi.org/10.1016/j.vetmic.2014.01.002] [PMID: 24480585]

[12] Daněk T, Kalous L, Vesel T, *et al.* Massive mortality of Prussian carp *Carassius gibelio* in the upper Elbe basin associated with herpesviral hematopoietic necrosis (CyHV-2). Dis Aquat Organ 2012; 102(2): 87-95.
[http://dx.doi.org/10.3354/dao02535] [PMID: 23269383]

[13] Groff JM, LaPatra SE, Munn RJ, Zinkl JG. A viral epizootic in cultured populations of juvenile goldfish due to a putative herpesvirus etiology. J Vet Diagn Invest 1998; 10(4): 375-8.
[http://dx.doi.org/10.1177/104063879801000415] [PMID: 9786531]

[14] Goodwin AE, Koo L, LaPatra SE, Bonar C. Goldfish hematopoietic necrosis herpesvirus (Cyprinid herpesvirus 2) in the USA: molecular confirmation of isolates from diseased fish. J Aquat Anim Health 2006; 18: 11-8. a
[http://dx.doi.org/10.1577/H05-007.1]

[15] Chang PH, Lee SH, Chiang HC, Jong MH. Epizootic of Herpes-like virus infection in goldfish, *Carassius auratus* in Taiwan. Fish Pathol 1999; 34: 209-10.
[http://dx.doi.org/10.3147/jsfp.34.209]

[16] Stephens FJ, Raidal SR, Jones B. Haematopoietic necrosis in a goldfish (*Carassius auratus*) associated with an agent morphologically similar to herpesvirus. Aust Vet J 2004; 82(3): 167-9.
[http://dx.doi.org/10.1111/j.1751-0813.2004.tb12650.x] [PMID: 15088985]

[17] Hine PM, Tham KM, Morrison R. Cyprinid herpesvirus 2 in New Zealand goldfish. Surveillance 2006; 33: 3-5.

[18] Jeffery KR, Bateman K, Bayley A, *et al.* Isolation of a cyprinid herpesvirus 2 from goldfish, *Carassius auratus* (L.), in the UK. J Fish Dis 2007; 30(11): 649-56.
[http://dx.doi.org/10.1111/j.1365-2761.2007.00847.x] [PMID: 17958608]

[19] Wu T, Ding Z, Ren M, *et al.* The histo- and ultra-pathological studies on a fatal diaease of Prussian

carp (*Carassius gibelio*) in mainland China associated with cyprinid herpesvirus 2 (CyHV-2). Aquaculture 2013; 412-413: 8-13. [http://dx.doi.org/10.1016/j.aquaculture.2013.07.004]

[20] Pretto T, Quaglio F, Rampazzo E, *et al.* Episodio di necrosi ematopoietica da Cyprinid Herpesvirus-2 in pesci rossi, *Carassius auratus* (L.) di importazione. Atti del Convegno Nazionale Società Italiana di Patologia Ittica, Ostuni (BR) 19-21 maggio. 2011.

[21] Fichi G, Cardeti G, Cocumelli C, *et al.* Detection of Cyprinid herpesvirus 2 in association with an *Aeromonas sobria* infection of *Carassius carassius* (L.), in Italy. J Fish Dis 2013; 36(10): 823-30. [PMID: 23488630]

[22] Fichi G, Susini F, Cocumelli C, *et al.* Nuovo episodio di mortalità da CyHV-2 in *Carassius carassius* in Toscana. Atti Convegno Nazionale Società Italiana di Patologia Ittica, Calci (PI) 18-19 settembre. 2014.

[23] Waltzek TB, Kurobe T, Goodwin AE, Hedrick RP. Development of a polymerase chain reaction assay to detect cyprinid herpesvirus 2 in goldfish. J Aquat Anim Health 2009; 21(1): 60-7. [http://dx.doi.org/10.1577/H08-045.1] [PMID: 19485127]

[24] OIE. Manual of diagnostic tests for aquatic animals. 2015.

[25] Lazar M, Vulpe V, Guguianu E, Lazar R, Rotaru A, Oprean OZ. An episode of bacterian haemorrhagic septicemia with supraacute course in Ciprinides in a farm in Iasi County, Romania. Bull Univ Agric Sci Vet Med Cluj-Napoca Vet Med 2011; 68: 203-8.

CHAPTER 15

Findings from a 16-year monitoring of Viral Notifiable Diseases in Salmonid Fish in Piedmont Region (Italy)

Maria Cristina Bona[1,*], **Arsieni P.**[2], **Righetti M.**[2], **Pastorino P.**[2,3], **Foglini C.**[2], **Burioli E.A.V.**, **Ru G.**[1] and **Prearo M.**[2]

[1] *Unit of Biostatistics, Epidemiology and Risk Analysis, Istituto Zooprofilattico Sperimentale del Piemonte, Liguria e Valle d'Aosta, via Bologna 148, 10154 Turin, Italy*

[2] *Fish Diseases Laboratory, Istituto Zooprofilattico Sperimentale del Piemonte, Liguria e Valle d'Aosta, via Bologna 148, 10154 Turin, Italy*

[3] *Department of Life Sciences, University of Trieste, via Giorgieri 10, 34127 Trieste, Italy*

Abstract: Infectious Hematopoietic Necrosis (IHN) and Viral Haemorrhagic Septicaemia (VHS) are two systemic infections of several species of salmonids; whose reporting is mandatory in accordance with Directive 67/1991 EC. In order to obtain and preserve the free status at farm level, the Italian legislation (D.L.vo 148/2008) adopted a surveillance program involving all freshwater farms raising salmonids to be released on watercourses. This study provides a description of the epidemiology of IHN and VHS in the Piedmont Region, during the period 2000-2015. In fact in the last 16 years the surveillance was extended to all fisheries. We firstly described the disease distribution by year and by geographical area., then the potential role of several farm charcteristics was tested using unvariate and multivariate analysis.

Keywords: *VHS, IHN, Epidemiology, Piedmont Region, Salmonid fish.*

INTRODUCTION

During the last years over-fishing led to progressive depletion of wild fish populations, meanwhile aquaculture was the most rapidly growing food producing sector and was often indicated as a chance to improve food security.

According to the latest FAO report on the "State of World Fisheries and Aqua-

* **Corresponding author Maria Cristina Bona:** Department of Biostatistics, Epidemiology and Risk Analysis, Istituto Zooprofilattico Sperimentale del Piemonte, Liguria e Valle d'Aosta, *via* Bologna 148, 10154 Turin, Italy; Tel: +39 0112686359; E-mail: cristina.bona@izsto.it

Panagiotis Berillis (Ed)

culture" [1], total world fisheries production reached 154 million tonnes in 2011, of which 90.4 million tonnes from capture fisheries and 63.6 million tonnes from aquaculture. The increase in production is mainly due to the aquaculture sector, which currently accounts for 48.6% of all fish consumed for food use (equivalent to 130.8 million tonnes) [1]. The Italian salmonid farming produces above 40,000 tonnes per year, mainly rainbow trout [2].

The strong growth in aquaculture products, farmed species and their trade are paralleled by an increase in the number and extent of infectious diseases outbreaks. Infectious diseases not only affect survival, but also the growth and zootechnical performance of farmed animals. The risk of parasites, fungi, bacteria, virus and protozoa infections in aquaculture is like as in animal husbandry, especially when population densities are high. In particular in intensively reared fish, viruses facilitate secondary infections of opportunistic microorganisms. Moreover viral diseases threaten aquaculture productions not only by their severity but also because of the unavailability of pharmacological treatments.

Infectious Hematopoietic Necrosis (IHN) and Viral Haemorrhagic Septicaemia (VHS) are two systemic infections caused by *Rhabdoviridae*, affecting several species of salmonids and causing a lot of damage and considerable economic losses. The disease peaks during winter season, as the viral replication happens in the coldest months, generally at temperatures below 14°C.

VHS is a highly infectious disease affecting rainbow trout (*Oncorhynchus mykiss*) in aquaculture. So far, it has been isolated from 82 freshwater fish species as well as marine fish species [3, 4], throughout North America, Asia and Europe [5, 6]. Juvenile fish are more severely affected, with mortality rates approaching 100%. Older fish may also develop the disease with a mortality of 25-7.5%, depending on the presence of stressors [7]. IHN is a disease affecting most species of salmonid fishes. The main clinical and economic consequences of IHN occur on farms rearing fry or fingerlings of rainbow trout in freshwater, where acute outbreaks can cause very high mortality.

Main risk factors for the diseases are: 1) introduction of fishes in farm; 2) breeding zone (mountain, hill and plain), as the water can act as a vector for the virus [8]; 3) water supply (source, well, stream, river), because wild fish, in open water, can act as a reservoir.

The reporting of both diseases is mandatory in accordance with Directive 67/1991 EC [9]. In order to obtain and preserve the free status at farm level, the Italian legislation [10] introduced a surveillance program of all freshwater farms of salmonid fish to be released on watercourses. Since the end of the '90s, given the presence of a large number of restocking salmonid farms and provincial fish hatcheries, the Piedmont Region has been extending its surveillance program to all fish farms.

Aim of our work is to provide data on the descriptive epidemiology of IHN and VHS in Piedmont Region during the period 2000-2015.

MATERIAL AND METHODS

The current study covers the years 2000 – 2015. The data were collected using a database created ad hoc.

The samples sent to the laboratory were reproductive fluids (ovarian fluid), organs (anterior kidney, spleen, heart and brain) or fish (fry, juveniles and trout), as required by the European legislation [11]. All samples were taken with sterile dissection tools and placed in sterile plastic tubes containing at least 4 ml of a transport medium (MEM-Earle antibiotic 5x).

Pools were completely homogenized (using a stomacher or a homogenizer) and then centrifuged in a refrigerated centrifuge at 3,750 rpm for 15 minutes; the supernatant was incubated in a refrigerated incubator at 15 ± 2°C for 4 hours or in a refrigerator at 5 ± 3°C overnight. The virological examination was conducted on cell monolayers using the EPC cell lines and BF2. For each pool, two 7-10 day serial passages were performed .

In case of positivity, based on cytopathic effects, viral recognition has been carried out by immunofluorescence using fluorescein-conjugated antisera (BIOX, Euroclone, Milan); positivity confirmation was provided by the National Reference Center for Fish Diseases, Istituto Zooprofilattico delle Venezie, Legnaro (PD).

The diseases spread was described by year and geographical area, at district (NUTS3) level. The occurrence of the viral agents was described in terms of prevalence, defined as the number of positive aquaculture farms per 100 investigated farms, associated with the 95% confidence interval (95%CI). Univariate and multivariate analyses were used to test the potential role as risk factors of a number of farm characteristics.

The hypothesised risk factors associated with an increase in the odds to develop infections were: 1) type of production (with or without introduction of fish, assuming a higher risk for breeding without a closed cycle); 2) breeding zone assuming a higher risk for farms located in hills and plains than those located in the mountains; 3) water supply, assuming a greater risk for farms that use surface water (rivers and streams) compared to farms that use only ground water (wells and/or springs).

The association between risk factor and disease was then calculated in terms of Odds Ratio (OR). A mixed effects logistic model was used to take into account repeated sampling in the same farm.

The analysis was performed using the statistical package Stata14.1 (StataCorp. 2015).

RESULTS

From 2000 to 2015 a total of 59,233 fish were collected and examined in 5,811 pools. 1,429 samplings were carried out from 186 aquaculture farms of the Piedmont Region (Fig. **1**). 15 outbreaks were detected, respectively 7 of IHN, 7 of VHS, and 1 coinfection.

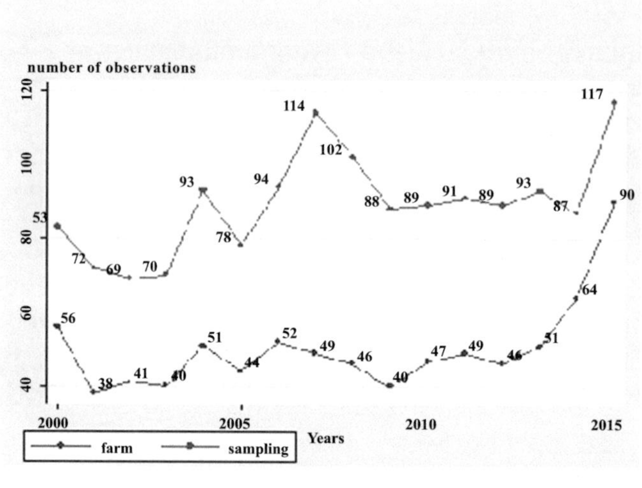

Fig. (1). Controlled fish farms and number of observations per year: 2000-2015.

Over the past 16 years, sporadic outbreaks have appeared, with a very low annual occurrence (1 or 2 cases per year); only in 2003 four cases of VHS were diagnosed with a prevalence of 9.1% (95%CI: 2.5 to 21.6).

Data were aggregated at the district level to describe the spread of disease; VHS outbreaks are present in 4 different districts, the highest number in Cuneo, whereas IHN outbreaks are limited to Torino and Cuneo districts (Figs. **2** and **3**). Disease mapping shows considerable heterogeneity in VHS and IHN prevalence by district likely due to the higher density of aquaculture in Torino and Cuneo (Figs. **4** and **5**).

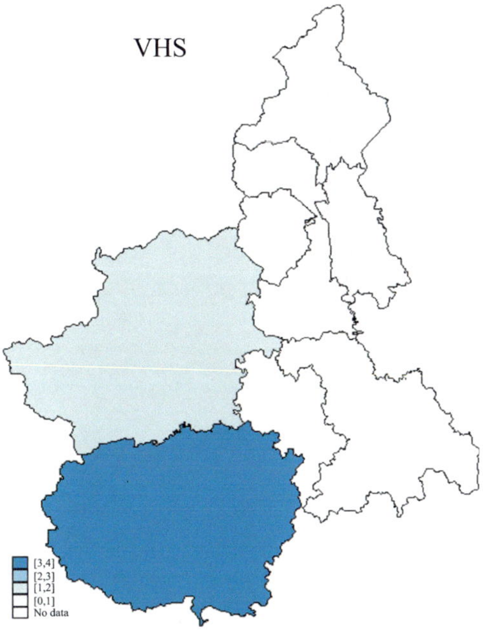

Fig. (2). Geographical distribution of outbreaks by district of VHS: years 2000-2015.

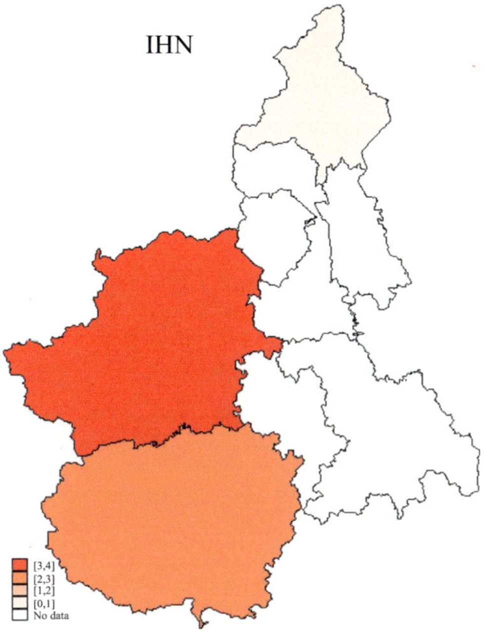

Fig. (3). Geographical distribution of outbreaks by district of IHN: years 2000-2015.

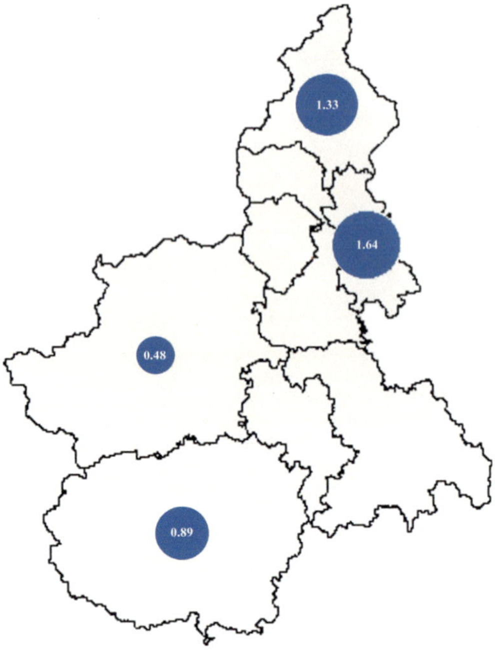

Fig. (4). Prevalence by district of VHS: years 2000-2015.

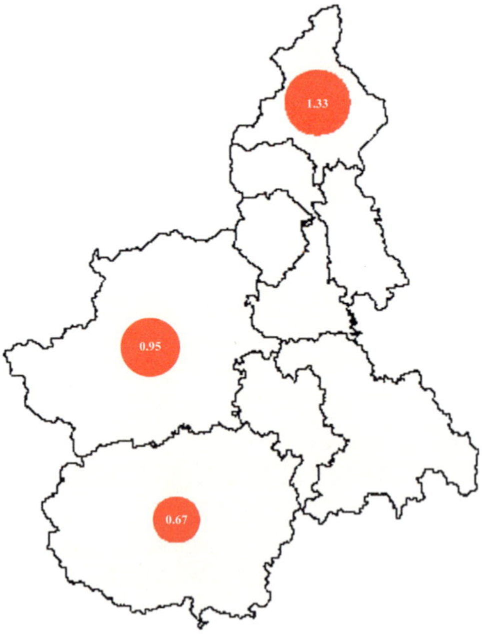

Fig. (5). Prevalence by district of IHN: years 2000-2015.

None of the considered risk factors (Fig. **6**) was statistically associated to disease:

the OR for farms exposed to fish introduction was 0.69 (95%CI 0.15-3.21); fish farms from hills or plains had a OR of 4.61 (CI 0.75 to 28.1) whereas an OR of 1.17 (CI 0.29 to 4.06) was associated to the supply of surface water.

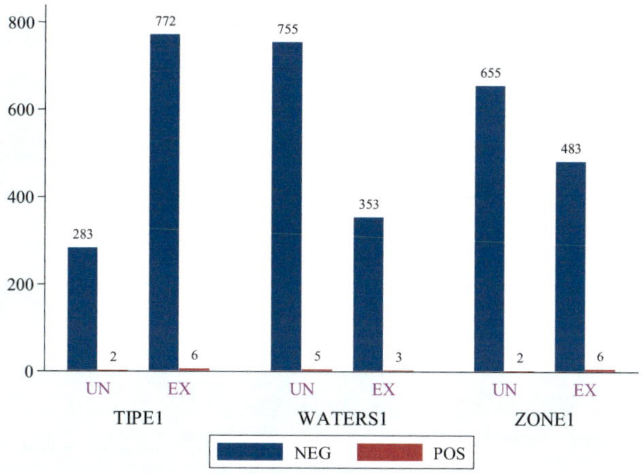

Fig. (6). VHS. Outcomes by risk category and exposure (UN unexposed, EX exposed).

Multivariate analysis for IHN showed an OR value of 0.78 (CI 0.21 to 2.84) for the farms introducing fishes, 0.44 (CI 0.08 to 2.21) for the farms from hills or plains and 1.06 (CI 0.28 to 3.92) for the farms using surface waters (Fig. **7**).

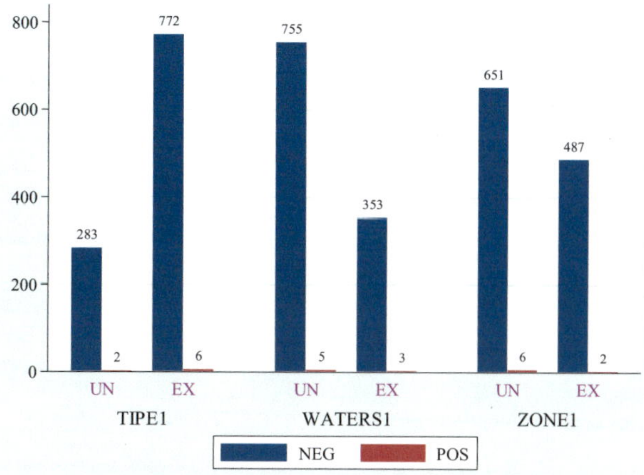

Fig. (7). IHN. Outcomes by risk category and exposure (UN unexposed, EX exposed).

DISCUSSION

This study is a first assessment of the epidemiological situation of IHN and VHS in Piedmont Region. Data shows that in our Region both the diseases have a very low prevalence despite and occur only sporadically in the region. Rainbow trout, the species most sensitive to viral salmonids diseases [3, 6] is the only species affected.

The study of the main risk factors does not allow the identification of a statistical association between any considered of fish farm characteristics and the two diseases. The only risk factor approaching statistical significance is plains/hills farming which would seem to favor the occurrence of VHS. Further investigations on risk factors is needed to increase the understanding of the diseases and to implement effective eradication strategies.

Only the description and monitoring of the epidemiological situation of both viral diseases can help in devising effective measures to prevent their spread in fish farms and impact on the aquaculture management. Our study confirms the need for a continuing surveillance as it plays a key role in effectively detecting and describing the real distribution of such diseases in our country.

CONFLICT OF INTEREST

The author (editor) declares no conflict of interest, financial or otherwise.

ACKNOWLEDGEMENT

Declared none.

REFERENCES

[1] FAO (2011) Fisheries & Aquaculture http://www.fao.org/fishery/aquaculture/en.

[2] Associazione Piscicoltori Italiani AP. 2012.http://www.api-online.it/index.cfm/en/home/

[3] OIE. (2009) Manual of Diagnostic Tests for Aquatic Animals. Chapter 2.3.9. VHS http://www.oie.int/eng/normes/fmanual/ 2.3.09_VHS.pdf.

[4] Skall HF, Olesen NJ, Mellergaard S. Viral haemorrhagic septicaemia virus in marine fish and its implications for fish farming--a review. J Fish Dis 2005; 28(9): 509-29. [http://dx.doi.org/10.1111/j.1365-2761.2005.00654.x] [PMID: 16266325]

[5] Bain MB, Cornwell ER, Hope KM, *et al.* Distribution of an invasive aquatic pathogen (viral hemorrhagic septicemia virus) in the Great Lakes and its relationship to shipping. PLoS One 2010; 5(4): e10156. [http://dx.doi.org/10.1371/journal.pone.0010156] [PMID: 20405014]

[6] Hill B, Reese A, Dixon P, *et al.* Epidemiology of different agents causing disease in aquatic animals: scientific review and database development Scientific/Technical Report submitted to EFSA; Question No EFSA-Q-2008-05009 (CFP/EFSA/AHAW/2008/01) 2010. [http://dx.doi.org/10.2903/sp.efsa.2010.EN-37]

[7] Meyers TR, Winton JR. Viral hemorrhagic septicemia virus in North America. Annu Rev Fish Dis 1995; 5: 3-24. [http://dx.doi.org/10.1016/0959-8030(95)00002-X]

[8] Renault T, Guichard B. Facteurs de risque d'apparition et d'émergence des maladies infectieuses en aquaculture. Prod Anim 2007; 20: 219-22.

[9] Council Directive 91/67/EEC https://ec.europa.eu/food/animals/animalproducts/aquaculture_en.

[10] Decreto Legislativo n. 148 del 04/08/08. Attuazionc della direttiva 2006/88/CE relativa alle condizioni di polizia sanitaria applicabili alle specie animali d'acquacoltura e ai relativi prodotti, nonch alla prevenzione di talune malattie degli animali acquatici e alle misure di lotta contro tali malattie. Gazzetta Ufficiale Serie Generale, 25 Settembre 2008, n. 225.

[11] Council Directive 91/67/EEC http://eur-lex.europa.eu/legal-content/en/ALL/?uri=CELEX.

CHAPTER 16

The Effects of Chronic Low Level Zinc (Zn) Exposure on the Hematological Profile of Tench, *Tinca tinca* L., 1758

Mehmet B. Ergönül* and **Sibel Atasagun**

Ankara University, Faculty of Science, Department of Biology, 06100-Tandoğan, Ankara, Turkey

Abstract: Fish, *Tinca tinca* were exposed to chronic low levels of zinc (2 mg/L, 10% of 96h LC_{50}) for 3, 10 and 20 days respectively and its hematological parameters were studied. It was found that Clotting time (CT), Thrombocyte count (TC), Red blood cell (RBC) counts, Hematocrit (Hct), Hemoglobin (Hb), Sedimentation rate (SR), White blood cell (WBC) count and Leucocrit (Lct) did not show a significant difference in 3 and 10 days chronic exposures. However, a significant increase in Clotting time, Sedimentation rate and White blood cell (WBC) count and significant decrease in Thrombocyte count and Hemoglobin were observed in 20 days exposure ($P<0.05$).

Keywords: Blood, Chronic, Exposure, Fish, Heavy metal, Stress, *Tinca tinca*, Zinc.

INTRODUCTION

The increase in the demands of human race gave rise to the industrialization which in turn lead to dramatic changes in the amount of pollutants released into environment especially to the aquatic environment. These detrimental changes in water quality have created a major thread for human health *via* food chain or direct use of those water resources. Among those pollutants existing in environment, heavy metals have a special place due to their long half-life. A large number of studies have investigated the acute toxicity and the effects of heavy metals on the aquatic organisms such as cladocerans and fishes [1 - 5]. Among heavy metals, zinc is less pronounced due to its essential role for the functioning and structure of many enzymes which participate in a variety of cellular metabolic processes [6]. Zinc is a rare metal in nature but it is commercially one of the most important metals in the world and it has been stated that the level of zinc com-

* **Corresponding author Mehmet B. ERGÖNÜL:** Department of Biology, Faculty of Science, Ankara University, 016100, Tandoğan, Ankara; Tel: +90 312 212 67 20; E-mail: borga@science.ankara.edu.tr

Panagiotis Berillis (Ed)

pounds increase in storm water, sewage effluents and sediments. Zinc concentration in runoff has been reported 1 to 15 mg/L [7] and in polluted waters up to 43 mg/L [8].

Although zinc is an essential element, it can be toxic at high levels [9], since as stated by Paracelsus "*the dose makes the poison*". However, in nature, most of the pollutants as well as heavy metals rarely exceed sub-lethal levels [10]. Thus aquatic biota frequently encounters sub-lethal levels of these pollutants. Although, there are a few papers on toxicity of zinc salts on aquatic organisms; toxicity data for tench (*Tinca tinca* L., 1758) which is known to be a tolerant fish species to unfavorable environmental conditions [11], is absent. Thus, in this study our aims were:

- Zn is deposited mainly in sediments through adsorption and precipitation [12] thus, we chose tench because of its benthic feeding properties; which makes it more vulnerable to Zn pollution.

- We wanted to test how a tolerant fish species responds to an essential heavy metal, zinc.

- To test the indicative capacity of blood parameters such as thrombocyte counts and clotting time in fish for monitoring programs.

For this purpose the blood parameters of fish exposed to chronic low levels of zinc, was investigated and compared to control.

MATERIALS AND METHODS

Test fish, tench (*Tinca tinca* L., 1758), were captured with fyke nets from Lake Mogan (Ankara, Turkey) and brought to laboratory in 50 L plastic cans. Fish were acclimatized to laboratory conditions for 2 weeks in 800 l tanks before they were used in the bioassays. A 12 hour photoperiod was applied during the maintenance of the fish. Mean weight and length of the fish were 209.5±12.3 g and 25.2±3.1 cm, respectively. The tanks were supplied with aerated, dechlorinated tap water (temperature 17.7±4.47 °C, pH 7.71±0.49, electrical conductivity (EC) 217.47±16.95 µS cm^{-1}, dissolved oxygen 6.39±0.45 mg l^{-1}, total hardness and total alkalinity 77.5 mg l^{-1} and 80 mg l^{-1} as $CaCO_3$, respectively). The zinc concentration of the tap water was determined with atomic absorption spectrometer (Varian AA240FS) and was below detection limits (0.01 mg/L). During the bioassays fish were fed until they do not respond to feeding attempts and commercial feed (45% crude protein; BioAqua.) was used at a rate of 2% body weight. Feeding was ceased 24 h prior to the blood sampling.

$ZnCl_2$ (Merck) was used to prepare stock solutions of zinc and the stock solution was diluted to the desired concentrations with local tap water. Fish was exposed to 2 mg/l Zn^{2+}, corresponding approximately to 10% of LC_{50} 96h for tench [13]. Tests were performed under semi static conditions and half of the aquaria water was renewed at 48 h intervals with the appropriate $ZnCl_2$ solution to keep the required concentration constant. Sub-lethal exposure tests were composed of two replicate tanks each containing 36 fish in 400 L tanks. Twelve of the fish were removed from each of the replicate tanks on days 3, 10 and 20 and seven of the fish bled from the caudal vein using heparinised needles and the rest 5 fish were used to determine clotting time using non-heparinized plastic syringes. The clotting time of blood was determined according to the methods described by Kawatsu and Sato [14]. Briefly a drop of blood was dropped on a clear watch glass and it was wappered. The time passed between the contact of blood on glass and the clot form was recorded as clotting time (seconds).

The red blood cell (RBC) and white blood cell (WBC) and thrombocyte (TC) counts were performed within 2 hours of after blood collection by Improved Neubauer hemocytometer, using Natt-Herrick as the diluent and stain [15]. Hemoglobin (Hb) levels were determined with a commercial kit (Roche) according to the instructions supplied with. Sedimentation rate of red blood cells were made according to Svobodova [16] in microcapillary tubes. Hematocrit (Hct) and leucocrit (Lct) measurements were made according to Blaxhall and Daisley [17] and McLeay and Gordon [18], respectively.

Data are presented as mean±standart deviation (SD) and were compared with One-Way ANOVA using the statistical packet program (SPPS v 9.0) at a significance level of $P<0.05$.

RESULTS

No mortality was observed in any of the exposure groups and control group through-out the study period. It was found that clotting time of fish exposed to sub-lethal level zinc, did not show a significant difference for 3 and 10 days. However, a significant increase was observed for fish in 20 days exposure group ($P<0.05$) (Fig. **1a**). Thrombocyte counts of fish did not show a significant difference in 3 and 10 days chronic low level exposure groups. However, a significant decrease was observed at 20 days sub-lethal exposure group (Fig. **1b**).

Red blood cell counts of fish did not show a significant difference in any of the exposure groups (Fig. **2a**). Hemoglobin levels of fish showed a similar pattern to RBC count; it did not show a significant difference in 3 and 10 days exposure groups. However, it showed a significant decrease in 20 days sub-lethal exposure group (Fig. **2b**). Hematocrit levels of fish did not show a significant difference in any of the exposure groups (Fig. **2c**). It was found that of sedimentation rates fish exposed to sub-lethal level of zinc, did not show a significant difference for 3 and 10 days sub-lethal and acute lethal exposure groups. However, a significant increase was observed for fish in 20 days exposure group ($P<0.05$) (Fig. **2d**).

White blood cell counts of fish showed a gradual decrease throughout exposure periods. Although, this decrease was not significant in 3 and 10 days, it became marked in 20 days (Fig. **3a**). Leucocrit levels of fish showed a similar pattern to white blood cell count (Fig. **3b**).

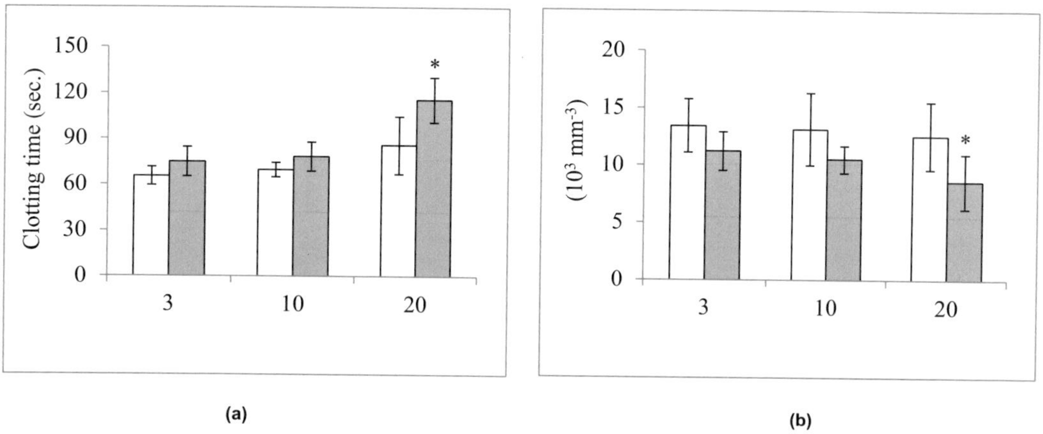

Fig. (1). Clotting time (**a**) and thrombocyte count (TC) (**b**) in *Tinca tinca* exposed to chronic low-level (2 mg l^{-1} Zn^{2+}) for 3, 10 and 20 days. Dark and light bars represent exposure and control groups respectively, * $P<0.05$.

DISCUSSION

The evaluation of blood chemistry of fish can be used as sensitive indicators of stress in fish exposed to pollutants such as heavy metals, biocides, pesticides, industrial effluents which in turn may reflect the environmental variables [5, 19]. Blood tests of fish from artificial conditions allow us to establish the hematopoietic regulation of fish, and therefore to assess deviations from the norm and nature of blood adaptations. Since fish is a good model for higher vertebrates including humans, these data can be also used for the interpretation of metals of human health. In recent decades, the acute effects of highly toxic metals such as Hg, Pb and Cd has been investigated thoroughly [2, 3, 5]. However, in most natural waters, the concentration heavy metals rarely exceed sub-lethal concentrations and less is known about the effects of chronic low level exposure in fish. However, understanding the responses of fish to chronic low level metal exposures such as zinc is further complicated by the fact that it is a required micronutrient [10].

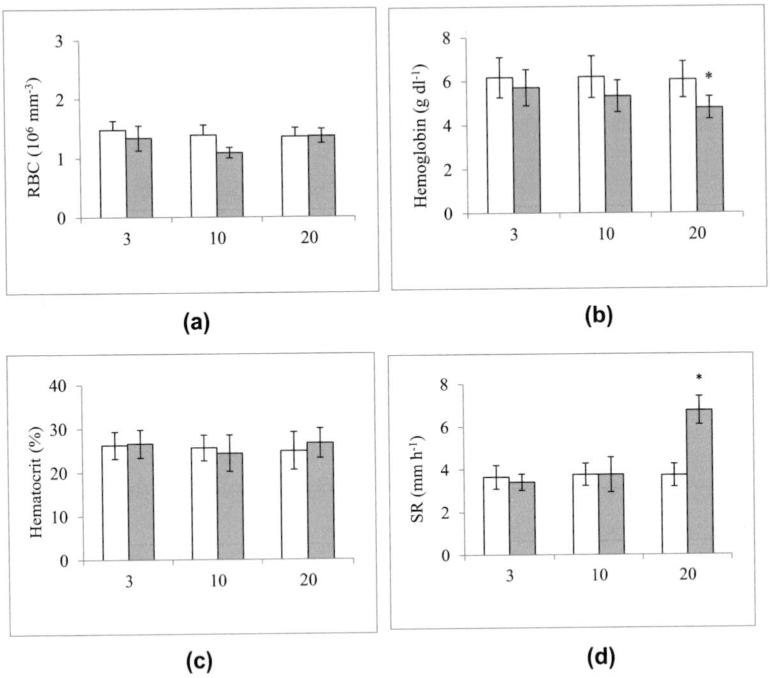

Fig. (2). Red blood cell (RBC) counts (**a**), hematocrit levels (Hct) (**b**), hemoglobin (Hb) (**c**) and sedimentation rate (SR) (**d**) of *Tinca tinca* exposed to chronic low level zinc (2 mg l^{-1} Zn^{2+}) for 3, 10 and 20 days. Dark and light bars represent exposure and control groups respectively, * $P<0.05$.

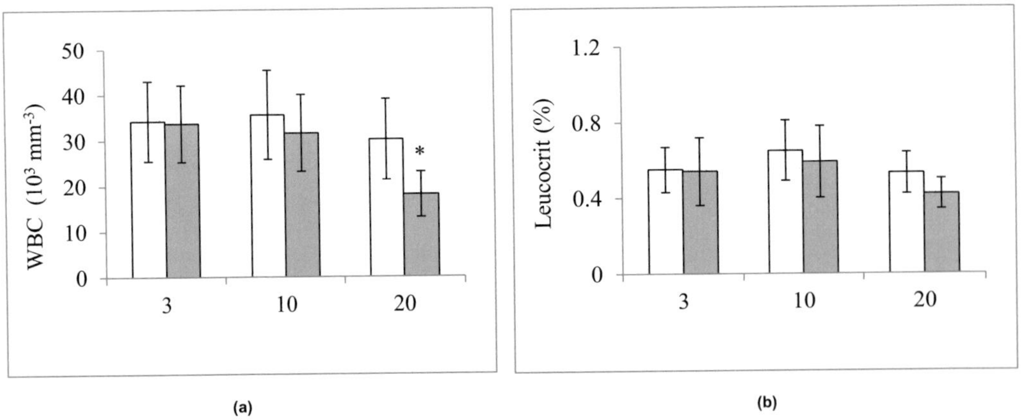

Fig. (3). White blood cell (WBC) counts (**a**) and leucocrit (**b**) levels of *Tinca tinca* exposed to chronic low level zinc (2 mg l^{-1} Zn^{2+}) for 3, 10 and 20 days. Dark and light bars represent exposure and control groups respectively, * $P<0.05$.

The hematological parameters of tench in the control group falls within the range of normal values reported for tench [2, 3] and there was no mortality during the exposure tests which shows that the chosen concentration was at a level that would induce physiological changes yet be just below the threshold for producing mortalities during chronic exposure.

Thrombocytes play a central role in the clotting of blood and there is a close relationship between clotting time and thrombocyte count and/or functions [14, 20, 21]. Our findings are in agreement with this assumption. A gradual but not significant increase was observed for clotting time of fish exposed to chronic zinc for 3 and 10 days, which became marked in 20 days ($P<0.05$). There is limited data on the effects of pollutants on thrombocyte counts and clotting time in fish. Witeska and Kosciuk [8] found that thrombocyte counts increased in carp (*C. carpio*) exposed to 20 g/dm^3 zinc for 3h, and recorded a sharp decrease at 96 h [22]. found that thrombocyte counts did not show a definite pattern in *Clarias gariepinus* exposed to sub-lethal levels of zinc; they observed both increases and decreases on a dose dependent manner. In the current study, the pattern observed for clotting time in fish could be a simply result of the decrease in thrombocyte counts; which could be a result of the direct toxic effects of heavy metal on cells or the cortisol released as a result of metal stress. It is known that the direct toxic effects of metal ions on cells or cortisol release due to metal stress could lead to a decrease in thrombocyte counts [23]. On the other hand, it is also reported that clotting time may change without a significant change in thrombocyte count which shows that there could be other processes that control clotting [21]. Nevertheless, it should be kept in mind that clotting time shows great variation both in inter- and intraspecies as well as other blood parameters in fish. Ranzani-Paiva [24] found that clotting time range between 43 and 530 seconds for *Cyprinus carpio*. Since there is not a widely acclaimed method for determination of clotting time in fish, variation may also arise from discrepancy among methods [21].

In this study, a slight decrease was observed in RBC counts for 3 and 10 days in sub-lethal exposure tests which turned back to control levels at 20 days. Hb levels of fish showed a gradual but not significant decrease throughout the chronic exposure tests for 3 and 10 days; however, it became marked in 20 days ($P<0.05$). Hematocrit levels of fish did not show any markable differences for sub-lethal exposure groups. It has been widely reported that heavy metals may alter red blood cell system and decreases or increases or both may be observed in red blood cell system parameters in many fish species on a dose and duration dependent manner [2, 3, 5, 8, 25]. Akahori [26] found that zinc caused a decrease in the hemolysis resistance of carp RBCs *in vitro*. Decreases caused by the reduced or inhibited RBC production in hematopoetic tissues and/or denaturation of RBCs because of altered cell permeability [2, 3] and/or the increase of the lysis of RBCs [3] are also reported for tench exposed to Cd, Pb and Hg. In the current study it is reasonable to speculate that sub-lethal zinc exposure has little effect on RBC system of tench. A parallel response was observed for Hb levels during 3 and 10 days exposure. The only significant alteration was observed for Hb at 20 days exposure test, which might be a result of the direct effects of zinc on iron metabolism. It is known that excess zinc may cause a decrease in iron levels in body, which in turn may lead to a decrease in Hb levels as a result of iron deficiency. Recently, it was shown that zinc exposure decreased Fe levels in liver of *Carassius auratus* [27].

Sedimentation rate which is a measure of the speed of red blood cells that settle down in heparinized blood that can be used to detect inflammation, disorders or infections [17, 28]. Sedimentation rate remained steady in chronic exposure groups for 3 and 10 days. However, a sharp increase was observed in 20 days ($P<0.05$) which may be related to the increase of fibrinogen which might decelerated the speed of red blood cells [29]. Similarly, Atamanalp [30] found that sedimentation rate of *Salmo trutta fario* exposed to cobalt for 4 weeks increased.

In this study WBC counts of fish exposed to sub-lethal zinc did not show any alterations at 3 days, but a slight decrease was observed at 10 days ($P>0.05$) which become significant at 20 days ($P<0.05$). Lct levels of fish did not show a distinct pattern; a steady level at 3 days and slight decrease at 10 and 20 days. However, none of those alterations were significant. Kori-Siakpere and Ubogu [25] recorded a slight decrease in WBC counts of *Heteroclarias* sp. exposed to 10 mg /L zinc for 15 days. Ololade and Ogini [22] found a gradual decrease in WBC counts of *Clarias gariepinus* exposed to 30-50 mg/L zinc for 96h. However, a few papers also report that WBC counts of *Catla catla* exposed to sub-lethal zinc concentrations for 25 days did not show an alteration, or show an increase in WBC counts of *O. mykiss* exposed to 2 ppm zinc for 72 h [31] or an initial increase and a gradual decrease in recovery period in *C. carpio* exposed to zinc for 3 hours [8]. Cortisol release as a response to stress caused by heavy metal exposure, may lead to apoptosis in white blood cells *via* inhibition of protein synthesis in lymphocytosis [32 - 34] and impaired proliferative capacity of cells [35] and decreasing the activities of WBCs [34, 36] and impairment of hematopoetic cells in the kidney [37]. It is also possible that the decrease in WBCs may also arise from direct toxic effects of heavy metal [38]. Leucorit is a measure of WBC and TC counts and there is a close relationship between them. A linear relationship between Lct and WBC-TC counts was also observed in this study.

Fish are highly sensitive to environmental variables and any disturbance of these variables result in a primer response in fish which includes hormonal changes that plays a central role in re-establishment of the homeostasis. However, if those unfavorable conditions prolongs fish may become unable to tolerate these changes and a seconder response takes place which in turn may lead to physiological and behavioural changes [39, 40]. Among stressors heavy metals have a special position due to their longevity besides their direct toxicity. In conclusion, the results clearly indicate that tench, known to be a relatively tolerant fish species, was unable to cope with the effects of zinc which is known to be an essential element. The most marked effect was observed on 20 days chronic exposure tests, thus it is reasonable to conclude that tench was unable to adapt heavy metal stress for prolonged periods. However, it should be kept in mind that understanding the response of fish to micronutrients such as zinc is more complicated as it is essential trace element required for normal growth and development of fish.

CONFLICT OF INTEREST

The author (editor) declares no conflict of interest, financial or otherwise.

ACKNOWLEDGEMENT

A part of this study was supported by The Scientific and Technological Council of Turkey (TUBİTAK) under grant TBAG HD 108T662.

REFERENCES

[1] Bagdonas E. Vosyliene MZ. A study of toxicty and genotoxicity of copper, zinc and their mixture to rainbow trout. Biologia 2006; 1: 8-13.

[2] Shah SL, Altındağ A. Hematological parameters of tench (*Tinca tinca* L.) after acute and chronic exposure to lethal and sublethal mercury treatments. Bull Environ Contam Toxicol 2004; 73(5): 911-8. [http://dx.doi.org/10.1007/s00128-004-0513-y] [PMID: 15669737]

[3] Shah SL, Altındağ A. Alterations in the immunological parameters of tench (*Tinca tinca*) after acute and chronic exposure to lethal and sub-lethal treatments with mercury, cadmium and lead. Turk J Vet Anim Sci 2005; 29: 1163-8.

[4] Witeska M, Jezierska B, Wolnicki J. Respiratory and hematological response of tench, *Tinca tinca* (L.) to a short-term cadmium exposure. Aquacult Int 2006; 14: 141-52. [http://dx.doi.org/10.1007/s10499-005-9020-3]

[5] Ergönül MB, Atasağun S, Kocatürk K. Alterations in the Hematological and Biochemical Parameters and Plasma Ion Concentrations of Common Carp, (*Cyprinus carpio* L., 1758) After Short Term Exposure to Sub-lethal Concentrations of Lead. J Fac Vet Med Kafkas Univ 2012; 18: 297-302.

[6] Vallee BL, Auld DS. Zinc coordination, function, and structure of zinc enzymes and other proteins. Biochemistry 1990; 29(24): 5647-59. [http://dx.doi.org/10.1021/bi00476a001] [PMID: 2200508]

[7] Golding S. A Survey of Zinc Concentrations in Industrial Stormwater Runoff. Washington State Department of Ecology, Environmental Assessment Program Publication 2006. No. 06-03-009.

[8] Witeska M, Kośiuk B. The changes in common carp blood after short-term zinc exposure. Environ Sci Pollut Res Int 2003; 10(5): 284-6. [http://dx.doi.org/10.1065/espr2003.07.161] [PMID: 14535641]

[9] Chukhlovi AB, Tokalov SV, Yagunov AS, Westendorf J, Reincke H, Karbe L. *In vitro* suppression of thymocyte apoptosis by metal-rich complex environmental mixtures: potential role of zinc and cadmium excess. Sci Total Environ 2001; 281(1-3): 153-63. [http://dx.doi.org/10.1016/S0048-9697(01)00843-9] [PMID: 11778948]

[10] McGeer JC, Szebedinszky C, Wood CM, Wood CM. Effects of chronic sublethal exposure to waterborne Cu, Cd or Zn in rainbow trout 2: tissue specific metal accumulation. Aquat Toxicol 2000; 50(3): 245-56. [http://dx.doi.org/10.1016/S0166-445X(99)00106-X] [PMID: 10958958]

[11] Svobodova Z, Kolarova J. A review of the diseases and contaminant related mortalities of tench. Vet Med 2004; 49: 19-34.

[12] Kadlec RH, Wallace S. Treatment Wetlands. 2nd ed. Halogens, Sulfur, Metals, and Metalloids 2008; pp. 403-80. [http://dx.doi.org/10.1201/9781420012514.ch11]

[13] Ergönül MB, Altındağ A. Çinko ve Bakırın Kadife Balığı (*Tinca tinca* L., 1758) için akut toksisitesinin belirlenmesi. KSU J Nat Sci 2011; 14: 19-24.

[14] Kawatsu H, Sato M. Effect of glass contact on the clotting time of carp blood. Nippon Suisan Gakkaishi 1987; 53: 1533-6. [http://dx.doi.org/10.2331/suisan.53.1533]

[15] Stoskopf MK. Fish Medicine. Philadelphia: Saunders 1993.

[16] Svobodova Z, Pravda D, Palackova J. Unified methods of hematological examination of fish. 1991.

[17] Blaxhall PC, Daisley KW. Routine haematological methods for use with fish blood. J Fish Biol 1973; 5: 771-81. [http://dx.doi.org/10.1111/j.1095-8649.1973.tb04510.x]

[18] McLeay DJ, Gordon MR. Leucocrit: A simple hematological technique for measuring acute stress in salmonid fish, including stressful concentrations of pulpmill effluent. J Fish Res Board Can 1977; 34: 2164-75. [http://dx.doi.org/10.1139/f77-284]

[19] Fernandes MN, Mazon AF. Environmental pollution and fish gill morphology.Fish adaptation. Enfield: Science 2003; pp. 203-31.

[20] Passantino L, Cianciotta A, Patruno R, Ribaud MR, Jirillo E, Passantino GF. Do fish thrombocytes play an immunological role? Their cytoenzymatic profiles and function during an accidental piscine candidiasis in aquarium. Immunopharmacol Immunotoxicol 2005; 27(2): 345-56. [http://dx.doi.org/10.1081/IPH-200067959] [PMID: 16114515]

[21] Tavares-Dias M, Oliveira SR. A review of the blood coagulation system of fish. Rev Bras Biocienc 2009; 7: 205-24.

[22] Ololade I, Ogini O. Behavioral and hematological effects of zinc on African Catfish, *Clarias gariepinus*. Int J Fish Aquacult 2009; 1: 22-7.

[23] Al-Akel AS, Shamsi MJ. Hexavalent chromium: toxicity and impact on carbohydrate metabolism and hematological parameters of carp (*Cyprinus carpio* L.) from Saudi Arabia. Aquat Sci 1996; 58: 24-30. [http://dx.doi.org/10.1007/BF00877638]

[24] Ranzani-Paiva MJ, Rodrigues EL, Veiga ML, Eiras AC, Campos BE. Differential leukocytes count in "dourado" *Salaminus maxilosus*, from the Mogi-Guacu River, Pirassununga. Braz J Biol 2003; 63: 517-25. [http://dx.doi.org/10.1590/S1519-69842003000300018] [PMID: 14758711]

[25] Kori-Siakpere O, Ubogu E. Sub-lethal hematological effects of zinc on the freshwater fish, *Heteroclarias* sp. (Osteichthyes: Clariidae). Afr J Biotechnol 2008; 7: 2068-73. [http://dx.doi.org/10.5897/AJB07.706]

[26] Akahori A, Jóźwiak Z, Gabryelak T, Gondko R. Effect of zinc on carp (*Cyprinus carpio* L.) erythrocytes. Comp Biochem Physiol C Pharmacol Toxicol Endocrinol 1999; 123(3): 209-15. [http://dx.doi.org/10.1016/S0742-8413(99)00031-6] [PMID: 10530892]

[27] Qu R, Feng M, Wang X, *et al.* Metal accumulation and oxidative stress biomarkers in liver of freshwater fish *Carassius auratus* following *in vivo* exposure to waterborne zinc under different pH values. Aquat Toxicol 2014; 150: 9-16. [http://dx.doi.org/10.1016/j.aquatox.2014.02.008] [PMID: 24632310]

[28] Sox HC Jr, Liang MH. The erythrocyte sedimentation rate. Guidelines for rational use. Ann Intern Med 1986; 104(4): 515-23. [http://dx.doi.org/10.7326/0003-4819-104-4-515] [PMID: 3954279]

[29] Malla FA, Sharma G, Singh S. Chlorpyrifos pesticide toxicity on erythrocyte sedimentation rate in fish, *Channa punctatus.* Biol Med 2009; 1: 54-5.

[30] Atamanalp M, Kocaman EM, Uçar A, Atak G. The alterations in the hematological parameters of brown trout exposed to cobalt chloride. J Anim Vet Adv 2010; 9: 2167-70. [http://dx.doi.org/10.3923/javaa.2010.2167.2170]

[31] Ateş B, Orun İ, Selamoğlu T, Durmaz Z, Yılmaz İ. Effects of sodium selenite on some biochemical and hematological parameters of rainbow trout exposed to Pb and Cu. Fish Physiol Biochem 2008; 34: 53-9. [http://dx.doi.org/10.1007/s10695-007-9146-5] [PMID: 18649023]

[32] Weyts FA, Flik G, Verburg-van Kemenade BM. Cortisol inhibits apoptosis in carp neutrophilic granulocytes. Dev Comp Immunol 1998; 22(5-6): 563-72. [http://dx.doi.org/10.1016/S0145-305X(98)00027-5] [PMID: 9877437]

[33] Verburg Van Kemenade BM, Nowak B, Engelsma MY, Wyets FA. Differential effects of cortisol on apoptosis and proliferation of carp B lymphocytes from head kidney, spleen and blood. Fish Shellfish Immunol 1999; 9: 405-15. [http://dx.doi.org/10.1006/fsim.1998.0197]

[34] Witeska M. Stress in fish - Hematological and immunological effects of heavy metals. Elec J Ichthyol 2005; 1: 35-41.

[35] Espelid S, Lokken GB, Steiro K, Bogwald J. Effects of cortisol and stress on the immune system in Atlantic Salmon (*Salmo salar* L.). Fish Shellfish Immunol 1996; 6: 95-110. [http://dx.doi.org/10.1006/fsim.1996.0011]

[36] Svobodova Z, Vykusova B, Machova J. The effects of pollutants on selected haematological and biochemical parameters in fish.Sublethal and chronic effects of pollutants on freshwater fish. London: Fishing New Books 1993.

[37] Kotsanis N, Georgudaki JL, Zoumbos KK. Changes in selected hematological parameters at early stages of the rainbow trout, *Oncorhynchus mykiss*, subjected to metal toxicants: arsenic, cadmium and mercury. J Appl Ichthyology 2000; 16: 276-8. [http://dx.doi.org/10.1046/j.1439-0426.2000.00163.x]

[38] Siwicki AK, Studnicka M. 1992.

[39] Barton BA, Iwama GK. Physiological changes in fish from stress in aquaculture with emphasis on the response and effects of corticosteroids. Annu Rev Fish Dis 1991; 1: 3-26. [http://dx.doi.org/10.1016/0959-8030(91)90019-G]

[40] Wendelaar Bonga SE. The stress response in fish. Physiol Rev 1997; 77(3): 591-625. [PMID: 9234959]

CHAPTER 17

Use of Acoustic Telemetry to Management of Fishery in Artificial Reefs

Aytaç Özgül* and **Altan Lök**

Ege University, Faculty of Fisheries, 35100 İzmir, Turkey

Abstract: Artificial reefs are to be used as a tool in fisheries management in world-wide since long time. Fishing practices in artificial reef areas should also be operated with a management model as it is in the other fishing zones. Unmanaged artificial reef system may be more harmful than its profits. Acoustic telemetry can be an extremely effective tool to understand the home range and movement models of the fish species for data of management fisheries in artificial reef areas. Acoustic telemetry technology has an important potential to determine strategy in fisheries management around artificial reefs and in construct more efficient artificial reef project.

Keywords: Artificial Reefs, Acoustic Telemetry, Fisheries, Management.

INTRODUCTION

The coastal areas have always performed a significant role in human activities, but their economic importance has been significantly developing in the past few decades [1]. Increasing of the population, economic activities carried out near or on the coast such as construction of cities, harbors, and tourism are main factors of this development. Nevertheless, the growing demand for marine fisheries resources is compounded by the increase in the world population, which is expected to reach 8.9 billion in 2050 [2, 3] implying serious pressure on fish stocks [4, 5]. According to Worm *et al.* [6], 63% of assessed fish stocks worldwide require rebuilding, and lower exploitation rates are needed to reverse the collapse of vulnerable species. Therefore, there is an urgent need for policies on the recovery and protect of fish stocks.

In the management models to be pursued in the coastline, one important step is undoubtedly protection, recovery and development of depleted or exploited marine sources. Artificial reefs (AR) have made positive contributions to the re-

* **Corresponding author Aytaç Özgül:** Ege University, Facuty of Fisheries, Department Fisheries and Fish Processing Technology, 35100, İzmir, Turkey; Tel: +90 232 3435225; E-mail: aytac.ozgul@ege.edu.tr

covery of the habitats lost in the coastal zone and to manage of fish stocks [6, 7]. It is defined that "a submerged structure deliberately placed on the seabed to mimic some functions of a natural reef, such as protecting, regenerating, concentrating and/or enhancing populations of living marine resources" [8]. They are used for coastal management purposes in many countries and regions across the world. Recently, artificial reefs have been used for several purposes such as enhancing marine bio-diversity, protecting sensitive habitats from illegal fishing activities, preventing coastal erosion, providing new substrates for algae and mollusk culture (mariculture), supporting small-scale fisheries, creating suitable areas for diving tourism, scientific research and educational activities on seas around the world [9 - 11].

MANAGEMENT OF FISHERY IN ARTIFICIAL REEFS

Increasing trend in artificial reef projects over the world in recent years has made it mandatory to develop a management model for industrial, recreational fishing and tourism activities [7 - 12]. Fishing practices in artificial reef areas should also be operated with a management model as it is in the other fishing zones. Unmanaged artificial reef system may be more harmful than its profits. Parallel with the traditional and ecosystem based fishery management model, applications such as closed season, closed area, fishing gear restriction should be adapted to artificial reef zones.

Variables such as the characteristics of the fishing fleet (number of boats, fishing power, fishing methods *etc.*), other stakeholders (scuba diving tourism, aquaculture *etc.*) and socio-economic conditions of coastal village as well as the effective area of the reefs, home ranges and movement patterns of the fish species are the major factors which have to be considered in the management strategies [13 - 15].

Most of the scientific studies related with artificial reefs to date have focused on environmental, ecological and physical factors influencing reef productivity. Nonetheless, very few studies have focused on how reef location, and specifically reef densities, influences of reef fish assemblages. In fact, reef placement remains one of the most neglected variables, and potentially one of the most important variable affecting artificial reef fish-assemblages [16]. Another important criteria in the fisheries management model is the interaction between fish, fisheries and the artificial reefs. In this respect, understanding the home range and side fidelity of the reef species during the day has a direct influence on the success of management of fishery in the artificial reefs. Most fish do not roam randomly. Instead, they exhibit site-fidelity and establish areas where they undertake routine activities such as feeding, resting and defending [17, 18]. Therefore, determining variance of reef fish behavior's in time and space is essential to the study of fish ecology and to the design of effective conservation and resource management strategies [19 - 21].

In determining the behavior of the fish species in the artificial reefs area, different methods are used such as visual census, experimental fishing, underwater video techniques and tag-monitoring [22 - 24]. Acoustic telemetry technology has been preferred over the last few years due to the fact that the methods mentioned above are inadequate for some reasons such as bad environmental conditions, diving limits, differences in fish behaviors. Among these methods, acoustic telemetry has been frequently preferred recently. In addition to improving the understanding of the value of artificial reefs, analyses of fine-scale movements of fish species also have been provided insight into the use of open habitats surrounding the reefs.

ACOUSTIC TELEMETRY

Acoustic telemetry has been used as an alternative to or as a complementary method for traditional mark and recapture studies to study fish movements [25, 26]. This technology has allowed researchers to quantify home ranges within MPAs to determine site fidelity and to infer temporal patterns of marine organisms [27 - 30].

Acoustic telemetry is the use of an acoustic transmitter attached to or implanted in an aquatic animal to locate and gather information about its presence, movements, and behavior in the underwater environment [31]. It uses high frequency sound (69 & 180 kHz) to transmit information. Typical systems consist of two main components: transmitters (tags) and receivers. A tag is an electronic device usually implanted or externally attached to a fish that transmits ultrasonic signals. Receivers are small, data-logging computers anchored near the bottom of a lake or stream or the ocean that "listen" for tagged fish (Fig. **1**). When a signal is identified, the tag's unique ID code is saved with the date and time. The data from any single receiver provide a record of each visit to that location by a tagged fish [32, 33].

It is of crucial importance to keep the fish alive which will be tagged in acoustic telemetry method. Tagged fish are generally caught by longline, hand line, pots or directly by divers [34, 35]. After fainting the fish with anesthetic materials (MS-222, Fenoksi etonol, clove oil), acoustic tags are surgically implanted into the peritoneal cavity through a dorso-ventral incision and sutured using non-absorbable sutures maximizing sterile conditions [36, 37].

Behaviours of fish species such as *Coryphaena hippurus* [40], *Thunnus albacares, Katsuwonus pelamis* [41], *Thunnus obesus* [42], *Diplodus vulgaris, Diplodus sargus* [43, 44], *Lutjanus campechanus* [45, 46], *Sebastes caurinus* ve *Ophiadon enlongatus* [47], *Cephalopholis taeniops* [48], *Gadus morhua* [49], *Sciaena umbra* [35] were researched by using acoustic telemetry technology in the artificial reef zones up to date.

Fig. (1). A main components of the acoustic telemetry systems.

Major benefits of acoustic telemetry over traditional sampling techniques (visual census, experimental fishing, standard tagging-monitoring *etc.*) include constant high resolution monitoring over large distances, and accuracy to within a few meters if receivers are arranged in closely-spaced grids. On its own, data from acoustic telemetry provides answers to basic questions, such as where and when fish move. When combined with other datasets, telemetry can also provide insights into how or why they move. Results from acoustic telemetry studies provide fishery managers with information about fish movement that improves management

Different methods are used in tracking and assessing the signals coming from the tagged fish with acoustic tag in studies of artificial reefs area (Fig. **2**). Active telemetry, once placing the transmitter, the fish is left into its natural environment and the signals coming from the tag is tracked by the hydrophone on the ship or boat [50, 51]. In the passive telemetry method, on the other hand, signals from the fish are detected by the transmitters placed under the water [31, 34]. Another method for monitoring tagged fish is acoustic positioning system (Vemco; VPS ve VRAP; Lotek; ALPS). System is composed of acoustic receivers, transmitters and analysis process. Tagged aquatic animals can be monitored with an accuracy of 1 m in coordination with the real time [52, 53]. This method works like GPS and satellites and positioning of an aquatic animal requires detection by at least 3 transmitters [54, 55]. Increasing the number of transmitters, correct deployment geometry and the depth of receivers increases the system performance and the success rate [56 - 58].

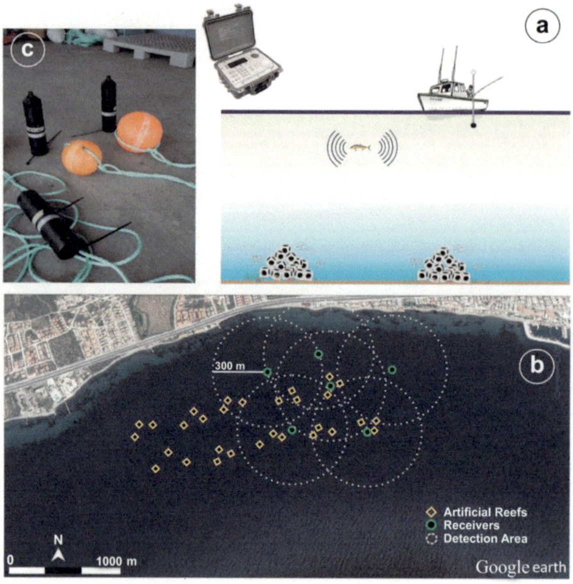

Fig. (2). Samples of acoustic telemetry methods for artificial reef areas. a: Active telemetry; b: VPS design c: Acoustic receivers and mooring system.

Basic telemetry data have at least three-dimensions (x and y in space and z in time), so that, date, time and some metric of horizontal position can be recorded as well as the tag number. It allows researchers to discover and explain site fidelity and habitat preferences. Furthermore, such variables as movement patterns and home range size of fish species have been used in artificial reefs studies especially. Home range is described as the area occupied by an animal during its normal activities of food gathering, mating, and caring for the young [59]. Accurate home range estimates are important to understand spatial and behavioral ecology of a species' [18], as well as the implementation of conservation and management programs [60, 61]. There are several methods of calculating home range, but minimum convex polygon (MCP) and kernel density estimation (KDE) are frequently used as estimates of a reef fish's home range [18, 61].

MCP calculates home range size by drawing a convex polygon around the location points of an individual. Although MCP has limitations, especially its in ability to distinguish between areas of high and low use by the subject animal, it is still one of the most widely used methods of calculating home range [18 - 62]. KDE have been considered to be more accurate than MCP. To calculate KDE, a nonparametric statistical method for estimating probability densities from a set of points is used. Generally, 90–95% volume contours are used to describe the boundaries of animal home ranges, and 50–60% contours are generated to represent "core" areas (Fig. **3**). In addition to geographic information system (GIS) software programs like ArcGIS and QGIS, some statistical programs such as MATLAB and R are also used to calculate home range parameters like MCP and KDE [63, 64].

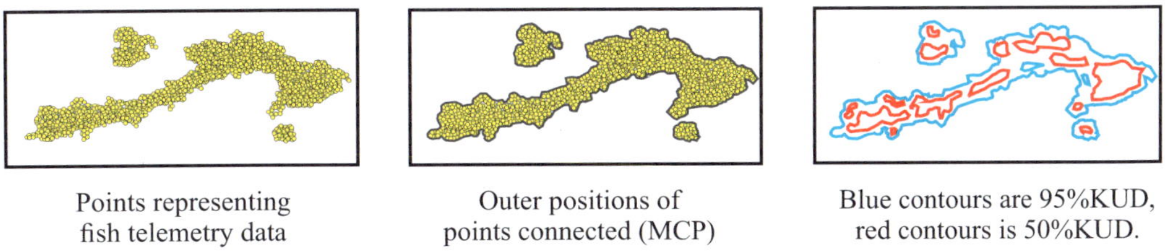

Fig. (3). Define of basic home range parameters. MCP: Minimum convex polygon, KDE: Kernel Density Estimation.

There are some limitations and problems of acoustic telemetry method in the artificial reef zones. Reflection and absorption by the artificial reef blocks may be decreased the system performance. Especially when the tagged fish hide in the reef blocks detection distance and range is fairly reduced [65]. Range test applications, increasing the number of transmitters, deployment of receivers with the correct geometry and depth are the methods in order to avoid this case [32, 56, 66, 67]. Another problem in the artificial reef zones is the loosing or replacement of transmitters by the waves and currents. Active tracking studies, where the subject is followed by boat with the assistance of a directional hydrophone, are very demanding of manpower and funding, generally limiting studies to less than a week and risking disruption by rough sea and weather conditions. Acoustic tracking is also acknowledged with some financial disadvantages. The equipment is substantially more expensive compared with simple external tags used in traditional mark and recapture studies; acoustics transmitters cost hundreds of dollars apiece, and hydrophones cost thousands [68].

DISCUSSION AND CONCLUSION

Acoustic telemetry can be an extremely effective tool to understand the home range and movement models of the fish species for data of management fisheries in artificial reef areas. In comparison with other methods (underwater visual census, video technics, experimental fishing methods and standard tag-recapture studies, *etc.*), acoustic telemetry may help obtain appropriate results causing minimum damage on marine ecosystem [69, 70].

Acoustic telemetry method is not used in artificial reef zones but also in marine protected areas. Information on home range size and movement patterns of target species is an important aspect of applied ecology, fisheries and fisheries management and for providing guidance to locally effective marine reserve design and assessment [71 - 73]. Telemetry data is also used to determine the geographical borders of marine protected areas in a more accurate way. Species sucs as *Epinephelus marginatus* [74], *Semicossyphus pulcher* [75], *Diplodus annularis* [76], *S. umbra* [77, 78] were studied by acoustic telemetry method in the marine protected areas.

The best example of acoustic telemetry method in artificial reef zone fishery management is the *Lutjanus campechanus* in the Gulf of Mexico [79]. Red snapper fishery is performed around the petrol platforms and home range and movement models of the species have been determined successfully by acoustic telemetry method [80, 81].

Another example of the using acoustic telemetry in management of fisheries in artificial reefs is the Mediterranean Sea. In Turkey, the first large-scale artificial reef area was created along the Northern Aegean Sea between 2010 and 2012 by The Ministry of Food, Agriculture and Livestock. There are 215 reef sets each consisting of 30 concrete blocks in seven different rectangular subareas in the artificial reef area, which aims to promote small-scale fisheries in the region. Three of subareas have been closed to fishing activity for management of fisheries in this zone [82, 83].

The movement behaviour of *D. vulgaris, Sparus aurata, Diplodus puntozzo, S. umbra, Scorpaena porcus, Scorpaeana scrofa* and *Pomatamus saltatrix* were investigated using by acoustic telemetry method [84, 85]. The movement pattern of *S. aurata* around artificial reefs in reproduction period was seen in Fig. (**4**). When we look at fisheries management, if we want to protect spawning population of *S. aurata*, it will not be enough to close only sub areas. Result of the study shows that *S. aurata* is to use not only artificial reef area but also other habitats such as sea grass and natural rocky reefs.

Acoustic telemetry studies in big and high volume artificial reef areas are very new and limited. Furthermore these limited studies show that this method is successfully performed in the artificial reef zones. In this frame we face with three situations:

- Using the fish movement data obtained from the acoustic telemetry may provide to the management of fisheries in artificial reef areas. If the aim is to increase fisheries, fishing can be carried out within the home range parameters of the targeted fish species. These data in particular can help prevent possible ghost fishing problems in the artificial reef areas.

- Moreover, use of artificial reefs as sea protection zones is frequently seen in the Mediterranean Sea. In the management plans designed to protect artificial reef areas, data to be obtained with acoustic telemetry may provide valuable information for decision makers concerning the location and scope of the fishing ban restrictions to be imposed in the area.
- Spatial and temporal data from acoustic telemetry can be used in assessing the performance of artificial reefs successfully. It can particularly contribute to the determination of distances between reef sets in the planning of the projects and to the efficient use of reef sets to be produced in this respect.

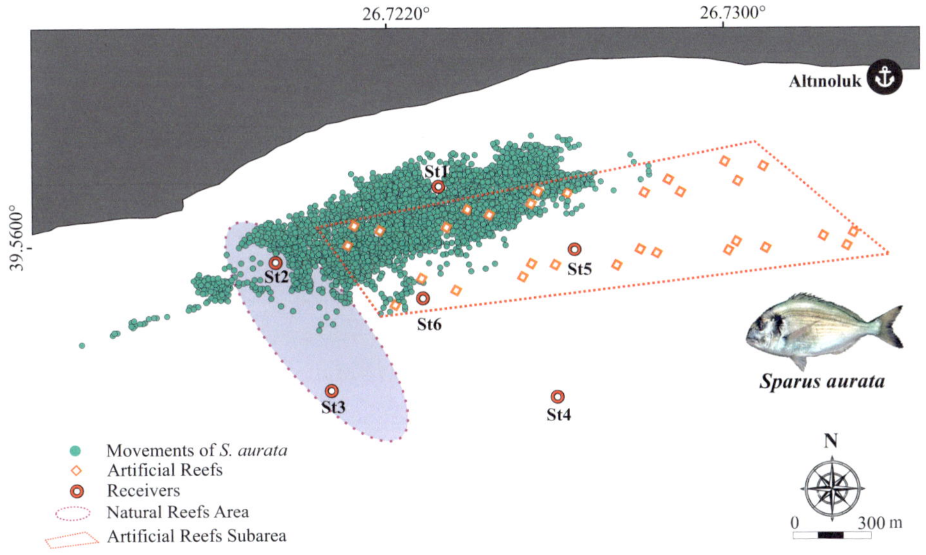

Fig. (4). Movement of *Sparus aurata* in artificial reefs area [86].

We suggest that acoustic telemetry technology will play an important role in fisheries management around artificial reefs and in construct more efficient artificial reef project in near feature.

CONFLICT OF INTEREST

The author (editor) declares no conflict of interest, financial or otherwise.

ACKNOWLEDGEMENTS

This study was part of the project (TOVAG-1120383), funded by TÜBİTAK and we would like to thank Ozan Soykan for revising the English text.

REFERENCES

[1] Arnouil DS. Shoreline Response for a Reef Ball TM Submerged Breakwater System Offshore of Grand Cayman Island 2006.

[2] UNFPA. State of the World Population 2004 The Cairo Consensus at Ten: Population. New York: Reproductive Health and the Global Effort to End Poverty 2004.

[3] Worm B, Barbier EB, Beaumont N, *et al.* Impacts of biodiversity loss on ocean ecosystem services. Science 2006; 314(5800): 787-90.
[http://dx.doi.org/10.1126/science.1132294] [PMID: 17082450]

[4] Cury P, Miserey Y. Une mer sans poisons. Paris: Calmann-Lévy 2008.

[5] Worm B, Hilborn R, Baum JK, *et al.* Rebuilding global fisheries. Science 2009; 325(5940): 578-85.
[http://dx.doi.org/10.1126/science.1173146] [PMID: 19644114]

[6] Clark W, Mosely J, Stone R. Foreword [sic]. In: Colunga L, Stone R, Eds. 3.Proceedings of the International Conference on Artificial Reefs. Houston. 1974; p.

[7] Pioch S, Raynal JC, Lasserre G, Aliaume C. An Integrated Coastal Area Management Strategy to Deploy Artificial Reefs.Artificial Reefs in Fisheries Management. London: CRC Press 2011; pp. 65-75.
[http://dx.doi.org/10.1201/b10910-6]

[8] FAO. Practical guidelines for the use of artificial reefs in the Mediterranean and the Black Sea. by Fabi G, Scarcella G, Spagnolo A, Bortone SA, Charbonnel E, Goutayer JJ, Haddad N, Lök A, Trommelen M. Studies and Reviews. GFCM, No. 96; Rome- Italy. 2015.

[9] Fabi G, Fiorentini L, Giannini S. Experimental shellfish culture on an artificial reef in the Adriatic Sea. Bull Mar Sci 1989; 44(21): 913-22.

[10] Clark S, Edwards AJ. An evaluation of artificial reef structures as tools for marine rehabilitation in the Maldives. Aquat Conserv 1999; 9: 5-21.
[http://dx.doi.org/10.1002/(SICI)1099-0755(199901/02)9:1<5::AID-AQC330>3.0.CO;2-U]

[11] Jakšić S, Stamenković I, Đorđević J. Impacts of Artificial Reefs and Diving Tourism. Turizam 2013; 17(4): 155-65.

[12] FAO. Report of GFCM Workshop on Artificial Reefs in the Mediterranean and Black Sea. GFCM Scientific Advisory Committee (SAC), Subcommittee on Marine Environment and Ecosystems (SCMEE); Izmir-Turkey.

[13] Kramer D, Chapman M. Implications of fish home range size and relocation for marine reserve function. Environ Biol Fishes 1999; 55: 65-79.
[http://dx.doi.org/10.1023/A:1007481206399]

[14] Pittman SJ, McAlpine CA. Movements of marine fish and decapod crustaceans: process, theory and application. Adv Mar Biol 2003; 44: 205-94.
[http://dx.doi.org/10.1016/S0065-2881(03)44004-2] [PMID: 12846043]

[15] Addis DT, Patterson WF, Dance MA. Implications of reef fish movement from unreported artificial reef sites in the northern Gulf of Mexico. Fish Res 2013; 147: 349-58.
[http://dx.doi.org/10.1016/j.fishres.2013.07.011]

[16] Bortone SA. Resolving the attraction-production dilemma in artificial reef research: Some yeas and nays. Fisheries (Bethesda, Md) 1998; 23: 6-10.
[http://dx.doi.org/10.1577/1548-8446(1998)023<0006:RTADIA>2.0.CO;2]

[17] Stamps J. Motor learning and the value of familiar space. Am Nat 1995; 146: 41-58.
[http://dx.doi.org/10.1086/285786]

[18] Powell RA. Animal home ranges and territories and home range estimators.Research techniques in animal ecology: Controversies and consequences. New York: Columbia University Press 2000; pp.

65-110.

[19] McCauley DE. Effects of population dynamics on genetics in mosaic landscapes.Mosaic Landscapes and Ecological Processes. London: Chapman & Hall 1995; pp. 178-98. [http://dx.doi.org/10.1007/978-94-011-0717-4_8]

[20] Hanski I, Gilpin ME. Metapopulation Biology: Ecology, Genetics and Evolution. San Diego: Academic Press 1997.

[21] Wiens JA. Ecological heterogeneity: an ontogeny of concepts and approaches.The Ecological Consequences of Environmental Heterogeneity. Oxford: Blackwell 2000; pp. 9-31.

[22] Bortone SA, Bohnsack JA. Sampling and studying fish on artificial reefs 1991.

[23] Addis DT, Patterson II, Dance MA. Site fidelity and movement of reef fishes tagged at unreported artificial reef sites off NW Florida. Proc Gulf Caribb Fish Inst 2008; 60: 297-304.

[24] Özgül A. The adaptation of floating artificial reefs for pelagic fisheries in the Aegean Sea. [PhD Thesis]. İzmir: Ege University, Graduate School of Natural and Applied Sciences 2010; 161.

[25] Voegeli FA, Smale MJ, Webber DM, Andrade Y, O'Dor RK. Ultrasonic telemetry, tracking and automated monitoring technology for sharks. Environ Biol Fishes 2001; 60: 267-82. [http://dx.doi.org/10.1023/A:1007682304720]

[26] Parsons D, Egli D. Fish movement in a temperate marine reserve: new insights through application of acoustic tracking. Mar Technol Soc J 2005; 39: 56-63. [http://dx.doi.org/10.4031/002533205787521703]

[27] Jadot C, Donnay A, Acolas ML, Cornet Y, Anras ML. Activity patterns, home-range size, and habitat utilization of *Sarpa salpa* (Teleostei: Sparidae) in the Mediterranean Sea. ICES J Mar Sci 2006; 63: 128-39. [http://dx.doi.org/10.1016/j.icesjms.2005.06.010]

[28] Jorgensen SJ, Kaplan DM, Klimley AP, Morgan SG, O'Farrell MR, Botsford LW. Limited movement in blue rockfish *Sebastes mystinus*: internal structure of home range. Mar Ecol Prog Ser 2006; 327: 157-70. [http://dx.doi.org/10.3354/meps327157]

[29] Collins AB, Heupel MR, Motta PJ. Residence and movement patterns of cownose rays *Rhinoptera bonasus* within a south-west Florida estuary. J Fish Biol 2007; 71: 1159-78. [http://dx.doi.org/10.1111/j.1095-8649.2007.01590.x]

[30] Abecasis D, Erzini K. Site fidelity and movements of gilthead sea bream (*Sparus aurata*) in a coastal lagoon (Ria Formosa, Portugal). Estuar Coast Shelf Sci 2008; 79: 758-63. [http://dx.doi.org/10.1016/j.ecss.2008.06.019]

[31] Pincock DG, Johnston SV. Acoustic telemetry overview.Telemetry techniques: A users guide for fisheries research. Maryland: American Fisheries Society 2012; pp. 305-38.

[32] Heupel MR, Semmens JM, Hobday AJ. Automated acoustic tracking of aquatic animals: Scales, design and deployment of listening station arrays. Mar Freshw Res 2006; 57: 1-13. [http://dx.doi.org/10.1071/MF05091]

[33] Cooke SJ, Midwood JD, Thiem JD, *et al.* Tracking animals in freshwater with electronic tags: past, present and future. Anim Biotelem 2013; 1: 1-5. [http://dx.doi.org/10.1186/2050-3385-1-5]

[34] Thorsteinsson V. Tagging Methods for Stock Assessment and Research in Fisheries 2002.

[35] Özgül A, Lök A, Ulaş A, Düzbastılar FO, Tanrıkul TT, Pelister C. Preliminary study on the use of the Vemco Positioning System to determine fish movements in artificial reef areas: a case study on *Sciaena umbra* Linnaeus, 1758. J Appl Ichthyology 2015; 31: 41-7. [http://dx.doi.org/10.1111/jai.12922]

[36] Holland KN, Lowe CG, Peterson JD. Gill, Tracking Coastal Sharks with Small Boats: Hammer-Head Shark Pups as a Case Study. Aust J Mar Freshwater Res 1992; 43: 61-6. [http://dx.doi.org/10.1071/MF9920061]

[37] Wagner GN, Cooke SJ. Methodological approaches and opinions of researchers involved in the surgical implantation of telemetry transmitters in fish. J Aquat Anim Health 2005; 17: 160-9. [http://dx.doi.org/10.1577/H04-037.1]

[38] Collins KJ, Jensen AC, Lockwood AP. Acoustic tagging of lobsters on the Poole Bay artificial reef. 1992.

[39] Szedlmayer ST. Ultrasonic telemetry of red snapper, *Lutjanus campechanus*, at artificial reef sites in the northeast Gulf of Mexico. Copeia 1997; 1: 846-50. [http://dx.doi.org/10.2307/1447304]

[40] Josse E, Dagorn L, Bertrand A. Typology and behaviour of tuna aggregations around fish aggregating devices from acoustic surveys in French Polynesia. Aquat Living Resour 2000; 13: 183-92. [http://dx.doi.org/10.1016/S0990-7440(00)00051-6]

[41] Doray M, Josse E, Gervain P, Reynal L, Chantrel J. Acoustic characterization of pelagic fish aggregations around moored fish aggregating devices in Martinique (Lesser Antilles). Fish Res 2006; 82: 162-75. [http://dx.doi.org/10.1016/j.fishres.2006.06.025]

[42] Hallier J, Gaertner D. Drifting fish aggregation devices could act as an ecological trap for tropical tuna species. Mar Ecol Prog Ser 2008; 353: 255-64. [http://dx.doi.org/10.3354/meps07180]

[43] D'Anna G, Giacalone VM, Pipitone C, Badalamenti F. Movement pattern of white seabream, *Diplodus sargus* (L., 1758) (Osteichthyes, Sparidae) acoustically tracked in an artificial reef area. Ital J Zool (Modena) 2011; 78(2): 255-63. [http://dx.doi.org/10.1080/11250000903464059]

[44] Koeck B, Gudefin A, Romans P, Loubet J, Lenfant P. Effect of intra-coelomic tagging on white seabream (*Diplodus sargus*) behavior and survival. J Exp Mar Biol Ecol 2013; 440: 1-7. [http://dx.doi.org/10.1016/j.jembe.2012.11.014]

[45] Szedlmayer ST, Schroepfer RL. Long-term residence of red snapper on artificial reefs in the northeastern Gulf of Mexico. Trans Am Fish Soc 2005; 134: 315-25. [http://dx.doi.org/10.1577/T04-070.1]

[46] Topping DT, Szedlmayer ST. Home range and movement patterns of red snapper (*Lutjanus campechanus*) on artificial reefs. Fish Res 2011; 112: 77-84. [http://dx.doi.org/10.1016/j.fishres.2011.08.013]

[47] Reynolds BF, Powers SP, Bishop MA. Application of acoustic telemetry to assess residency and movements of rockfish and lingcod at created and natural habitats in Prince William Sound. PLoS One 2010; 5(8): e12130. [http://dx.doi.org/10.1371/journal.pone.0012130] [PMID: 20730090]

[48] Lino PG, Bentes L, Oliveira MT, Erzini K, Santos MN. The African Hind's (*Cephalopholis taeniops*, Serranidae) use of artificial reefs off Sal Island (Cape Verde): A preliminary study based on acoustic telemetry. Braz J Oceanogr 2011; 59: 69-76. [http://dx.doi.org/10.1590/S1679-87592011000500009]

[49] Reubens JT, Pasotti F, Degraer S, Vincx M. Residency, site fidelity and habitat use of Atlantic cod (*Gadus morhua*) at an offshore wind farm using acoustic telemetry. Mar Environ Res 2013; 90: 128-35. [http://dx.doi.org/10.1016/j.marenvres.2013.07.001] [PMID: 23937893]

[50] Carey FG, Olson RJ. Sonic Tracking Experiments with Tunas. ICCAT Collect Coll Vol Sci Pap 1982; 17: 458-66.

[51] Holland KN, Brill RW, Chang RK. Horizontal and vertical movements of yellowfin and bigeye tuna associated with fish aggregating devices. Fish Bull 1990; 88: 493-07.

[52] Espinoza M, Farrugia TJ, Webber DM, Smith F, Lowe CG. Testing a new acoustic telemetry technique to quantify long-term, fine-scale movements of aquatic animals. Fish Res 2011; 108: 364-71.
[http://dx.doi.org/10.1016/j.fishres.2011.01.011]

[53] Özgül A. Use of the acoustic positioning system in artificial reef sites. EgeJFAS 2016; 33(4): 405-11.
[http://dx.doi.org/10.12714/egejfas.2016.33.4.15]

[54] Klimley A, Le Boeuf B, Cantara K, Richert J, Davis S, Van Sommeran S. Radio acoustic positioning as a tool for studying site-specific behavior of the white shark and other large marine species. Mar Biol 2001; 138(2): 429-46.
[http://dx.doi.org/10.1007/s002270000394]

[55] Andrews KS, Tolimieri N, Williams GD, Samhouri JF, Harvey CJ, Levin PS. Comparison of fine-scale acoustic monitoring systems using home range size of a demersal fish. Mar Biol 2011; 158: 2377-87.
[http://dx.doi.org/10.1007/s00227-011-1724-5]

[56] Lacroix GL, Voegeli FA. Development of automated monitoring systems for ultrasonic transmitters.

[57] Welsh JQ, Fox RJ, Webber DM, Bellwood DR. Performance of remote acoustic receivers within a coral reef habitat: implications for array design. Coral Reefs 2012; 31: 693-02.
[http://dx.doi.org/10.1007/s00338-012-0892-1]

[58] Smedbol SJ, Smith F, Webber DM, Vallée RE, King TD. Using underwater coded acoustic telemetry for fine scale positioning of aquatic animals 20th Symposium of the International Society on Biotelemetry Proceedings.

[59] Burt WH. Territoriality and home range concepts as applied to mammals. J Mammal 1943; 24: 346-52.
[http://dx.doi.org/10.2307/1374834]

[60] Caro T. The significance of behavioral ecology for conservation biology.Behavioral ecology and conservation biology. Oxford: Oxford University Press 1998; pp. 3-26.

[61] Rodgers AR, Carr AP. HRE: The home range extension for Arc View Thunder Bay. Ontario, Canada: Centre for Northern Forest Ecosystem Research, Ontario Ministry of Natural Resources 2002.

[62] Row JR, Blouin-Demers G. Kernels are not accurate estimators of home range size for herpetofauna. Copeia 2006; 1: 797-802.
[http://dx.doi.org/10.1643/0045-8511(2006)6[797:KANAEO]2.0.CO;2]

[63] Hooge PN. Using GIS to analyze animal movements in the marine environment. In: Kruse GH, Bez N, Booth A, Dorn MW, Hills S, Lipcius RN, Pelletier D, Roy C, Smith SJ, Witherell D, editors. Spatial pro-cesses and management of marine populations, University of Alaska Sea Grant College. 2001; pp. 37-51.

[64] Vander-Wal E, Rodgers AR. An individual based quantitative approach for delineating core areas of animal space use: A function of maximum use and minimum area. Ecol Modell 2012; 224: 48-53.
[http://dx.doi.org/10.1016/j.ecolmodel.2011.10.006]

[65] Giacalone VM, D'Anna G, Garofalo G, Collins K, Badalamenti F. Estimation of positioning error from an array of automated omni-directional receivers in an artificial reef area.

[66] Voegeli FA, Pincock DG. Overview of underwater acoustics as it applies to telemetry. In: Baras E, Philippart Jc, Eds. Proceedings of the First Conference and Workshop on Fish Telemetry in Europe. 235-41.

[67] Simpfendorfer CA, Heupel MR, Collins AB. Variation in the performance of acoustic receivers and its implication for positioning algorithms in a riverine setting. Can J Fish Aquat Sci 2008; 65(3): 482-92.

[http://dx.doi.org/10.1139/f07-180]

[68] Klimley AP, Voegeli F, Beavers SC, Le Boeuf BJ. Automated listening stations for tagged marine fishes. Mar Technol Soc J 1998; 32: 94-01.

[69] Wolcott TG. New options in physiological and behavioural ecology through multichannel telemetry. J Exp Marine Bio 1995; 193: 257-75.
[http://dx.doi.org/10.1016/0022-0981(95)00121-2]

[70] Zeller DC. Home range and activity patterns of the coral trout *Plectropomus leopardus* (Serranidae). Mar Ecol Prog Ser 1997; 154: 65-77.
[http://dx.doi.org/10.3354/meps154065]

[71] Bolden SK. Using ultrasonic telemetry to determine home range of a coral reef fish.Electronic tagging and tracking in marine fisheries. The Netherlands: Kluwer Academic Publishers 2001; pp. 167-88.
[http://dx.doi.org/10.1007/978-94-017-1402-0_8]

[72] O'Dor RK, Aitken JP, Babcock RC, *et al.* Using radio-acoustic positioning and telemetry (RAPT) to define and assess marine protected areas (MPAs).Electronic Tagging and Tracking in Marine Fisheries. The Netherlands: Kluwer Academic Publishers 2001; pp. 147-66.
[http://dx.doi.org/10.1007/978-94-017-1402-0_7]

[73] Pittman SJ, Monaco ME, Friedlander AM, *et al.* Fish with chips: tracking reef fish movements to evaluate size and connectivity of Caribbean marine protected areas. PLoS One 2014; 9(5): e96028.
[http://dx.doi.org/10.1371/journal.pone.0096028] [PMID: 24797815]

[74] Spedicato M, Carbonara P, Lembo G.

[75] Topping DT, Lowe CG, Caselle JE. Home range and habitat utilization of adult California sheephead, *Semicossyphus pulcher* (Labridae), in a temperate no-take marine reserve. Mar Biol 2005; 147: 301-11.
[http://dx.doi.org/10.1007/s00227-005-1573-1]

[76] March D, Alós J, Palmer M, Grau A. Short-term residence and movement patterns of the annular seabream *Diplodus annularis* in a temperate marine reserve. Estuar Coast Shelf Sci 2011; 92: 581-7.
[http://dx.doi.org/10.1016/j.ecss.2011.02.015]

[77] Picciulin M, Umani M, Costantini M, Spoto M, Ferrero EA.

[78] Alós J, Cabanellas-Reboredo M. Experimental acoustic telemetry experiment reveals strong site fidelity during the sexual resting period of wild brown meagre, *Sciaena umbra.* J Appl Ichthyology 2012; 28(4): 606-11.
[http://dx.doi.org/10.1111/j.1439-0426.2012.01955.x]

[79] Peabody MB, Wilson CA. Fidelity of red snapper (*Lutjanus campechanus*) to petroleum platforms and artificial reefs in the northern Gulf of Mexico: final report. Coastal Fisheries Institute, School of the Coast and Envi-ronment, Louisiana State University, U.S. Dept. of the Interior, Minerals Management Service, Gulf of Mexico OCS Region; New Orleans 2006.

[80] Szedlmayer ST, Schroepfer RL. Long-term residence of red snapper on artificial reefs in the north eastern Gulf of Mexico. Trans Am Fish Soc 2005; 134: 315-25.
[http://dx.doi.org/10.1577/T04-070.1]

[81] Williams-Grove LJ, Szedlmayer ST. Acoustic positioning and movement patterns of red snapper *Lutjanus campechanus* around artificial reefs in the northern Gulf of Mexico. Mar Ecol Prog Ser 2016; 553: 233-51.
[http://dx.doi.org/10.3354/meps11778]

[82] Lök A, Ünal V, Özgül A, Tunca S. Biological and socio-economic effects of artificial reefs: A case study of Edremit bay coastal fishery Turkey. 3rd International Symposia on Integrated Coastal Zone Management; October 2014; Antalya, Turkey.

[83] Lök A, Düzbastılar FO, Gül B, Özgül A, Ulaş A. The Role of Artificial Reefs in Fisheries

Management in Turkey.Artificial Reefs in Fisheries Management. London: CRC Press 2011; pp. 155-66.
[http://dx.doi.org/10.1201/b10910-11]

[84] Lök A, Özgül A. Movements of *Pomatamus saltatrix* in artificial reef area: preliminary results. 3rd International Conference on Fish Telemetry. Halifax, Canada. 2015.

[85] Özgül A, Lök A. Long-term residency of fish species on artificial reefs in the northern Aegean Sea, Turkey. 2nd International Congress on Applied Ichthyology & Aquatic Environment. Messolonghi, Greece. 2016; pp. 259-63.

[86] Özgül A, Lök A, Tanrıkul TT, Ulaş A, Düzbastılar FO. Determining movement patterns of the fish in artificial reefs and their use for small scale fisheries 2016.

SUBJECT INDEX

Made in the USA
Las Vegas, NV
03 November 2026

57424113R00217